DATE DUE FOR RETURN

Methods in Enzymology

Volume 146
PEPTIDE GROWTH FACTORS
Part A

METHODS IN ENZYMOLOGY

EDITORS-IN-CHIEF

Sidney P. Colowick   Nathan O. Kaplan

Methods in Enzymology

Volume 146

Peptide Growth Factors

Part A

EDITED BY

David Barnes

DEPARTMENT OF BIOCHEMISTRY
AND BIOPHYSICS
OREGON STATE UNIVERSITY
CORVALLIS, OREGON

David A. Sirbasku

DEPARTMENT OF BIOCHEMISTRY
AND MOLECULAR BIOLOGY
UNIVERSITY OF TEXAS MEDICAL SCHOOL
HOUSTON, TEXAS

1987

ACADEMIC PRESS, INC.
Harcourt Brace Jovanovich, Publishers

Orlando   San Diego   New York   Austin
Boston   London   Sydney   Tokyo   Toronto

ACADEMIC PRESS, INC.
Orlando, Florida 32887

United Kingdom Edition published by
ACADEMIC PRESS INC. (LONDON) LTD.
24–28 Oval Road, London NW1 7DX

LIBRARY OF CONGRESS CATALOG CARD NUMBER: 54-9110

ISBN 0–12–182046–7 (alk. paper)

PRINTED IN THE UNITED STATES OF AMERICA

87 88 89 90   9 8 7 6 5 4 3 2 1

Table of Contents

Section I. Epidermal Growth Factor

Section II. Transforming Growth Factors

v

Section III. Somatomedin/Insulin-Like Growth Factors

Contributors to Volume 146

Article numbers are in parentheses following the names of contributors.
Affiliations listed are current.

SUSAN AITKEN (31), *Medical Breast Cancer Section, Medicine Branch, Division of Cancer Treatment, National Cancer Institute, National Institutes of Health, Bethesda, Maryland 20892*

RICHARD K. ASSOIAN (14), *Department of Biochemistry and Molecular Biophysics (Center for Reproductive Sciences), Columbia University, New York, New York 10032*

JULIA H. BALDWIN (15), *Molecular Biology Laboratory, The Salk Institute, San Diego, California 92138*

DAVID BARNES (8), *Department of Biochemistry and Biophysics, Environmental Health Sciences Center, Oregon State University, Corvallis, Oregon 97331*

DAVID J. BAYLINK (27), *Departments of Medicine, Biochemistry, and Orthopaedics, Loma Linda University, and the Mineral Metabolism Laboratory, Jerry L. Pettis Veterans Hospital, Loma Linda, California 92357*

GRAEME I. BELL (23), *Howard Hughes Medical Institute and Department of Biochemistry and Molecular Biology, The University of Chicago, Chicago, Illinois 60637*

A. G. BROWNELL (28), *Department of Surgery, Division of Orthopedics, UCLA Bone Research Laboratory, Los Angeles, California 90024*

GRAHAM CARPENTER (5), *Division of Dermatology, Vanderbilt University School of Medicine, Nashville, Tennessee 37232*

J. J. CHANG (28), *Department of Surgery, Division of Orthopedics, UCLA Bone Research Laboratory, Los Angeles, California 90024*

SIDNEY P. COLOWICK[1] (38), *Department of Microbiology, Vanderbilt University School of Medicine, Nashville, Tennessee 37232*

A. JOSEPH D'ERCOLE (21), *Department of Pediatrics, Division of Endocrinology, University of North Carolina, Chapel Hill, North Carolina 27514*

WILLIAM H. DAUGHADAY (24), *Metabolism Division, Department of Medicine, Washington University School of Medicine, St. Louis, Missouri 63110*

SUSAN DAVIDSON (4), *Department of Surgery, The Children's Hospital, Boston, Massachusetts 02115*

R. J. DELANGE (28), *Department of Surgery, Division of Orthopedics, UCLA Bone Research Laboratory, Los Angeles, California 90024*

JOSEPH E. DE LARCO (9), *Otsuka Pharmaceutical Co., Ltd., Rockville, Maryland 20850*

ROBERT B. DICKSON (31), *Medical Breast Cancer Section, Medicine Branch, Division of Cancer Treatment, National Cancer Institute, National Institutes of Health, Bethesda, Maryland 20892*

CHARLES A. FROLIK (9), *Biochemistry Division, Lilly Corporate Center, Indianapolis, Indiana 46285*

RICHARD W. FURLANETTO (20), *University of Pennsylvania, School of Medicine, Philadelphia, Pennsylvania 19104*

LYNNE A. GAYNES (25), *Metabolism Branch, National Cancer Institute, National Institutes of Health, Bethesda, Maryland 20892*

GORDON N. GILL (7), *Department of Medicine, School of Medicine, University of California, San Diego, La Jolla, California 92093*

G. YANCEY GILLESPIE (19), *Division of Neurosurgery, Department of Pediatrics, University of North Carolina School of Medicine, Chapel Hill, North Carolina 27514*

SYBIL GREENFIELD (15), *Molecular Biology Laboratory, The Salk Institute, San Diego, California 92138*

[1] Deceased

LAWRENCE A. GREENSTEIN (25), *Metabolism Branch, National Cancer Institute, National Institutes of Health, Bethesda, Maryland 20892*

RICHARD N. HARKINS (26), *Department of Protein Chemistry, Triton Biosciences, Alameda, California 94501*

ROBERT A. HARPER (1), *Connective Tissue Research Institute, University of Pennsylvania, Philadelphia, Pennsylvania 19104*

RAYMOND L. HINTZ (22), *Division of Pediatric Endocrinology, Department of Pediatrics, Stanford University, Stanford, California 94305*

YUJI HIRAKI (29), *Department of Biochemistry and Calcified-Tissue Metabolism, Faculty of Dentistry, Osaka University, Osaka 565, Japan*

ROBERT W. HOLLEY (15), *Molecular Biology Laboratory, The Salk Institute, San Diego, California 92138*

Y. K. HUO (28), *Department of Surgery, Division of Orthopedics, UCLA Bone Research Laboratory, Los Angeles, California 90024*

SHIGERU IKUTA (2), *Research Institute for Biochemistry, Toyo Jozo Co. Ltd., Ohito, Shizuoka 410–23, Japan*

WENDELYN H. INMAN (38), *Department of Medicine, Division of Dermatology, Vanderbilt University School of Medicine, Nashville, Tennessee 37232*

JOHN C. JENNINGS (27), *Department of Medicine, Loma Linda University Medical School, and Research and Mineral Services, Jerry L. Pettis Veterans Hospital, Loma Linda, California 92357*

YUKIO KATO (29), *Department of Biochemistry and Calcified-Tissue Metabolism, Faculty of Dentistry, Osaka University, Osaka 565, Japan*

TOMOYUKI KAWAMOTO (6), *Department of Biochemistry, Okayama University Dental School, Okayama City 700, Japan*

MICHAEL KLAGSBRUN (4, 30), *Departments of Surgery and Biological Chemistry, The Children's Hospital and Harvard Medical School, Boston, Massachusetts 02115*

DANIEL J. KNAUER (26), *Department of Developmental and Cell Biology, University of California-Irvine, Irvine, California 92717*

ANH D. LE (6), *Department of Chemistry, University of California, San Diego, La Jolla, California 92093*

LILLY LEE (25), *Metabolism Branch, National Cancer Institute, National Institutes of Health, Bethesda, Maryland 20892*

JULIA E. LEVER (36), *Department of Biochemistry and Molecular Biology, The University of Texas Medical School, Houston, Texas 77225*

CHOH HAO LI (18), *Laboratory of Molecular Endocrinology, University of California, San Francisco, California 94143*

A. LIETZE (28), *Department of Surgery, Division of Orthopedics, UCLA Bone Research Laboratory, Los Angeles, California 90024*

MARC E. LIPPMAN (31), *Medical Breast Cancer Section, Medicine Branch, Division of Cancer Treatment, National Cancer Institute, National Institutes of Health, Bethesda, Maryland 20892*

RUSSETTE M. LYONS (26), *Department of Cell Biology, School of Medicine, Vanderbilt University, Nashville, Tennessee 37232*

JEAN M. MARINO (20), *Department of Endocrinology, The Children's Hospital of Philadelphia, Philadelphia, Pennsylvania 19104*

JORGE MARTIN-PÉREZ (35), *Department of Biology, Harvard University, Cambridge, Massachusetts 02138*

JOAN MASSAGUÉ (10, 13, 17), *Department of Biochemistry, University of Massachusetts Medical Center, Worcester, Massachusetts 01605*

DANA N. MCKINLEY (39), *Division of Immunology and Cell Biology, Department of Pathology, Health Sciences Center, University of Utah, Salt Lake City, Utah 84132*

STANLEY A. MENDOZA (37), *Department of Pediatrics, School of Medicine, University of California, San Diego, La Jolla, California 92093*

SUBBURAMAN MOHAN (27), *Departments of Medicine and Physiology, Loma Linda*

University, and Research Service, Jerry L. Pettis Veterans Hospital, Loma Linda, California 92357

KATSUZO NISHIKAWA (2), Department of Biochemistry, Kanazawa Medical University, Uchinada, Ishikawa 920-02, Japan

S. PETER NISSLEY (25), Metabolism Branch, National Cancer Institute, National Institutes of Health, Bethesda, Maryland 20892

JAMES PIERCE (1), Department of Biochemistry, Temple University School of Medicine, Philadelphia, Pennsylvania 19140

LINDA J. PIKE (33), Howard Hughes Medical Institute, Department of Biological Chemistry, Washington University, St. Louis, Missouri 63110

LESLIE B. RALL (23), Chiron Corporation, Emeryville, California 94608

MATTHEW M. RECHLER (25), Molecular, Cellular, and Nutritional Endocrinology Branch, National Institute of Arthritis, Diabetes, and Digestive and Kidney Diseases, National Institutes of Health, Bethesda, Maryland 20892

ANN RICHMOND (11), Veterans Administration Medical Center (Atlanta), and Department of Medicine, Emory University, Atlanta, Georgia 30322

ANGIE RIZZINO (32), Eppley Institute for Research in Cancer and Allied Diseases, University of Nebraska Medical Center, Omaha, Nebraska 68105

JOYCE A. ROMANUS (25), Molecular, Cellular, and Nutritional Endocrinology Branch, National Institute of Arthritis, Diabetes, and Digestive and Kidney Diseases, National Institutes of Health, Bethesda, Maryland 20892

ROBERT G. B. ROY (11), Veterans Administration Medical Center (Atlanta), and Department of Medicine, Division of Endocrinology, Emory University, Atlanta, Georgia 30322

ENRIQUE ROZENGURT (37), Imperial Cancer Research Fund, Lincoln's Inn Fields, London WC2A 3PX, United Kingdom

JOACHIM SASSE (30), Immunology Section, Shriner's Hospital, Tampa, Florida 33612

J. DENRY SATO (6), W. Alton Jones Cell Science Center, Lake Placid, New York 12946

C. RICHARD SAVAGE, JR. (1), Department of Biochemistry, Temple University School of Medicine, Philadelphia, Pennsylvania 19140

JAMES SCOTT (23), Molecular Medicine Group, MRC, Clinical Research Centre, Harrow, Middlesex HA1 3UJ, United Kingdom

YUEN SHING (4), Departments of Surgery and Biological Chemistry, The Children's Hospital and Harvard Medical School, Boston, Massachusetts 02115

MICHEL SIEGMANN (34), Friedrich Miescher-Institut, CH-4002 Basel, Switzerland

GARY L. SMITH (26), School of Biological Sciences, University of Nebraska, Lincoln, Nebraska 68588

ANN MANGELSDORF SODERQUIST (5), Department of Biochemistry, Vanderbilt University School of Medicine, Nashville, Tennessee 37232

ROBERT SULLIVAN (30), Department of Surgical Research, Children's Hospital, Boston, Massachusetts 02115

FUJIO SUZUKI (29), Department of Biochemistry and Calcified-Tissue Metabolism, Faculty of Dentistry, Osaka University, Osaka 565, Japan

MARJORIE E. SVOBODA (19), Department of Surgery and Division of Pediatric Endocrinology, Department of Pediatrics, University of North Carolina School of Medicine, Chapel Hill, North Carolina 27514

JAMES P. TAM (12), The Rockefeller University, New York, New York 10021

GEORGE THOMAS (34, 35), Friedrich Miescher-Institut, CH-4002 Basel, Switzerland

H. GREG THOMAS (11), Veterans Administration Medical Center (Atlanta), and Department of Medicine, Division of Endocrinology, Emory University, Atlanta, Georgia 30322

LOUIS E. UNDERWOOD (19, 21) Department of Pediatrics, Division of Endocrinology,

University of North Carolina, Chapel Hill, North Carolina 27514

MICKEY S. URDEA (3), *Chiron Research Laboratories, Chiron Corporation, Emeryville, California 94608*

MARSHALL R. URIST (28), *Department of Surgery, Division of Orthopedics, UCLA Bone Research Laboratory, Los Angeles, California 90024*

JUDSON J. VAN WYK (19), *Department of Surgery and Division of Pediatric Endocrinology, Department of Pediatrics, University of North Carolina School of Medicine, Chapel Hill, North Carolina 27514*

LALAGE M. WAKEFIELD (16), *Laboratory of Chemoprevention, National Cancer Institute, Bethesda, Maryland 20205*

WOLFGANG WEBER (7), *Physiologisch-Chemisches Institut, Universität Hamburg, 2000 Hamburg 20, Federal Republic of Germany*

H. STEVEN WILEY (39), *Division of Immunology and Cell Biology, Department of Pathology, University of Utah Health Sciences Center, Salt Lake City, Utah 84132*

DONALD YAMASHIRO (18), *Laboratory of Molecular Endocrinology, University of California, San Francisco, California 94143*

YOSHINO YOSHITAKE (2), *Department of Biochemistry, Kanazawa Medical University, Uchinada, Ishikawa 920-02, Japan*

Dedication

Sidney Colowick and Nate Kaplan were personally involved to a great degree in the development of Peptide Growth Factors, Parts A and B, Volumes 146 and 147 of *Methods in Enzymology.* Although their scientific interests and research directions were varied, they directed their interests toward hormone-related research from time to time throughout their careers. (See, for instance, the chapter on Growth Factor Stimulation of Sugar Uptake by Inman and Colowick in Volume 146.) From the early discussions regarding these volumes, Sidney's and Nate's comments, advice, and suggestions made our task easier, and their wisdom contributed greatly to the quality of the final product. Sadly, neither lived to see the completion of the work.

Other tributes to these extraordinary men have appeared in *Methods in Enzymology,* and we could add little that has not been expressed. The series itself may be the best demonstration of their remarkable scientific insight and comprehension. However, those who knew them realize that in spite of their enormous intellectual contributions it is the personal impact they had on colleagues, students, and friends that is their greatest legacy. These men and their families touched our lives profoundly. We dedicate these volumes to Sidney and Maryda, Nate and Goldie, and their families.

DAVID BARNES
DAVID A. SIRBASKU

Preface

Applications of new techniques and proliferation of investigators in the area of peptide growth factors have led to enormous expansion in understanding the mechanisms of action of these molecules in the past few years and have, in turn, raised new and intriguing questions concerning the relationship of peptide growth factors, receptors, and related molecules to normal and abnormal regulation of cell division and differentiation *in vivo.* Volumes 146 and 147 of *Methods in Enzymology* are a reflection of the recent advances. In these volumes we have attempted to cover methods specific for each growth factor in the areas of purification, immunoassay, radioreceptor assay, biological activity assay, and receptor identification and quantitation. Also included in both volumes are general methods for the study of mechanisms of action that are applicable to many of the factors.

Volume 146 includes techniques concerning epidermal growth factor, transforming growth factors alpha and beta, somatomedin C/insulin-like growth factors, and bone and cartilage growth factors. It also contains methods for quantitative cell growth assays and techniques for the study of growth factor-modulated protein phosphorylation and cell surface membrane effects. Volume 147 covers techniques concerning platelet-derived growth factor, angiogenesis, endothelial and fibroblast growth factors (heparin-binding growth factors), nerve and glial growth factors, transferrin, erythropoietin, and related factors. Also included are genetic approaches to growth factor action and additional methods for the study of biological effects of these molecules.

Procedures related to the basic aspects of research concerning epidermal, nerve, platelet-derived, and insulin-like growth factors have appeared in the Hormone Action series of *Methods in Enzymology,* and methods concerning most aspects of growth factors for lymphoid cells have appeared in the Immunochemical Techniques series. We have attempted to avoid major duplication of material appearing elsewhere in this series in the belief that volumes of moderate size providing recent and complementary methodology are of more practical value to the researcher. For example, we have not attempted to cover growth factors for lymphoid cells, with the exception of transferrin, which appears to be a requirement for optimal growth of many cell types. Similarly, some approaches that are finding current intensive use in the field, such as recombinant DNA, hybridoma, nucleic acid, or peptide synthesis techniques, are covered per se in other volumes of *Methods in Enzymology.* Detailed techniques in these areas are included only in situations in which the authors or editors felt that they were sufficiently novel or specific to the areas of growth factor research.

We thank the many contributors to this project, and hope these volumes will be useful to investigators in the field.

DAVID BARNES
DAVID A. SIRBASKU

METHODS IN ENZYMOLOGY

EDITED BY

Sidney P. Colowick and Nathan O. Kaplan

VANDERBILT UNIVERSITY
SCHOOL OF MEDICINE
NASHVILLE, TENNESSEE

DEPARTMENT OF CHEMISTRY
UNIVERSITY OF CALIFORNIA
AT SAN DIEGO
LA JOLLA, CALIFORNIA

I. Preparation and Assay of Enzymes
II. Preparation and Assay of Enzymes
III. Preparation and Assay of Substrates
IV. Special Techniques for the Enzymologist
V. Preparation and Assay of Enzymes
VI. Preparation and Assay of Enzymes (*Continued*)
   Preparation and Assay of Substrates
   Special Techniques
VII. Cumulative Subject Index

METHODS IN ENZYMOLOGY

EDITORS-IN-CHIEF

Sidney P. Colowick and Nathan O. Kaplan

VOLUME 74. Immunochemical Techniques (Part C)
Edited by JOHN J. LANGONE AND HELEN VAN VUNAKIS

VOLUME 75. Cumulative Subject Index Volumes XXXI, XXXII, and XXXIV–LX
Edited by EDWARD A. DENNIS AND MARTHA G. DENNIS

VOLUME 76. Hemoglobins
Edited by ERALDO ANTONINI, LUIGI ROSSI-BERNARDI, AND EMILIA CHIANCONE

VOLUME 77. Detoxication and Drug Metabolism
Edited by WILLIAM B. JAKOBY

VOLUME 78. Interferons (Part A)
Edited by SIDNEY PESTKA

VOLUME 79. Interferons (Part B)
Edited by SIDNEY PESTKA

VOLUME 80. Proteolytic Enzymes (Part C)
Edited by LASZLO LORAND

VOLUME 81. Biomembranes (Part H: Visual Pigments and Purple Membranes, I)
Edited by LESTER PACKER

VOLUME 82. Structural and Contractile Proteins (Part A: Extracellular Matrix)
Edited by LEON W. CUNNINGHAM AND DIXIE W. FREDERIKSEN

VOLUME 83. Complex Carbohydrates (Part D)
Edited by VICTOR GINSBURG

VOLUME 84. Immunochemical Techniques (Part D: Selected Immunoassays)
Edited by JOHN J. LANGONE AND HELEN VAN VUNAKIS

VOLUME 85. Structural and Contractile Proteins (Part B: The Contractile Apparatus and the Cytoskeleton)
Edited by DIXIE W. FREDERIKSEN AND LEON W. CUNNINGHAM

VOLUME 86. Prostaglandins and Arachidonate Metabolites
Edited by WILLIAM E. M. LANDS AND WILLIAM L. SMITH

VOLUME 87. Enzyme Kinetics and Mechanism (Part C: Intermediates, Stereochemistry, and Rate Studies)
Edited by DANIEL L. PURICH

VOLUME 88. Biomembranes (Part I: Visual Pigments and Purple Membranes, II)
Edited by LESTER PACKER

VOLUME 89. Carbohydrate Metabolism (Part D)
Edited by WILLIS A. WOOD

VOLUME 90. Carbohydrate Metabolism (Part E)
Edited by WILLIS A. WOOD

VOLUME 91. Enzyme Structure (Part I)
Edited by C. H. W. HIRS AND SERGE N. TIMASHEFF

VOLUME 92. Immunochemical Techniques (Part E: Monoclonal Antibodies and General Immunoassay Methods)
Edited by JOHN J. LANGONE AND HELEN VAN VUNAKIS

VOLUME 93. Immunochemical Techniques (Part F: Conventional Antibodies, Fc Receptors, and Cytotoxicity)
Edited by JOHN J. LANGONE AND HELEN VAN VUNAKIS

VOLUME 94. Polyamines
Edited by HERBERT TABOR AND CELIA WHITE TABOR

VOLUME 95. Cumulative Subject Index Volumes 61–74 and 76–80
Edited by EDWARD A. DENNIS AND MARTHA G. DENNIS

VOLUME 96. Biomembranes [Part J: Membrane Biogenesis: Assembly and Targeting (General Methods; Eukaryotes)]
Edited by SIDNEY FLEISCHER AND BECCA FLEISCHER

VOLUME 97. Biomembranes [Part K: Membrane Biogenesis: Assembly and Targeting (Prokaryotes, Mitochondria, and Chloroplasts)]
Edited by SIDNEY FLEISCHER AND BECCA FLEISCHER

Section I

Epidermal Growth Factor

[1] Purification of Human Epidermal Growth Factor by Monoclonal Antibody Affinity Chromatography

By ROBERT A. HARPER, JAMES PIERCE, and C. RICHARD SAVAGE, JR.

Introduction

Epidermal growth factor (EGF) is a single chain polypeptide containing 53 amino acid residues that exhibits potent mitogenic activity for a variety of cell types both *in vivo* and *in vitro*. Many reviews are available concerning the mechanism of action and biological effects of EGF.[1-5] This molecule was first described by Cohen,[6] who isolated EGF from the submaxillary glands of adult male mice (mEGF). Since then EGF has been isolated from the rat,[7] the guinea pig prostate,[8] and human urine (hEGF).[9-11]

At the time Cohen[9] first reported the amino acid composition of hEGF in 1975, Gregory[12] simultaneously reported the structure of the antisecretory agent "urogastrone" which is also found in urine. Gregory demonstrated that urogastrone and EGF has similar amino acid structures as well as biological properties. From this early work of Cohen and Gregory it was soon clear that these two molecules were one and the same. A comparison of the primary structure of mouse EGF[13] and human EGF[12] is shown in Fig. 1. There exist a remarkable degree of sequence homology between hEGF and mEGF with 37 of the 53 amino acid residues comprising the two molecules being identical and with the three disulfide linkages located in the same relative positions. Only 2 of the remaining 16 nonhomologous

[1] G. Carpenter, *in* "Handbook of Experimental Pharmacology" (R. Baserga, ed.), p. 89. Springer-Verlag, New York, 1981.
[2] G. Carpenter and S. Cohen, *Annu. Rev. Biochem.* **48**, 193 (1979).
[3] M. Das, *Int. Rev. Cytol.* **78**, 233 (1982).
[4] D. Gospodarowicz, *Annu. Rev. Physiol.* **43**, 251 (1981).
[5] H. Haigler, *in* "Growth and Maturation Factors" (G. Guroff, ed.), p. 117. Wiley, New York, 1983.
[6] S. Cohen, *J. Biol. Chem.* **237**, 1555 (1962).
[7] J. B. Moore, Jr., *Arch. Biochem. Biophys.* **189**, 1 (1978).
[8] J. A. Rubin and R. A. Bradshaw, *in* "Methods for Preparation of Media, Supplements and Substrata for Serum-Free Animal Cell Culture" (D. W. Barnes, O. A. Sirbasku, and G. H. Sato, eds.), p. 139. Liss, New York, 1984.
[9] S. Cohen and G. Carpenter, *Proc. Natl. Acad. Sci. U.S.A.* **72**, 1317 (1975).
[10] H. Gregory and I. R. Willshire, *Hoppe-Seyler's Z. Physiol. Chem.* **356**, 1765 (1975).
[11] C. R. Savage, Jr., and R. A. Harper, *Anal. Biochem.* **111**, 195 (1981).
[12] H. Gregory, *Nature (London)* **257**, 325 (1975).
[13] C. R. Savage, Jr., T. Inagami, and S. Cohen, *J. Biol. Chem.* **247**, 7612 (1972).

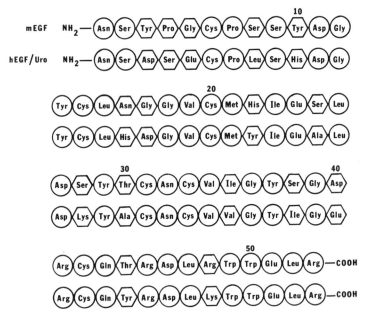

FIG. 1. Primary structures of mouse and human EGF. Amino acid residues shown in circles are identical; amino acid residues shown in hexagons are nonidentical.

amino acid residues cannot be interconverted by a single base change in the triplet code. It is also estimated that guinea pig EGF has a 70% homology with the sequences of mouse and human EGF.[8]

The most abundant source of EGF known is the submaxillary glands of the mouse. It is present in the gland and at levels up to 0.5% of the dry weight. Milligram quantities of EGF can be purified readily from the mouse[14]; however, this is not the case for hEGF. In 1975 Cohen and Carpenter[9] published a procedure for the purification of hEGF using concentrates of urinary proteins (not raw urine). That same year Gregory and Willshire[10] reported a procedure which involved the fractionation of up to 3000 liters of human urine utilizing a 12-step procedure which resulted in relatively low yields of hEGF (3–5%). In 1981[11] our laboratory reported a purification procedure for hEGF using 20 liters of raw urine which was less time consuming and resulted in good yields of highly purified hEGF. We have subsequently refined this method to incorporate the use of a rabbit polyclonal antibody affinity column.[15]

[14] C. R. Savage, Jr., and S. Cohen, *J. Biol. Chem.* **247**, 7609 (1972).

[15] C. R. Savage, Jr., and R. A. Harper, *in* "Methods for Preparation of Media, Supplements and Substrata for Serum-Free Animal Cell Culture" (D. W. Barnes, O. A. Sirbasku, and G. H. Sato, eds.), p. 147. Liss, New York, 1984.

In this chapter we would like to summarize part of our standard published procedure and present a modification that employs the use of monoclonal antibody affinity chromatography. We will discuss the use of monoclonal antibody affinity chromatography in the light of other published reports using this technology. (See also Nishikawa *et al.*, this volume [2].)

Materials and Methods

Human Epidermal Growth Factor or Urogastrone

Five liters of adult human urine is collected from the first morning void, acidified with 250 ml of glacial acetic acid, and the pH adjusted to 3.0 with concentrated HCl. This represents the starting material for the purification procedure described in this chapter.

Monoclonal Antibody to hEGF, 836.D4

The monoclonal antibody (836.D4) used for the purification of hEGF from urine is obtained by fusing murine myeloma cells with BALB/c mouse splenocytes sensitized to hEGF. The full characterization of the monoclonal antibody has been published previously.[16] The antibody does not react with either rat or mouse EGF or with 11 other polypeptide hormones tested as shown by solid phase radioimmunoassay and immunoprecipitation followed by sodium dodecyl sulfate–polyacrylamide gel electrophoresis. Scatchard analysis of the antibody binding to purified hEGF reveals an apparent equilibrium dissociation constant of 1×10^{-8} M. The antibody blocks both the binding of hEGF and hEGF stimulation of [^{3}H]thymidine incorporation into DNA by $>90\%$ in confluent cultures of human foreskin fibroblasts.

Purification of Monoclonal Antibody by DEAE Affi-Gel Blue Chromatography

Ascites fluid is centrifuged at 2000 g for 5 min to remove cells and then centrifuged at 100,000 g for 30 min. The fluid is then dialyzed overnight at 4° against 100 vol of 0.02 M Tris–HCl, pH 7.2. Purification of the monoclonal antibody is carried out according to Bruck *et al.*[17] All procedures are carried out at 4°. One milliliter of ascites fluid is applied to a 7-ml column (a 10-ml disposable CoStar plastic pipet) of DEAE-Affi-Gel Blue

[16] D. M. Moriarity, R. A. Harper, B. B. Knowles, and C. R. Savage, Jr., *Hybridoma* 2, 321 (1983).
[17] C. Bruck, D. Portetelle, C. Glineur, and A. Bollen, *J. Immunol. Methods* 53, 313 (1982).

(Bio-Rad, Richmond, CA). The column is washed with 3 column volumes of 0.02 M Tris–HCl, pH 7.2, and then with 3 column volumes of the column buffer containing 25 mM NaCl. The monoclonal antibody is then eluted with 3 column volumes of buffer containing 50 mM NaCl. The flow rate is approximately 30 ml/hr and 4-ml fractions are collected. The eluted fractions are tested for monoclonal antibodies directed toward hEGF by our previously published solid phase radioimmunoassay.[16]

Radioreceptor Assay for hEGF

The assay is conducted in 24-multiwell tissue culture plates using confluent monolayers of human skin fibroblasts as previously described.[11] Briefly, a standard curve is generated in a final volume of 0.2 ml of binding medium containing 0.5 mg of ^{125}I-labeled mEGF (~1 × 10^5 cpm) and 0 to 10 ng of nonlabeled mEGF. After 40 min at 37° the binding medium is removed and the cells washed four times with ice-cold buffer. The plates are drained, the cells are dissolved in 1.0 ml of 10% NaOH for 10 min, and the amount of radioactivity bound is then determined. Nonspecific binding is determined in the presence of 1000-fold excess of nonlabeled mEGF. This value, which is less than 6% of the total amount bound, is subtracted from each point. Using this assay, we can routinely measure 0.5 ng of hEGF.

Preparation of the Monoclonal Antibody Affinity Gel

Monoclonal antibody 836.D4 is coupled to Affi-Gel 10 (Bio-Rad; Richmond, CA) according to the manufacturer's protocol. In short, 2 ml of gel is washed three times with ice-cold isopropanol and then three times with ice-cold distilled water. Purified monoclonal antibody (20 mg) is added to the gel in 0.12 M HEPES, pH 7.5, containing 0.15 M NaCl and the mixture is rotated overnight at 4°. The Affi-Gel 10 is sedimented by centrifugation and treated at room temperature for 1 hr with 1 M ethanolamine, pH 8.0, in order to block any active esters which might remain. The Affi-Gel is then washed extensively with 0.05 M phosphate buffer, pH 8.0, containing 0.15 M NaCl, and stored at 4° in the same buffer containing 0.1% NaN$_3$ as preservative.

Results

Premonoclonal Antibody Affinity Column Purification Steps

Step 1: Batch Bio-Rex 70. Five liters of adult human urine contain approximately 150–200 μg of hEGF as determined by the radioreceptor

TABLE I
PURIFICATION OF HUMAN EGF FROM URINE

Procedure	Total volume (ml)	Concentration of hEGF (μg/ml)[a]	Total hEGF (μg)	Average yield (%)
Urine	5000	0.03–0.04	150–200	100
Batch Bio-Rex 70	12[b]	12–18	144–216	100
Ethanol precipitation	10	15–20	150–200	100
Passage over DE-52 Cellulose	10[b]	14–18	140–180	>90
Monoclonal antibody affinity column	1	80–103	80–103	52

[a] Determined by the radioreceptor assay, assuming that hEGF and mEGF compete equally for binding of [125] I-labeled mEGF to the membrane receptor on human skin fibroblasts.

[b] Indicates volumes after lyophilization and resuspension of the dry residue. This was necessary since the presence of high concentrations of ammonium acetate interfere with the radioreceptor assay.

assay (Table I). At the outset, we attempted to adsorb the EGF directly from the urine using the monoclonal antibody coupled to Affi-Gel 10. We were unsuccessful in getting the EGF to bind to the affinity matrix. We therefore decided to use the first three steps in our previously published procedure in order to remove large amounts of brown – black pigments and protein.

Briefly, Bio-Rex 70 ion-exchange resin (Bio-Rad; Richmond, CA) is suspended in water and the pH is adjusted to 3.0 with glacial acetic acid. Sixty milliliters of settled resin is added to the 5 liters of urine and the mixture is stirred for 18 hr at 4°. The resin is then allowed to settle and the supernatant fraction discarded. The resin is transferred into a 4-cm-diameter glass column and washed with 3.6 liters of 10^{-3} N HCl at 25°. The EGF is then eluted with 240 ml of 1.0 M ammonium acetate, pH 8.0. The eluate, containing 144–216 μg of EGF, is then lyophilized (Table I).

Step 2: Ethanol Precipitation. Pepstatin (0.5 mg) is added and the dry residue is suspended at 25° in 10 ml of a 2 mM aqueous solution containing 50 mg of crystalline bovine serum albumin. To this mixture 125 ml of absolute ethanol is added. The resulting precipitate is collected by centrifugation and the clear brown supernatant fraction is discarded. The pellet is suspended in 5 ml of 2 mM arginine and the pH is adjusted to 3.0. The mixture is recentrifuged and the clear dark brown supernatant fraction (10 ml) containing 150–200 μg of EGF (Table I) is kept.

Step 3: Passage over DE-52 Cellulose. A 4 × 5 cm column of DE-52 cellulose (Whatman) is prepared and equilibrated at 25° with 0.05% formic acid, pH 3.0. The supernatant fraction from the previous step is applied to the column and the column washed with 135 ml of 0.05% formic acid, pH 3.0. Under these conditions the EGF is not adsorbed to the cellulose. The column effluent is collected on ice and then lyophilized. The residue is suspended in 10 ml of 0.05 M phosphate-buffered saline, pH 8.0. The mixture is centrifuged and the brown supernatant fraction containing 140–180 μg EGF is retained. Through this stage of purification greater than 90% of the starting amount of EGF is recovered (Table I).

Monoclonal Antibody Affinity Column. All procedures are carried out at room temperature. A 2-ml column (a 5-ml disposable Costar plastic pipet) of monoclonal antibody Affi-Gel 10 is prepared. The column is washed successively with 3 column volumes of 0.05 M phosphate-buffered saline, pH 8.0 (column buffer), 1 M acetic acid, and column buffer. Ten milliliters of the preparation from the above purification step is applied to the column. The column is then washed with 10 ml of column buffer followed by 6 ml of 0.01 M ammonium acetate, pH 6.5, to remove the salt that is present in the column buffer. The EGF is then stripped from the column with 6 ml of 1 M acetic acid and 1-ml fractions are collected.

The pattern of electrophoretic mobility of the two major forms of hEGF together with mouse and rat EGF is shown in Fig. 2. It can be seen that monoclonal antibody affinity chromatography yields a mixture of the two major forms of hEGF in a highly purified state.[11]

The preparation of hEGF from the affinity column is assayed for biological activity by measuring the growth stimulation of human foreskin fibroblasts in culture. At 10 ng/ml hEGF stimulates growth of these cells approximately 2-fold (Table II). At 1 ng/ml a 30% increase in cell growth is observed.

Discussion

The procedure we have described in this chapter permits the isolation of approximately 100 μg of hEGF from 5 liters of raw urine with an average yield of 50%. As shown in Fig. 2, the hEGF elutes from the monoclonal antibody affinity column in a highly purified state as two major forms, hEGF-1 and hEGF-2.[11] These two forms of hEGF can be easily separated by DEAE–cellulose chromatography as described previously.[11] Interestingly, the minor form of hEGF from urine, hEGF-A, found previously utilizing a rabbit antibody affinity column,[15] was not observed in preparations using the monoclonal antibody affinity column. It could be that this minor form of hEGF has a weak binding affinity for

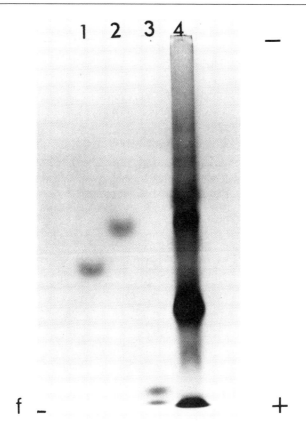

FIG. 2. Polyacrylamide gel electrophoresis of mouse, rat, and human EGF. Polyacrylamide gel electrophoresis was carried out in standard 7.5% slab gels at pH 9.5.[18] The samples were electrophoresed at 25° for 3.5 hr at 20 mA per gel and the gels were subsequently stained for 10 min with Coomassie brilliant blue R250 (0.24% w/v solution in methanol/H_2O/acetic acid, 5/5/1). The gel was destained by soaking in methanol/H_2O/acetic acid, 4/33/3. Lanes 1 and 2 contained 5 μg of mouse EGF and rat EGF, respectively [isolated by the method of R. P. Schaudies and C. R. Savage, Jr., Comp. Biochem. Physiol. **84B,** 497 (1986)]; lane 3 contained about 8 μg of a mixture of hEGF-1 (slower migrating) and hEGF-2[11] (faster migrating); and lane 4 contained 20 μl of the hEGF preparation just prior to affinity chromatography. Running dye front is labeled "f."

the monoclonal antibody or the antibody is reactive with an epitope of hEGF-1 and hEGF-2 which is not present on hEGF-A. We believe hEGF-1 is the intact molecule, whereas, hEGF-2 lacks the carboxy-terminal Arg or Leu-Arg residues.[11] The structure of hEGF-A is not known.

[18] B. J. Davies, Ann. N.Y. Acad. Sci. **121,** 404 (1964).

TABLE II
EFFECT OF EGF ON HUMAN FORESKIN FIBROBLAST
GROWTH[a]

Additions	Number of cells/dish × 10^6 (day 11)
None	1.82
mEGF; 10 ng/ml	3.68
hEGF ; 10 ng/ml	3.64
hEGF ; 1 ng/ml	2.37

[a] Human foreskin fibroblast (passage #5), 4×10^5 cells, were seeded into 100-mm CoStar tissue culture dishes in 5 ml of Dulbecco's modified Eagle's essential medium containing 100 U/ml penicillin, 100 μg/ml streptomycin, and 5% fetal calf serum. The next day EGF was added to the cultures in 50 μl Hanks' BSS containing 0.2% bovine serum albumin and 25 mM HEPES, pH 7.4. Control dishes received only buffer. On the fourth, seventh, and ninth days after seeding, the culture fluid was removed from all dishes and replaced with 5 ml of fresh culture medium plus EGF. On the eleventh day cell numbers were determined using a hemocytometer and trypan blue dye exclusion to monitor viability. The values in the table represent the average between two dishes. The difference in cell counts between duplicate dishes was < 15%.

An attempt was made to use the immobilized monoclonal antibody at the beginning of the purification procedure, thereby hoping to eliminate the first three steps of our purification protocol. Addition of immobilized antibody in a batch-wise fashion directly to neutralized urine with stirring overnight was unsuccessful due to poor absorption of the hEGF to the antibody. Likewise, passage of neutralized urine directly over a monoclonal antibody affinity column was not effective when large volumes (20–40 liters) of urine were used. Thus, we find it necessary to remove large amounts of the insoluble salts, denatured protein, and pigments before using the monoclonal antibody affinity column. Our first three steps consisting of batch Bio-Rex 70, ethanol precipitation, and DE-52 cellulose chromatography are rapid, highly efficient (a 500-fold concentration of hEGF with a >90% recovery), and can easily be used to work up large volumes of raw urine (40–80 liters).

In a recent paper by Hissey *et al.*[19] the authors reported the purification of human epidermal growth factor/urogastrone from urine utilizing affinity chromatography with a monoclonal antibody raised against purified recombinant human urogastrone. Their yields of the two major forms of β- and γ urogastrone, were similar to the yields we obtained (~20 μg/liter of urine). However, they used 0.2 M glycine · HCl buffer to elute the hEGF from the column, which upon lyophilization results in salts being present in the preparation. Our use of 1 M acetic acid to strip the hEGF from the affinity column allows, upon lyophilization, a pure preparation of hEGF.

The utilization of monoclonal antibody affinity chromatography for the purification of hEGF/urogastrone has many obvious advantages over conventional purification techniques. However, it still remains a formable task to obtain relatively large quantities of first morning void (i.e., 40–80 liters) in order to obtain 1–2 mg of highly purified hEGF. Although in the future the major source of hEGF will likely be recombinant derived material, monoclonal antibody affinity chromatography will still be a necessary tool in isolating native material from raw urine for comparative studies.

[19] P. H. Hissey, K. J. Thompson, and L. Bawden, *J. Immunol. Methods* **78**, 211 (1985).

[2] Derivation of Monoclonal Antibody to Human Epidermal Growth Factor

By KATSUZO NISHIKAWA, YOSHINO YOSHITAKE, and SHIGERU IKUTA

Köhler and Milstein first developed techniques for the production of monoclonal antibody (MAb) by hybridization of spleen cells from immunized mice with myeloma cell lines, and many investigators have since used these techniques for various studies.[1] Monoclonal antibodies are useful in studies of growth factors because they are specific for various antigens, homogeneous, and easy to produce.

Epidermal growth factor (EGF) is a strong stimulator of numerous cell types in culture.[2] Mouse EGF (mEGF), isolated from mouse submaxillary gland, and human EGF (hEGF), isolated from human urine, have nearly identical biological activities. These polypeptides both consist of single chains of 53 amino acids and their peptide sequences show 70% homology with 3 disulfide bonds in the same relative positions. This Chapter de-

[1] G. Galfré and C. Milstein, this series, Vol. 73, p.3.
[2] S. Cohen, this series, Vol. 37, p. 425.

scribes techniques for preparation of diverse monoclonal antibodies to hEGF and for studies on the properties of these antibodies.

Preparation of hEGF

A preparation of hEGF (20% purity) from human urine (Japan Chemical Research Co., Kobe, Japan) in 0.15 M NaCl solution is used for immunization of mice. For assay of anti-hEGF activity in the serum of immunized mice and medium of cultured hybridoma cells and for study of the properties of monoclonal antibodies, hEGF is purified further by the method of Savage and Harper[3] (see also Harper *et al.*, this volume [1]) and iodinated as follows: Partially purified hEGF (4.5 mg hEGF equivalent) in 0.15 M NaCl solution is dialyzed against 0.02 M ammonium acetate buffer (pH 5.3) in Spectrapor dialysis tubing with a molecular weight cut-off of 3500. The dialyzate is applied to a column of DE-52 cellulose (2 × 10 cm) equilibrated with the same solution at room temperature, and material is eluted with a linear gradient formed with each 160 ml of 0.02 and 0.3 M ammonium acetate buffer (pH 5.3). Most of the hEGF, determined by assay of activity for stimulating DNA synthesis (see below), appears in fractions eluted with 0.12 and 0.2 M ammonium acetate buffer in a ratio of 1:3. As suggested by Savage and Harper[3] for mEGF, the first and the second peaks appear to contain the native form of hEGF and that lacking one or two C-terminal amino acids, respectively. The first fraction is lyophilized, dissolved in H_2O, and used for radioiodination.

When a MAb to hEGF has been obtained, hEGF can be purified by affinity chromatography on a column of Sepharose-coupled MAb. The purified MAb (30 mg; HA, see below) is coupled to cyanogen bromide-activated Sepharose 4B (2.5 g) by the method of Mescher *et al.*[4] Partially purified hEGF (1.8 mg hEGF equivalent) in 0.15 M NaCl solution is diluted 10 times with 0.1 M sodium phosphate buffer (pH 7.4) and applied to a column (8 ml) equilibrated with 0.1 M sodium phosphate buffer (pH 7.4) at room temperature. The column is washed with 40 ml of 0.2 M ammonium formate and hEGF is eluted with 1 M acetic acid. This fraction is lyophilized, and the residue is dissolved in H_2O. The Sepharose-coupled MAb can be regenerated by washing with 0.1 M sodium phosphate buffer (pH 7.4) and reused at least five times without destroying binding capacity. Although the preparation of hEGF purified by this chromatography can be used for studies of the properties of MAb, it is further

---

[3] C. R. Savage, Jr., and R. Harper, *Anal. Biochem.* **111**, 195 (1981).
[4] M. F. Mescher, K. C. Stallcup, C. P. Sullivan, A. P. Turkewitz, and S. H. Herrmann, this series, Vol. 92, p. 86.

separated into at least five distinct species by reversed-phase HPLC. Five protein peaks, which show equal activities for stimulating DNA synthesis in BALB/c 3T3 cells and equal immunnoreactivities (see below), are separated when the preparation is chromatographed on an ODS column (Dupon, 0.46 × 25 cm) at room temperature in a linear gradient of 0.1 M sodium phosphate buffer (pH 2.3) to that containing 50% acetonitrile. The relationship between these hEGF species and the mEGF species described by Burgess et al.[5] and Petrides et al.[6] is unclear.

Radioiodination of hEGF

Purified hEGF is labeled with ^{125}I by the chloramine-T method of Aharonov et al.[7] with slight modifications. Two micrograms of hEGF in 30 μl of 0.3 M sodium phosphate buffer (pH 7.4), 5 μl of 20 mCi/ml solution of Na^{125}I and 20 μl of 1 mg/ml of chloramine-T solution in 0.05 M sodium phosphate buffer (pH 7.4) are mixed and incubated for 1 min at room temperature. The amounts of hEGF and Na^{125}I can be scaled up 5- to 10-fold, if necessary. The reaction is stopped by adding 60 μl of a solution of sodium metabisulfite at 1 mg/ml in 0.05 M sodium phosphate buffer (pH 7.4), followed by successive additions of 130 μl of KI at 10 mg/ml and 200 μl of bovine serum albumin (BSA) at 20 mg/ml in phosphate-buffered saline (PBS). The reaction mixture is passed through a column of Sephadex G-25 (0.7 × 28 cm) in PBS. The iodinated hEGF (^{125}I-labeled hEGF, 220 Ci/mmol) thus separated can be stored at −20° and is used for assays after appropriate dilution. ^{125}I-labeled hEGF and unlabeled hEGF have identical biological activities as judged by stimulation of DNA synthesis in BALB/c 3T3 cells (see below).

Production of Monoclonal Antibodies[8,9]

Media and Cell Lines

Media
RPMI-1640 + 10% FCS for maintenance of myeloma and hybridoma cell lines: 90% RPMI-1640 (Flow Laboratories) supplemented with

[5] A. W. Burgess, J. Knesel, L. G. Sparrow, N. A. Nicola, and E. C. Nice, Proc. Natl. Acad. Sci. U.S.A. **79**, 5753 (1982).
[6] P. E. Petrides, P. Böhlen, and J. E. Shively, Biochem. Biophys. Res. Commun. **125**, 218 (1984).
[7] A. Aharonov, R. M. Pruss, and H. R. Herschman, J. Biol. Chem. **253**, 3970 (1978).
[8] S. Ikuta, Y. Yamada, Y. Yoshitake, and K. Nishikawa, Biochem. Int. **10**, 251 (1985).
[9] Y. Yoshitake and K. Nishikawa, in "Growth and Differentiation of Cells in Defined Environment" (H. Murakami, I. Yamane, D. W. Barnes, J. P. Mather, I. Hayashi, and G. H. Sato, eds.), p. 135. Kodansha/Springer-Verlag, Tokyo/Berlin, 1985.

2 g/liter of sodium bicarbonate, 100 U/ml of penicillin, 100 μg/ml of streptomycin and 15 mM N-hydroxyethylpiperazine-N'-2-ethanesulfonic acid buffer (HEPES) (pH 7.3); 10% fetal calf serum (FCS) (Flow Laboratories)

RPMI-1640 + 10% FCS + HAT for selection of hybridoma cells: RPMI-1640 + 10% FCS supplemented with HAT (10^{-4} M hypoxanthine, 4×10^{-7} M aminopterin, and 1.6×10^{-5} M thymidine). Before use, 100 × stock solutions of these reagents are added to RPMI-1640 medium

RPMI-1640 + 10% FCS + HT for growth of hybridoma cells: RPMI-1640 + 10% FCS + HAT, but without aminopterin

Cell Lines

Mouse myeloma P3-NSI-IAg4-1 (NS-1) and P3-X63-Ag8-U1 (P3U1) cells are kept in RPMI-1640 + 10% FCS and subcultured every 3–4 days

Immunization of Mice

Mice are readily immunized with hEGF. Female BALB/c mice of 1–2 months old are immunized by four injections of partially purified hEGF. On days 0 and 7, the mice are subcutaneously injected with 20 μg of hEGF emulsified with Freund's complete adjuvant. On day 14, they are subcutaneously injected with the same amount of hEGF emulsified with Freund's incomplete adjuvant. On day 21, a drop of blood is taken from the retroorbital venous plexus and the serum is separated and tested for activity to bind to labeled hEGF as described below. If the titer is high enough, the mice are given 20 μg of hEGF dissolved in 200 μl of 0.15 M NaCl solution intraperitoneally. Three days later, they are killed and their spleen is excised.

Generation of Hybridomas

Since the detailed procedures for cell fusion to generate monoclonal antibodies have been described in this series,[1] the procedures for hybridization and cloning are mentioned here only briefly.

A. Hybridization and Culture of Hybridoma Cells

1. Mixtures of 10^8 spleen cells from immunized mouse and 5×10^7 NS-1 cells are fused in 0.5 ml of 50% (w/v) polyethylene glycol 1540 (J. T. Baker Chemicals, B. V., Deventer, Holland) in DME containing 5% (v/v) dimethyl sulfoxide for 90 sec at 37°. P3U1 cells can also be used instead of NS-1 cells.

2. The fused cells are washed with DME by centrifugation and resuspended in RPMI-1640 + 10% FCS + HAT medium.

3. A suspension of the cells in 0.5 ml of medium (equivalent to about 5×10^5 original spleen cells) is distributed into the wells of eight 24-well multiwell plates (Falcon, 2 cm^2) in which mouse peritoneal macrophage cells (about 1.5×10^4 cells/well) or spleen cells (about 10^6 cells/well) have been cultured for 1–2 days in 0.5 ml of the same medium as a feeder layer.

4. The cells are cultured at 37° in a humidified atmosphere of 5% CO$_2$ in air.

5. Half the medium of each well is replaced by fresh medium on days 4 and 8.

6. The medium is replaced by RPMI-1640 + 10% FCS + HT medium on days 11 and 15.

7. The medium is replaced by RPMI-1640 + 10% FCS on days 19 and 21.

8. Hybridoma cell growth should be observable in all wells if conditions have been correct.

9. A sample of 0.1 ml of the conditioned medium is taken for assay of antibody activity.

B. Assay of Anti-hEGF Activity in Culture Medium
Reaction Mixture
0.3 ml of 0.1 M sodium phosphate buffer (pH 7.5)
0.05 ml of calf serum (CS)(Flow Laboratories)
0.05 ml of 10 ng/ml of ^{125}I-labeled hEGF (20,000–30,000 cpm/ng) in PBS containing 0.1% BSA (PBS/BSA)
0.1 ml of culture medium

1. The reaction mixture in a minisorp tube (Nunc) is incubated for 12 hr at 4°.

2. Then 1 ml of 25% polyethylene glycol 6000 (w/v) in 0.1 M sodium phosphate buffer (pH 7.5) is added to the reaction mixture.

3. The mixture is incubated for 30 min at 4°.

4. The mixture is centrifuged at 1200 g for 30 min and the supernatant is discarded.

5. The radioactivity in the precipitate is counted in a gamma counter.

The value for nonspecific binding of radioactivity (about 7% of the total tracer added) in the absence of the medium is subtracted. When the corrected value for specific binding of ^{125}I-labeled EGF is above 10% of the total tracer added, the well can be judged to contain antibody.

C. Cloning
1. The cells in the wells showing antibody activity are suspended in, and diluted with, RPMI-1640 + 10% FCS.

2. A suspension of the cells in 0.1 ml of medium (50 cells/ml) is distributed into the wells of 96-well multiwell plates (Falcon, 0.32 cm^2) in

which thymocytes from normal mice (about 10^6 cells/well) have been cultured in 0.1 ml of the same medium for 1–2 days as a feeder layer.

3. On day 10, when cell growth can be observed, 0.1 ml of the conditioned medium is taken for assay of the antibody activity.

4. The cells in wells in which the medium shows activity are used for recloning by limiting dilution (1 cell/well) by the same method as described above.

5. The cells in wells in which a single colony is observed at an early stage (about 7 days after inoculation) and in which the medium shows antibody activity are subcultured to obtain enough cells for injection into mice.

Determination of Immunoglobulin Classes

A large amount of MAb can be purified from the ascites fluid of mice inoculated with hybridoma cell lines (see below). However, it is desirable to determine the class and subclass of the MAb in the culture medium of the hybridoma cells, because immunoglobulins in ascites fluid from host mice sometimes interfere with the determination.

The classes and subclasses can be determined by the double immuno-diffusion technique (micro-Ouchterlony method). Culture medium (5 ml) from confluent cultures is centrifuged at 1000 g for 10 min, and the supernatant is lyophilized. The dried material is dissolved in 0.5 ml of H_2O.

Micro-Ouchterlony Method

1. Microscope slides (2.6 × 7.6 cm) are covered with 3 ml of 1.2% agar in PBS containing 0.1% NaN_3.

2. One central well and four to six surrounding wells (3–4.5 mm in diameter) 7–9 mm from the central well are cut out with a puncher.

3. Then 10–25 μl of concentrated medium is put in the central well, and 10–25 μl of rabbit or goat antiserum to mouse immunoglobulin heavy chains, IgM, IgA, IgG_1, IgG_{2a}, IgG_{2b}, IgG_3 and light chains, κ and λ (Miles and Nordic Immunological Laboratories) are put in surrounding wells.

4. Precipitin lines are developed by incubating the plate for about 12 hr at room temperature in a humidified box.

5. The slides are put in 30 ml of 0.87% NaCl solution containing 0.02% NaN_3 and the agar gels are stripped from the slide glasses. The gels are washed in 30 ml of the same solution by gentle shaking with four changes of solution over a 6-hr period.

6. The gels are stained with 0.25% Coomassie Brilliant Blue R250 solution in 50% methanol containing 10% acetic acid for 30 min at room temperature.

7. The gels are washed with 10% methanol containing 10% acetic acid.
8. The stained gels are put on new slide glasses and dried.

Purification of Monoclonal Antibodies

Monoclonal antibodies can be purified from either conditioned medium of the hybridoma cells or ascites fluid of mice bearing the hybridoma. Comparison of these two methods has been discussed in detail in this series.[1] In the present study we chose ascites fluid, since the concentration of the MAb in the fluid is usually about 500 times higher than that in conditioned medium.

Hybridoma cells (3×10^6) are injected intraperitoneally into a female BALB/c mouse that has received an intraperitoneal injection of 0.5 ml of pristane 4–10 days previously. The ascites fluid is obtained 10 days later. Usually 2–5 ml of fluid can be obtained from one mouse. When ascites fluid is collected without killing the animal, remaining hybridoma cells in the abdomen grow again. Therefore, it is possible to obtain ascites fluid from the same mouse 2–3 times. The method to produce MAb from ascites fluid is also described in this series.[1,10]

Monoclonal antibodies of the IgG class can be purified by the method as described in this series,[10] except that the step of Sephadex G-200 chromatography is omitted. This procedure involves precipitation with 50% saturated ammonium sulfate and DE-52 cellulose column chromatography, which is suitable for all IgG subclasses of MAb that usually arise. The purity of the final preparation can be evaluated by SDS–polyacrylamide gel (12.5%) electrophoresis. The reduced sample of the purified MAb gives two sharp bands of protein corresponding to heavy and light chains of immunoglobin. The antibody concentration is conventionally determined by measuring the absorbancy at 280 nm, assuming $A_{1\%} = 14$. The yield of MAb from 25 ml of ascites fluid is 75–450 mg. The NaCl concentration in the pooled active fractions is adjusted to 0.15 M and the solution can be stored at $-20°$.

Properties of Monoclonal Antibodies[8,9,11]

Determination of Specificity and Dissociation Constant

Reaction Mixture
0.3 ml of 0.1 M sodium phosphate buffer (pH 7.5)
0.05 ml of CS

[10] P. Parham, this series, Vol. 92, p. 110.
[11] Y. Yoshitake, K. Nishikawa, and K. Adachi, *Cell Struct. Funct.* 7, 229 (1982).

  0.05 ml of 10 ng/ml of ^{125}I-labeled hEGF (20,000–30,000 cpm/ng) in PBS/BSA

  0.05 ml of unlabeled hEGF or mEGF in PBS/BSA to give a final concentration of 0.5–500 ng/ml

  0.05 ml of 0.08–1.2 μg/ml of purified MAb in PBS/BSA

The procedure for this competitive binding experiment is the same as that for assay of anti-hEGF activity in the culture medium (see Generation of Hybridomas, part B). Values of B (specific binding of ^{125}I-labeled hEGF in the presence of various concentrations of unlabeled EGF) divided by B_0 (specific binding in the absence of unlabeled EGF), that is B/B_0, are plotted against the logarithm of the concentration of unlabeled EGF. The value for nonspecific binding (usually 5–7% of the total tracer added regardless of the presence and absence of unlabeled EGF) in the absence of MAb should be subtracted. The concentration of MAb in the reaction mixture is adjusted to give a B_0 value of 20–30% of the total tracer added. The apparent K_d value can be calculated by Scatchard plot analysis of the displacement curves of hEGF.

Competitive Binding of Various Monoclonal Antibodies to hEGF

Competitive antibody binding to the antigen (solid phase antibody binding assay) is useful for determining whether different MAb's are specific for different epitopes of hEGF molecule. A strategy for distinction of epitopes by MAb's is described in detail in this series.[12] Here a method using hEGF-coated wells of microtiter plates and MAb's labeled with ^{125}I is described.

A. Radioiodination of Monoclonal Antibodies

Purified MAb's are labeled with ^{125}I by the method described above (see Radioiodination of hEGF); 8–10 μg of MAb is treated with Na[125]I in the presence of chloramine-T in 60 μl of reaction mixture. The specific activity of the ^{125}I-labeled monoclonal antibodies fractionated by Sephadex G-25 is 5–10 mCi/mg.

B. Preparation of hEGF-Coated Wells

1. Wells of a Micro Test III flexible assay plate (Falcon 3912) are coated for 2 hr at room temperature with 50 μl of 2 μg/ml of purified hEGF in PBS.

2. The wells are filled with 3% BSA solution in PBS, and left to stand for 15 min.

[12] C. Stähli, V. Miggiano, J. Stocker, Th. Staehelin, P. Häring, and B. Takács, this series, Vol. 92, p. 242.

3. The liquid is pipetted off.
4. The wells are washed three times with 0.3 ml of PBS.

The coated plates can be stored at 4° for several weeks.

C. Competitive Binding

1. Samples of 50 μl of various concentrations of each unlabeled MAb (0.2–100 μg/ml) in PBS containing 1% BSA are added to the coated wells. Diluted ascites fluid or concentrated culture medium can also be used.

2. Then 50 μl of each labeled MAb diluted with PBS containing 1% BSA (2 μg/ml, $5 \times 10^5 - 10^6$ cpm/μg) is added.

3. The wells are incubated for 2 hr at 37°.

4. The liquid is sucked off.

5. The wells are washed four times with 0.3 ml of PBS.

6. The plates are cut up and the radioactivities of individual wells are counted in a gamma counter.

The value for nonspecific binding of radioactivity (about 0.2% of the total tracer added) with hEGF-noncoated well is subtracted. The specific binding of radioactivity decreases with increase in the concentration of homologous MAb. It also decreases in the presence of heterogeneous MAb that binds to a common domain of the hEGF molecule or to a neighboring domain. Heterogeneous unlabeled MAb that binds to a distant domain but prevents the binding of the labeled MAb may also cause competition. Nevertheless, when an unlabeled MAb does not compete with another labeled MAb, it is clear that the two MAb's bind to different domains.

Effect of Monoclonal Antibodies on Biological Activities of hEGF

Addition of hEGF to cultured quiescent BALB/c 3T3 cells stimulates DNA synthesis in the cells. hEGF also binds to a receptor of the cells. Therefore, it is possible to determine whether the MAb's inhibit these biological activities of hEGF.

A. Cell Lines and Medium

BALB/c 3T3 cells are kept in standard growth medium of the following composition: 90% DME (Dulbecco's modified Eagle's medium, Flow Laboratories) supplemented with 1.2 g/ml of sodium bicarbonate, 15 mM HEPES, 100 U/ml of penicillin, 100 μg/ml of streptomycin; 10% CS. The cells are subcultured every 3–4 days and grown at 37° in a humidified atmosphere (5% CO_2 in air). The cells should be checked for contamination with *Mycoplasma.*

B. Assay of DNA Synthesis

1. BALB/c 3T3 cells in subconfluent culture in standard growth medium are trypsinized.

2. The cells are suspended in DME containing 3% CS and inocula of 10^5 cells are introduced into 60-mm Falcon plates with 5 ml of the medium.

3. After 5 hr, the medium is replaced by 5 ml of DME containing 0.2% CS.

4. After 24 hr, 5 μl of 0.5 μg/ml of hEGF in PBS/BSA (final concentration, 0.5 ng/ml) and purified MAb (1–3 mg/ml) at a final concentration of 2–25 μg/ml are added.

5. After 16 hr, 50 μl of 2×10^5 M [^{3}H]thymidine (Radiochemical Centre, Amersham) (20 μCi/ml) is added.

6. After 3 hr, the medium is sucked off and the cells are washed twice with 3 ml of PBS and then three times with 3 ml of 10% cold trichloroacetic acid (w/v).

7. DNA is extracted with 1 ml of 0.5 N NaOH.

8. A volume of 0.5 ml of the extract is mixed with 6 ml of toluene-based scintillator containing 33% Triton X-100 (v/v) and 0.83% trichloroacetic acid (w/v).

9. Radioactivity is counted in a liquid scintillation counter.

All measurements should be done at least in duplicate. Since DNA synthesis starts at about 10 hr and ceases at about 24 hr after adding hEGF, almost maximal incorporation can be observed from 16 to 19 hr. The incorporation of [^{3}H]-thymidine in the presence of 0.5 ng/ml of hEGF, which gives the maximal stimulation, is about 200,000 dpm/plate/3 hr. The control value in the absence of hEGF is 2000–8000 dpm/plate/3 hr. The purified MAb itself has no effect on DNA synthesis. We prefer this assay system using quiescent BALB/c 3T3 cells in sparse culture to that of cells in confluent culture, since the concentration of hEFG for maximal stimulation is about 20 times less than that for confluent cells.

C. Assay of [^{125}I]EGF Binding

1. Inocula of 2×10^4 BALB/c 3T3 cells harvested as described above are introduced into 24-well multiwell plates (Falcon, 2 cm^2) with 1 ml of DME containing 10% CS and grown to confluency, which takes 4–6 days.

2. The confluent monolayer is washed twice with 1.5 ml of PBS.

3. A volume of 0.4 ml of DME containing 0.1% BSA (binding medium) is added to each well.

4. A volume of 50 μl of 0.1 μg/ml ^{125}I-labeled hEGF in PBS/BSA (20,000–30,000 cpm/ng) is added.

5. A volume of 50 μl of MAb (5–100 μg/ml) in binding medium is added.

6. The wells are incubated for 90 min at 4°.

7. The wells are washed three times with 1 ml of cold PBS.

8. The material in the well is solubilized by incubation with 1 ml of 0.5 N NaOH for 10 min at room temperature.

9. Radioactivity is counted in a gamma counter.

All measurements should be done at least in duplicate. The value for nonspecific binding in the presence of excess unlabeled hEGF (usually 0.08–0.2% of the total tracer added) is subtracted. Specific binding of radioactivity in the absence of MAb amounts to 0.3–1% of the total tracer added. ^{125}I-labeled hEGF binding can also be assayed by the same procedure with A431 cells, which express about 50 times more receptors than BALB/c 3T3 cells.

Example

We produced four different monoclonal antibodies to hEGF having the properties shown in Table I. In competitive binding assay of these MAb's with labeled hEGF , displacement is observed with unlabeled hEGF at concentrations of 0.5–50 ng/ml. Little or no displacement is observed with unlabeled mEGF even at 500 ng/ml, indicating that these MAb's are specific for hEGF. These MAb's are named HA, LA', LB and LC, where H and L indicate high and low affinity to hEGF, respectively, and A, A', B, and C indicate different binding domains on the hEGF molecule. The properties of these MAb's are deduced from the results of competitive binding experiments with hEGF. The unlabeled MAb's competed with the respective labeled MAb's. Neither HA nor LA' competed with labeled LB,

TABLE I

PROPERTIES OF MONOCLONAL ANTIBODIES TO hEGF

Monoclonal antibody	Myeloma	Class	$K_d/(M)$	Competitive binder to hEGF[a]	Inhibition	
					DNA synthesis[b]	hEGF receptor[c]
HA	P3U1	G_1, κ	1.7×10^{-10}	(HA), LA', LC	++	++
LA'	P3U1	G_1, κ	4.2×10^{-9}	(LA'), HA, LC	+	+
LB	NS-1	G_1, κ	1.4×10^{-9}	(LB), LC	++	+
LC	NS-1	G_{2a}, κ	2.4×10^{-9}	(LC), HA, LA', LB	+	+

[a] Competitive curves for binding of labeled MAb with unlabeled MAb's including the homologous and heterogeneous ones indicated are similar. The concentrations of unlabeled MAb's for half maximal replacement are 0.5–1 μg/ml.

[b] Concentration of MAb for half-maximal inhibition: ++, 0.1–1 μg/ml; +, 10–15 μg/ml.

[c] Concentration of MAb for half-maximal inhibition: ++, 0.2 μg/ml; +, 0.5–1 μg/ml.

and LB did not compete with either labeled HA or labeled LA'. On the other hand, LC competed with all the MAb's. From these results, there seem to be three different binding domains on the hEGF molecule: HA and LA' bind to a common domain or neighboring domains (A); LB binds to domain (B), which is too far from domain (A) to cause competition; LC binds to another domain (C), but covers both domain (A) and (B) and so prevents the binding of other MAb's.

These MAb's inhibit the biological activities of hEGF, including its binding to the receptor of BALB/c 3T3 cells and its stimulation of DNA synthesis in the cells, but do not inhibit similar activities of mEGF. The extents of the inhibitory effects of these MAb's appear to be correlated with the K_d values. When hEGF is reduced by dithiothreitol, it loses both its biological activities and its reactivity with all these MAb's, indicating that the conformation of the hEGF molecule, which is maintained by three disulfide bonds, is necessary for both reactivities. These MAb's should be useful for studies on EGF, including the structure–function relationship of this factor in binding to the receptor molecule.

Acknowledgments

This work was supported in part by a grant-in-aid for cancer research by the Ministry of Education, Science and Culture of Japan, and the Science Research Promotion Fund of the Japan Private School Promotion Foundation. We thank Hiromi Nakamura for technical assistance.

[3] Design, Chemical Synthesis, and Molecular Cloning of a Gene for Human Epidermal Growth Factor

By MICKEY S. URDEA

The chemical synthesis of oligodeoxyribonucleotides pioneered by Khorana and co-workers[1] has become an integral and essential part of recombinant DNA methodology. The commercial availability of automated synthesizers and reliable chemical reagents permits synthetic oligonucleotides to be routinely produced in many nonchemically oriented laboratories.

---

[1] E. L. Brown, R. Belagaje, M. J. Ryan, and H. G. Khorana, this series, Vol. 68, p. 109.

Since the first successful construction and genetic expression of a chemically synthesized gene coding for somatostatin,[2] many total human gene syntheses have been reported including growth hormone,[3] insulin,[4] leukocyte interferon,[5] insulin-like growth factors I and II,[6] and lymphotoxin.[7] Given the amino acid sequence of a peptide hormone or growth factor, it is probably easier to synthesize the gene than to isolate it by cDNA cloning (if it contains less than about 350 amino acids). Also, chemical synthesis permits gene design. Codon usage and restriction sites can be tailored for a specific expression system.

This chapter includes the methods used for the chemical synthesis, enzymatic assembly, and expression in yeast of human epidermal growth factor (hEGF).[8,9] A simple manual protocol for rapid oligonucleotide synthesis by a solid supported phosphoramidite method using commercially available reagents is presented for those without access to automated synthesizers. The design of synthetic genes for facile subcloning into expression systems is also described.

Using hEGF as an example, a general method for the one-step enzymatic ligation and subsequent gel purification of a full-length double-stranded gene constructed from several single-stranded synthetic oligomers is described. The techniques have been used to construct several genes and gene derivatives.[6,10]

[2] K. Itakura, T. Hirose, R. Crea, A. D. Riggs, H. L. Heyneker, F. Bolivar, and H. W. Boyer, *Science* **198**, 1056 (1977).

[3] M. Ikehara, E. Ohtsuka, T. Tokunaga, Y. Taniyama, S. Iwai, K. Kitano, S. Miyamoto, T. Ohgi, Y. Sakuragawa, K. Fujiyama, T. Ikari, M. Kobayashi, T. Miyake, S. Shibahara, A. Ono, T. Ueda, T. Tanaka, H. Baba, T. Miki, A. Sakurai, T. Oishi, O. Chisaka, and K. Matsubara, *Proc. Natl. Acad. Sci. U.S.A.* **81**, 5956 (1984).

[4] R. Crea, A. Kraszewski, T. Hirose, and K. Itakura, *Proc. Natl. Acad. Sci. U.S.A.* **75**, 5765 (1978).

[5] M. D. Edge, A. R. Greene, G. R. Heathcliffe, P. A. Meacock, W. Schuch, D. B. Scanlon, T. C. Atkinson, C. R. Newton, and A. F. Markham, *Nature (London)* **292**, 756 (1981).

[6] G. T. Mullenbach, Q. L. Choo, M. S. Urdea, P. J. Barr, J. P. Merryweather, A. J. Brake, and P. Valenzuela, *Fed. Proc. Fed. Am. Soc. Exp. Biol.* **42**, 434 (1983) (Abstr.).

[7] P. W. Gray, B. B. Aggarwal, C. V. Benton, T. S. Bringman, W. J. Henzel, J. A. Jarrett, D. W. Leung, B. Moffat, P. Ng, L. P. Svedersky, M. A. Palladino, and G. E. Nedwin, *Nature (London)* **312**, 721 (1984).

[8] M. S. Urdea, J. P. Merryweather, G. T. Mullenbach, D. Coit, U. Heberlein, P. Valenzuela, and P. J. Barr, *Proc. Natl. Acad. Sci. U.S.A.* **80**, 7461 (1983).

[9] A. J. Brake, J. P. Merryweather, D. Coit, U. Heberlein, F. R. Masiarz, G. T. Mullenbach, M. S. Urdea, P. Valenzuela, and P. J. Barr, *Proc. Natl. Acad. Sci. U.S.A.* **81**, 4642 (1984).

[10] M. S. Urdea, unpublished results.

Chemical Synthesis of Oligodeoxyribonucleotides

Principles of Method

The two most popular techniques for DNA synthesis are the phosphate triester[11] and phosphite triester[12a-c] methods. Both can be conducted on solid supports either manually or by automated synthesizers. Although commercial[13] or "home-made"[14] instruments are extensively utilized, the cost or availability of them may be prohibitive to some potential users of synthetic DNA. For these persons, the following manual phosphite triester (phosphoramidite[15]) method is easily and reliably performed and can be used to produce oligomers up to about 50 bases. A 20mer can be synthesized in about 3.5 hr.

The method involves four basic steps: detritylation, condensation, oxidation, and capping (see Fig. 1). First, detritylation is the 5'-deprotection of the solid supported nucleoside or growing synthetic oligomer with dichloroacetic acid dissolved in dichloromethane.[16] The 5'-(4,4'-dimethoxytrityl) protecting group (DMT) is the only acid-sensitive blocking functionality used in the synthesis. Second, after washing and drying, a condensation of the appropriate fully protected nucleoside 3'-phosphoramidite is conducted in an acetonitrile solution using 1H-tetrazole as an activator. The third step involves the oxidation of the condensed phosphite nucleoside (3+ oxidation state of phosphorus) to the more stable phosphate form (5+ oxidation state of phosphorus). Fourth, any 5'-hydroxyl functions remaining due to an incomplete condensation are acylated or "capped" with acetic anhydride in order to prevent their future participation in condensation reactions, thereby increasing the overall yield of the desired product and simplifying the purification process. The steps are repeated until the desired sequence of nucleotides is produced. Subsequently, the synthetic material is cleared from the support, and the phosphate and exocyclic nitrogen blocking functionalities are quantitatively removed in base.

[11] S. A. Narang, H. M. Hsiung, and R. Brousseau, this series, Vol. 68, p. 90 and references therein; B. S. Sproat and M. J. Gait, in "Oligonucleotide Synthesis: A Practical Approach" (M. J. Gait, ed.), p. 83. IRL Press, Oxford, England, 1984.

[12a] R. L. Letsinger and W. B. Lunsford, *J. Am. Chem. Soc.* **98**, 3655 (1976).

[12b] M. D. Matteucci and M. H. Caruthers, *J. Am. Chem. Soc.* **103**, 3185 (1981).

[12c] For an alternative method for manual synthesis of oligonucleotides by phosphite triester (phosphoramidite) coupling, see T. Atkinson and M. Smith, in "Oligonucleotide Synthesis: A Practical Approach" (M. J. Gait, ed.), p. 35. IRL Press, Oxford, England, 1984.

[13] J. A. Smith, *Am. Biotech. Lab.* **1**, 15 (1983).

[14] B. D. Warner, M. E. Warner, G. A. Karns, L. Ku, S. Brown-Shimer, and M. S. Urdea, *DNA* **3**, 401 (1984).

[15] S. L. Beaucage and M. H. Caruthers, *Tetrahedron Lett.* **22**, 1859 (1981).

[16] S. P. Adams, K. S. Kavka, E. J. Wykes, S. B. Holder, and G. R. Gallupi, *J. Am. Chem. Soc.* **105**, 661 (1983).

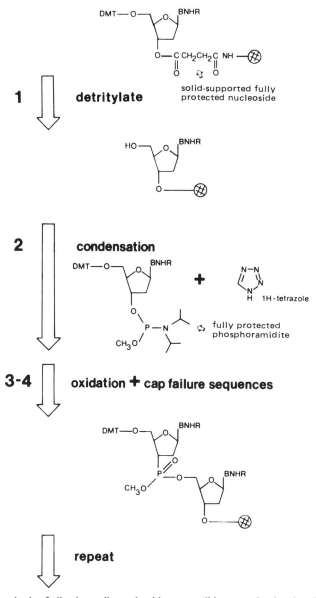

FIG. 1. Synthesis of oligodeoxyribonucleotides on a solid support by the phosphite triester method. The procedure is shown with a *N,N*-diisopropylmethylphosphoramidite intermediate.

Note that the synthesis proceeds from the 3' to 5' direction; therefore, the solid support initially contains the 3'-most nucleoside of the desired product.

Chemicals

The amounts given are meant as reasonable quantities for laboratory stocks and not the quantities needed for each synthesis. Suggested suppliers are also given.

Dichloromethane (CH_2Cl_2), tetrahydrofuran (THF), and acetonitrile (CH_3CN), one 4-liter bottle each of HPLC grade (EM Omnisolv or J. T. Baker)

4-Dimethylaminopyridine, 250 g (DMAP)

Distilled 2,6-lutidene, 100 ml (LUT; Cruachem, Bend, OR)

Dichloroacetic acid, 250 ml (DCA; 99+%, Aldrich)

Sublimed tetrazole, 5 g [Cruachem, American Bionuclear, Emeryville, CA (ABN) or Applied Biosystems, Foster City, CA (ABS)]

Dry, distilled acetonitrile, 500 ml (ABN, ABS, or Aldrich 99+%, Gold Label)

Reagent grade acetic anhydride, 250 ml (Ac_2O, J. T. Baker)

Iodine crystals, 100 g (Aldrich, Gold Label)

Pyrophosphoric acid (City Chemical Corp., New York, NY)

One gram each of the four fully protected nucleoside phosphoramidites[17] (5'-dimethoxytrityl form of N^6-benzoyladenosine, N^4-benzoylcytidine N,N^2-isobutyrylguanosine, and thymidine 2'-deoxyribose-3'-nucleoside, N-diisopropylmethylphosphoramidites; referred to as A, C, G, and T phosphoramidites; ABN or ABS)

One gram each fully blocked nucleoside derivatized solid support [5'-dimethoxytrityl form of the blocked nucleosides as above covalently attached by the 3'-hydroxyl to a solid supported carboxylate; use Vydac silica from Cruachem or long-chain alkylamino control pore glass (LCAA-CPG) from Cruachem, ABN, or ABS. Although LCAA-CPG supports typically give a better overall yield, Vydac silica is more easily handled in manual syntheses]

Thiophenol, 50 g (Aldrich 99%)

1,4-Dioxane, 500 ml (Aldrich, HPLC grade)

Triethylamine, 50 g (Aldrich 99%, Gold Label)

---

[17] At this time the 3'-N,N-diisopropylmethylphosphoramidite derivatives [Ref. 16 and L. J. McBride and M. H. Caruthers, *Tetrahedron Lett.* **24,** 245 (1983)] are favored, although the N,N-diisopropyl-β-cyanoethylphosphoramidite intermediates [N. D. Sinha, J. Biernat, J. McManus, and H. Koster, *Nucleic Acids Res.* **12,** 4539 (1984)] are gaining popularity.

Preparation of Reagents

The following stock solutions can be kept at room temperature for several weeks unless otherwise indicated. (Work in a fume hood with a lab coat, gloves, and safety glasses.)

1. DCA. Mix 8 ml of DCA and 242 ml of CH_2Cl_2 in a 250-ml Gibco bottle.

2. DMAP. Dissolve 6.5 g of DMAP into 100 ml of 1:9 (v/v) LUT/THF in a 250-ml beaker. Filter the amber solution with a 50-ml sintered glass funnel into a vacuum flask. Store in a 100-ml Gibco bottle.

3. Ac_2O. Fill a 100-ml Gibco bottle with acetic anhydride.

4. THF/LUT/H_2O. Fill a 250-ml wash bottle with 125 ml THF, 62.5 ml LUT, and 62.5 ml distilled water (2:1:1, v/v/v).

5. I_2 solution. Dissolve 2.5 g of iodine into 100 ml of 2:1:1 THF/LUT/H_2O and filter the dark brown solution as for DMAP. Put in a 250-ml wash bottle (0.1 M).

6. w/CH_2Cl_2 (w/ = wash). Fill a 250-ml wash bottle with CH_2Cl_2.

7. w/CH_3CN. Fill a 500-ml wash bottle with HPLC grade CH_3CN.

8. Syringe wash vials. Fill two 40-ml vials (#13219 Pierce Chemicals) with HPLC grade CH_3CN and one 40-ml vial with dry, distilled CH_3CN. Seal with the Teflon–rubber septa and cap. The dry CH_3CN vial is used for the final syringe wash.

9. d/CH_3CN (d/ = dry). Fill and seal three 40-ml vials with dry, distilled CH_3CN.

10. Tetrazole. Add 817 mg of tetrazole to a 40-ml vial, add 35 ml of dry CH_3CN, and seal. Shake vigorously to dissolve (0.5 M solution).

11. Pyrophosphoric acid. Place about 100 μl of pyrophosphoric acid into a 10-ml beaker. Although pyrophosphoric acid is a solid, it is very hygroscopic and typically the top of the bottle contains a layer of hydrated liquid. This very viscous acid is used for the capillary test (see below).

12. Phosphoramidite vials. The A, C, G, and T phosphoramidite powders are dissolved in acetonitrile, aliquoted into 1.5-ml vials in quantities sufficient for one condensation (30 μmol), evaporated, and dried to form very stable "glasses" as follows. Using a different color tape for each base (i.e., green for A, red for G, blue for C, and yellow for T), label 40 1.5-ml Wheaton vials (#224800 Pierce Chemicals) for each A, C, and G and 45 vials for T. While in the manufacturer's cardboard storage box, fill the 165 vials with w/CH_3CN and quickly aspirate off as much liquid as possible. Repeat the wash. Make a solution containing 1 g of the phosphoramidites of each A, C, and G in 20 ml and 1 g of T in 22.5 ml of dry, distilled CH_3CN in a 50-ml beaker. Aliquot 500 μl of the clear to yellowish solutions into each of the labeled 1.5-ml vials with a Pipetman. Place the

box of filled vials into a large desiccator and affix to a vacuum pump trapped with a dry ice trap. Carefully and slowly (full vacuum in about 1 min) open the desiccator to the vacuum source. *Avoid bumping.* (This can be a little tricky. You may wish to practice with a few vials filled with wash CH_3CN first.) Although the vials will be relatively dry in a few hours, leave them under vacuum overnight. In the morning, back flush the desiccator with argon and cap the vials with the Teflon–rubber septa. Using a syringe needle (19-gauge, 3-in. stainless steel), blow a little argon into each vial just before capping. Although the phosphoramidites can be stored in this form for weeks at room temperature, it is best to keep them at $-20°$ where they should remain active for at least 6 months. When bringing the vials to room temperature, wipe the tops off if any beads of condensation form.

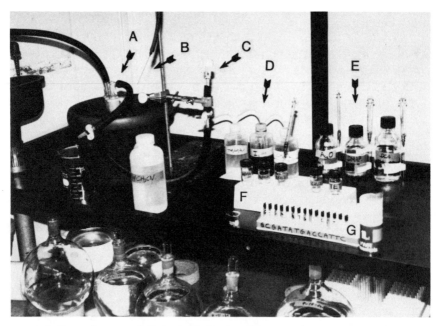

FIG. 2. Set-up for manual synthesis of oligodeoxyribonucleotides. (A) Ice trap in bucket. To the lower left of the bucket is a three-way stopcock that can be opened to a waste beaker during the dry acetonitrile washings. (B) Argon line attached to a needle. (C) Sintered glass funnel reactor (medium, 2 ml) with septum cap (24/40 rubber septum; 12.6 mm I.D. bottom with upper flap cut off) held in place by a clamp and ring stand. A two-way stopcock sits just below the clamp. (D) Reagent squeeze wash bottles. (E) Reagent Gibco bottles with syringes (BD multifit with stainless steel Iver-lock needles) sitting in glass test tubes taped to the sides. (F) A plastic centrifuge tube rack containing the syringe wash vials (left side), d/CH_3CN and tetrazole solution (right side). (G) A plastic test tube rack containing the 1.5-ml Wheaton vials of phosphoramidite glasses placed in order of addition.

13. Thiophenol reagent. Add, to a 40-ml vial, 7.5 ml of thiophenol, 7.5 ml of triethylamine, and 15 ml of reagent grade dioxane. Gently mix and seal with a cap and septum. This should be done in a fume hood. Thiophenol is toxic and very smelly. Spills can be deodorized with a small amount of commercial bleach. (Alternatively, the thiophenol reagent can be purchased from ABN or ABS.)

Set-Up

Place the reactor, filled bottles and vials, stopcocks, dry ice trap, vacuum lines, and argon line in a fume hood as shown in Fig. 2. For storage of syringes, a glass test tube can be attached to the side of the Gibco bottles. Be sure not to fasten the reactor to the clamp directly but instead by the vacuum tube just below the reactor in order to avoid breakage and to permit hand agitation. Be certain that the argon needle is securely fastened to the tubing. Since many steps are involved in the synthesis, progress should be recorded with a flow chart as in Fig. 3. (Gloves, safety glasses, and a lab coat are highly recommended.)

Oligonucleotide Synthesis

Start-Up. Place 30–60 mg (0.75–1.5 μmol) of the appropriate support in a clean glass funnel reactor. Set out the phosphoramidite vials in order of addition. In case of a poor condensation, keep an extra vial of each base on hand. Give the vials at least 30 min to come to room temperature. Be sure that you have filled the reagent bottles and vials.

Precap. Wash two times with w/CH$_3$CN.[18] With the syringes add approximately 1 ml of DMAP solution and then 0.1–0.2 ml of Ac$_2$O to the reactor. Don't cover with the septum. Agitate once by hand and wait 1.5 min. This step dries the support and blocks any free amines that may not have been protected during its synthesis. Wash two times with w/CH$_3$CN.

Detritylation. Wash one time with w/CH$_2$Cl$_2$. Close the stopcock below the reactor, add 1 ml of DCA solution, wait 1.5 min. A bright orange color develops indicating deprotection. Don't cover with the septum. Open the stopcock to evacuate the reactor, then repeat the DCA addition and wait another 1.5 min. Evacuate and wash one time with w/CH$_2$Cl$_2$.

---

[18] All squeeze bottle washes are conducted by flowing addition as follows: With the stopcocks open to the vacuum source, fill to about one-half volume (about 1 ml). Try to get the sides of the reactor. There should be no problem if the support surface dries slightly after the wash. You may find that a partial opening of the stopcock permits better control. If additional washes are necessary (i.e., 2 w/CH$_3$CN rinses or THF/LUT/H$_2$O followed by I$_2$ solution) just as the reactor empties, fill again.

DATE:

SEQUENCE:

SUPPORT:

	phosphoramidites →	2	3	4	5	6	7	8	9	10	11	12	13	14	15	16	17	18	19	20
Detritylation																				
	w/CH$_2$Cl$_2$, 1X																			
	DCA, 2X, 1.5 min each																			
	w/CH$_2$Cl$_2$, 1X																			
Condensation																				
	w/CH$_3$CN, 2X																			
	w/CH$_3$CN, 2X (syringes)																			
	Tetrazole																			
	Phosphoramidite, 1 min																			
	w/CH$_3$CN, 2X																			
	CH$_2$Cl$_2$, 1X																			
	Capillary Test																			
Oxidation																				
	LUT/THF/H$_2$O																			
	I$_2$ solution, 1X																			
	LUT/THF/H$_2$O, 2X																			
Cap	Precap																			
	w/CH$_3$CN, 2X																			
	DMAP solution																			
	Ac$_2$O, 1.5 min																			
	w/CH$_3$CN, 2X																			

Fig. 3. Flow chart for manual oligonucleotide synthesis.

Condensation. Wash two times with w/CH$_3$CN. Close the stopcock. Adjust the three-way stopcock leading to the dry ice trap so that it is open to the waste beaker. Affix the rubber septum to the top of the reactor. With a clean 1-ml syringe (BD Yale tuberculin affixed with a 1.5 inch, 20 gauge Iver-lock needle), add about 1 ml of d/CH$_3$CN from a 40-ml vial through the septum to the support. Open the two-way stopcock below the reactor and push the solution through the reactor by putting the argon needle through the rubber septum, thereby blowing the solution into the beaker. Don't dry the support too thoroughly or it will blow around and deposit on the reactor walls and the underside of the septum. Repeat the wash and close off the stopcock below the reactor. Readjust the three-way stopcock back to the ice trap and close the argon line (use a pinch clamp). Using the same syringe, add 0.5 ml of tetrazole solution to the support. Then take up 0.5 ml of d/CH$_3$CN and push it through the septum into the appropriate phosphoramidite vial. Take up some argon from the atmosphere above the liquid in the vial, then dissolve the phosphoramidite glass by rapidly pulling and pushing the syringe plunger (about 10 sec). Add the solution to the reactor through the septum and quickly hand agitate one time. Let the reaction stand 1 min. Put the empty vial back into the rack upside down as a bookkeeping measure. While waiting for the condensation, wash the syringe by pushing and pulling of the plunger first three times in w/CH$_3$CN vial 1, then vial 2, and finally d/CH$_3$CN vial 3. Leave the syringe in the septum of vial 3 until the next condensation.

Capillary Test. Remove the septum and open the stopcock to the vacuum to dry the support. Wash three times with w/CH$_3$CN, one time with w/CH$_2$Cl$_2$, and air dry. Take up as small a portion as possible of the support in a micropipet capillary tube. Dip the edge of the tube into the beaker of pyrophosphoric acid. If a bright orange color develops in the tip of the capillary, indicating a good condensation, you should then proceed with the oxidation. If there is little or no color, repeat the condensation. For comparison, you may wish to save the capillaries from previous couplings. Very poor condensations due to moisture can be corrected by employing this simple test. Since a capping step is used in the synthesis, if the condensation is not repeated at this point, the synthesis will be lost. Poor condensation should be a rare event.

Oxidation. Wash one time with THF/LUT/H$_2$O, one time with I$_2$ solution, and two times with THF/LUT/H$_2$O. Oxidation occurs very rapidly.

Capping. Wash two times with w/CH$_3$CN. Add 1 ml of DMAP solution, then 0.1 ml of Ac$_2$O. Shake and leave for 1.5 min. Wash two times with w/CH$_3$CN.

Repeat the detritylation, condensation (and capillary test), oxidation, and capping steps for each nucleotide addition. Each nucleotide coupling requires about 8 min.

Skip the capping step after the last condensation. After the last THF/ LUT/H_2O addition of the final oxidation, wash two times with w/CH_3CN, one time with w/CH_2Cl_2, and air dry.

It is convenient to stop during a synthesis (even overnight) after capping and resume at detritylation.

After completion of the synthesis, blow a little argon through the tetrazole solution and d/CH_3CN vials, then cover with Parafilm. If properly handled, the materials will remain dry for 2 days at room temperature. If no synthesis will be carried out for 3 days or more, replace the punctured septa and cover with Parafilm. Discard the vials of CH_3CN for syringe washing after 2 days or 100 condensations.

It is not necessary to start the protective group removal immediately upon completion of the synthesis. The dry, solid supported, fully blocked product can be kept in a sealed 1.5-ml Wheaton vial at room temperature indefinitely.

Several (up to four) syntheses can be carried out at the same time by simply adding more reactors, clamps, stopcocks, etc., to the synthesis set-up. Beyond four simultaneous syntheses, the time saving achieved is questionable.

Degenerate pools of oligonucleotides where more than one base is added at a specific position, can be produced by using phosphoramidite glasses consisting of equimolar mixes of the desired combination to give a final 30-μmol sample.

Protective Group Removal and Gel Purification of Synthetic Oligonucleotides[19]

To the solid support in the fritted funnel, add about 2 ml of thiophenol reagent.[12b] Cap, gently shake, then let the material sit for 1 hr. Open the stopcock to remove the thiophenol, then wash five times with reagent grade methanol and once with CH_2Cl_2.[20]

The air-dried product is transferred to a 1.5-ml Wheaton vial, approximately 1 ml of concentrated ammonium hydroxide is added, the vial is fitted with a septum and cap, and then hand shaken and left at room temperature for 3 hr (or at 37° for 1 hr) to remove the product from the support. After this period, the vial is hand shaken, cooled in ice for a few seconds (to avoid bumping upon opening), transferred to a new Wheaton vial, and placed at 60° for 10–12 hr to remove phosphate and exocyclic

---

[19] R. Sanchez-Pescador and M. S. Urdea, *DNA* **3**, 339 (1984).
[20] The thiophenol step can be eliminated if β-cyanoethyl phosphoramidites are used.

nitrogen blocking groups.[21] The solution is cooled on ice, uncapped, and dried in a Speed Vac concentrator (Savant Instruments, Hicksville, NY; rotor model #RH40-12 will accommodate the Wheaton vials). Removal of the 5'-dimethoxytrityl functionality is achieved by adding 100 μl of 80% acetic acid (v/v; glacial acetic acid/distilled water), vortex mixing, and setting the tube at room temperature for 30 min. The vial is then filled with ethanol, vortexed, and spun in the Speed Vac without the vacuum for 1 min to collect the white precipitate. The supernatant is removed with a Pasteur pipet and the precipitate is dried for 1 min. [Occasionally (~ 1 in 300 syntheses) no precipitate is apparent, in which case dry the entire sample, add 10 μl of ammonium hydroxide (to neutralize any remaining acid), and reevaporate.] To the pellet, 50 μl of 90% deionized formamide containing 0.01% (w/v) bromphenol blue (BPB) is added, vortex mixed, heated to 95° for 2 min on a heating block, and vortexed again. For sequences 5 to 35 bases or 36 to 120 bases, a 20 or 10% denaturing polyacrylamide gel, respectively, is used [three lanes, 0.15 × 20 × 40 cm gel, 19:1 acrylamide:bisacrylamide, 90 mM Tris, 90 mM boric acid, 2.7 mM EDTA, pH 8.3 (TBE), 8.3 M urea]. After flushing the lanes with a pipet, one-fifth to one-half of the samples are loaded with a Pipetman onto the gel (which has been prerun for 15 min) and run at 30 mA for 2.5 to 4 hr (BPB two-thirds to three-quarters down gel) or overnight at 7–10 mA. The remaining unpurified formamide solutions of crude DNA can be stored indefinitely at room temperature.

Upon completion, the gel is carefully removed from the plates and set upon a Saran Wrap-covered pair of 20 × 20 cm Baker F-254 silica 60 TLC plates taped to a 20 × 40 cm glass plate. Bands are visualized as UV shadows by illumination from above with a hand-held short wavelength UV lamp and cut out with a razor blade. Photographic records of the gel can be obtained with a Polaroid MP-4 camera system fitted with a Kodak No. 59 green filter (f stop 8; 1- to 3-sec exposure).

The uncrushed gel band is placed into 10 ml polypropylene Bio-Rad Econo column, 3–8 ml of elution buffer is added (0.1 M Tris–HCl, pH 8.0, 0.5 M NaCl, 5 mM EDTA) and left at room temperature overnight without mixing (or at 60° for 1 hr with occasional vortexing). The bottom seal of the Econo column is broken and the eluted DNA is filtered into a 20-ml disposable syringe fitted with a C_{18} Sep-Pak (Waters Associates) that has been washed with 5 ml of reagent grade methanol and 10 ml of distilled water.[22] The DNA solution is slowly applied to the cartridge

---

[21] Overnight deblocking can be conducted at home by placing a commercial heating pad (General Electric, Cat. No. B3P55, 115 V, 55 W, high setting) inside a terrycloth towel in a porcelain bowl. The temperature setting remains at approximately 60–65°.

[22] Push the solutions through the Sep-Pak, disconnect the cartridge, reset the syringe, and reaffix the Sep-Pak for all washes and applications.

($\sim$ 20 sec), washed with 20 ml of water to desalt, and eluted with $\sim$2.5 ml of 1:1 (v/v) 0.1 M triethylamine acetate, pH 7.3, and methanol into two 1.5-ml Eppendorf tubes. [A 10$\times$ triethylamine acetate solution (1 M) is prepared by adding 139.4 ml of triethylamine distilled from tosyl chloride (alternatively, use Aldrich 99% Gold Label) to 52.7 ml of glacial acetic acid made up to a total of 1 liter distilled water.] (Ammonium acetate can be used instead but is somewhat more difficult to fully remove in the subsequent drying step.) The filled Eppendorf tubes are evaporated to dryness in a Speed Vac, the pellets are dissolved and combined into a total of 100 μl of distilled water, and stored at $-20°$.

For some applications, purification of the oligomers is not necessary.[19]

Analysis of Purified Oligonucleotides

A 1-μl sample of the purified product (1% of the total) is 5'-^{32}P-labeled in a 10-μl solution containing 50 mM Tris–HCl, pH 7.6, 10 mM MgCl$_2$, 1 μg/ml spermidine (1$\times$ KB buffer), 1 mM DTT, 1–2 U T$_4$ polynucleotide kinase, and 1 μl (10 μCi) of [γ-^{32}P]ATP for 30 min at 37°. The solution is then evaporated to dryness, solubilized in 10 μl of 90% formamide, 0.01% BPB, and loaded on a prerun denaturing polyacrylamide gel (12 lanes, 0.05 $\times$ 20 $\times$ 40 cm, 15–20 mA, 10–20% as above, TBE as running buffer). Size markers from previous syntheses (or available from American Bionuclear, Emeryville, CA) should be labeled and run alongside the samples. One plate is removed and the gel is covered with Saran Wrap. Autoradiography without an intensifying screen requires 10–60 min. Mobility shift due to composition can significantly affect the relative position of oligomers in the gel and should be considered in interpretation of the results.[23]

Alternatively, unlabeled oligonucleotides can be analyzed in the same gel system. Approximately 0.05 OD$_{260}$ U is applied in 3–5 μl of 90% formamide solution to an 8- to 15-lane 0.15 $\times$ 20 $\times$ 40 cm denaturing gel, run and visualized by UV shadowing as described above (see Fig. 4).

If oligomers are greater than 99% pure and run on gels as anticipated for the composition, gene synthesis can be conducted without sequence determination of the individual oligonucleotides.

Gene Construction and Purification

Principles of Method

Several criteria were considered in the design of the synthetic oligonucleotide sequences. In order to facilitate molecular expression in yeast, the

---

[23] R. Frank and H. Koster, *Nucleic Acids Res.* **6**, 2069 (1979).

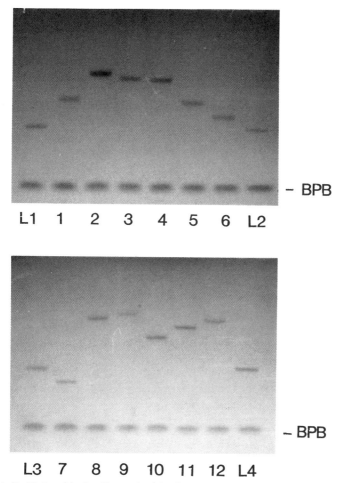

FIG. 4. Purified synthetic oligonucleotides for construction of hEGF analyzed without labeling by polyacrylamide gel. Fragments L1, 1, 2, 3, 4, 5, 6, L2, L3, 7, 8, 9, 10, 11, 12, and L4 contain 15, 23, 33, 30, 29, 23, 17, 15, 15, 12, 30, 33, 23, 27, 30, and 15 bases, respectively. Their relative positions and overlaps in the synthetic gene are given in Fig. 5. Visualization is by UV shadowing as described in text.

gene was constructed with the set of preferred codons inferred from the usage bias observed in highly expressed yeast genes.[24] Although visual inspection was used to remove internal repeats of eight bases or greater, computer-assisted analysis such as the Repeats Program[25] is recom-

---

[24] J. L. Bennetzen and B. D. Hall, *J. Biol. Chem.* **257**, 3026 (1982).
[25] H. Martinez, Department of Biochemistry and Biophysics, University of California, San Francisco, personal communication.

FIG. 5. Design of *Hga*I linker derivative of the hEGF synthetic gene. At the ends of the duplex are *Eco*RI adhesive ends. The boxes indicate the *Hga*I recognition sequences while the arrows show the cleavage points upon *Hga*I restriction. The lengths of the synthetic oligomers are given in Fig. 4.

mended. Also, undesired restriction sites were removed by conservative replacement of nucleotides in the third position of the codons.

Five bases away from either end of the coding segment of the synthetic gene, an HgaI recognition site was incorporated (Fig. 5). Since the HgaI restriction endonuclease cleaves 5 and 10 bases away from the GACGC recognition sequence,[26] it was possible to specifically remove the complete hEGF-coding sequence from the cloning vector. The resulting duplex possesses two 5' nonpalindromic overhangs that can be used to transfer the coding sequence to any desired expression vector with the aid of specific synthetic linkers. The method is illustrated below for the hEGF gene cloned into pBR328 then transferred to an α factor-derived plasmid for expression and secretion in yeast. This HgaI linker approach has also been applied to construction of insulin-like growth factors I and II.[6]

Construction of the gene involved 5'-phosphorylation of all but the 5'-terminal oligomers followed by annealing, ligation, and gel purification of the full-length duplex.[1] The purified synthetic gene was then 5'-phosphorylated for cloning.

Preparative 5'-Phosphorylation of Synthetic Fragments

Based on 20 OD_{260} U/1 mg DNA, 1000 pmol of each synthetic oligomer (except the two 5'-terminal fragments) were 5'-phosphorylated individually in 19 μl total volume containing 1× KB, 1 mM DTT, 1 mM ATP, 4–6 U T_4 kinase for 1 hr at 37°. After the addition of 1 μl of 250 mM EDTA, the stock solutions were stored at −20°. These solutions can be kept for several months.

Ligation and Purification

For the HgaI linker derivative of hEGF, 16 oligomers were employed ranging from 12 to 33 bases in length (see Fig. 5). Except for the 5'-hydroxyl terminal HgaI linkers 1 and 4, 1 μl (50 pmol) of each fragment was combined (14 μl total), from the 5'-phosphorylated stock solutions, then 1.4 μl of 1 M sodium acetate and 5 μl of 1 mg/ml poly(A) were added and the solution was heated to 65° for 5 min. After 50 pmol of the two 5'-terminal fragments was added, 3 vol (66 μl) of 100% ethanol was added. The solution was vortexed, then chilled for 20 min at −80°, centrifuged at room temperature in an Eppendorf centrifuge for 2 min, decanted, washed twice with −20° 100% ethanol, and dried in a Speed Vac.

[26] N. L. Brown and M. Smith, *Proc. Natl. Acad. Sci. U.S.A.* **74**, 3213 (1977).

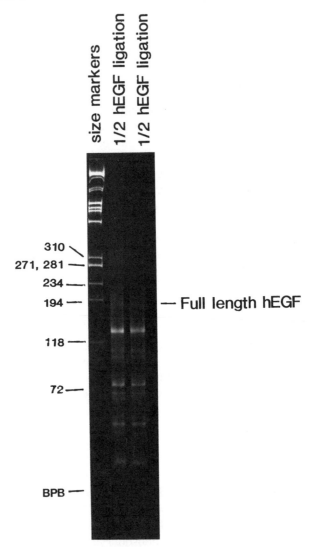

FIG. 6. Gel purification of the ligated full-length *Hga*I–hEGF synthetic duplex. All 16 oligomers were used in a one-step ligation reaction. The size markers contain 1 μg of *Hae*III-digested φX174 and *Hin*dIII-digested λ.

The pellet was dissolved in 18 μl of water to which 3 μl of 10× KB buffer was added. After 1 hr at 37°, 3 μl each of 10 mM DTT, 10 mM ATP, and T$_4$ DNA ligase (New England Biolabs, 4 × 10^5 U/ml) were added (total 30 μl) and the reaction was placed at 14° for 1 hr.

A 7% native polyacrylamide gel (TBE as running buffer,

0.15 × 20 × 20 cm, 10 lanes) was prerun for 15 min at 60 mA. The ligation reaction was quenched with 10 μl of stop mix (25% glycerol, 0.05% bromphenol blue, 0.5% SDS, 25 mM EDTA) and loaded into two lanes (20 μl per lane). Size markers were run in an adjacent lane. The gel was run until the bromphenol blue was approximately 1 cm from the bottom, then removed from the glass plates, put into a glass dish containing 500 ml of 1 μg/ml ethidium bromide, stained for 10 min in the dark, and washed for 5 min in distilled water. The bands were visualized on a short wave length (254 nm) light box and a photographic record was made with an MP-4 Polaroid camera fitted with an orange filter (see Fig. 6). The appropriate band was cut out with a razor blade.

The gel slab was placed into a small dialysis bag (Spectrapor 1000 MW cut off), filled with 10 mM Tris, pH 8.0, 1 mM EDTA (TE), clipped off, and electroeluted at 100 mA for 3 hr in an electroelution box containing approximately 200 ml of 20 mM Tris–acetate, 0.5 mM EDTA ($\frac{1}{2}$× TEA). The current was reversed for 90 sec and the contents of the dialysis bag were filtered through an Econo column into a 1.5-ml Eppendorf tube. The bag was washed with a small portion of water and the washing was also filtered into the tube. (If the volume is greater than 400 μl, the sample can be concentrated by n-butanol extraction.) The solution was made 0.1 M in sodium acetate, then 5 μg of poly(A) and 3 vol of 100% ethanol were added. The mixture was vortexed, left at $-80°$ for 20 min, centrifuged at room temperature, decanted, washed twice with $-20°$ 70% ethanol, and dried in a Speed Vac.

The pellet was dissolved in 20 μl of 1× KB, 1 mM DTT, 1 mM ATP containing 5 U of T_4 kinase. After 1 hr at 37°, the solution was extracted with an equal volume of equilibrated phenol then chloroform isoamyl alcohol (24:1, v/v). Then 20 μl of 0.2 M sodium acetate was added and the 5′-phosphorylated synthetic gene was ethanol precipitated, washed, and dried as described above.

To check the recovery, the pellet was dissolved in 20 μl of distilled water and 2 μl was applied to a 7% native polyacrylamide gel, run, and stained as described above.

For synthetic genes greater than 300 bases, a 1.5% agarose gel is more useful for purification. After electroelution a phenol extraction should be performed.[10]

Molecular Cloning and Expression

Principles of Method

The scheme for cloning the synthetic gene[8] and transferring it to an α factor-expression plasmid[9] is presented in Fig. 7.

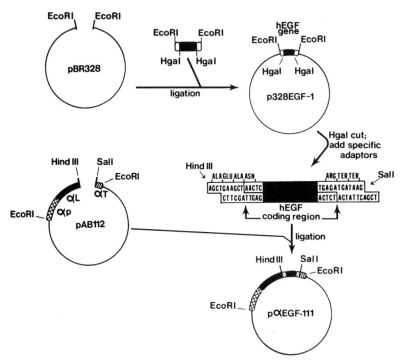

FIG. 7. Scheme for synthetic *Hga*I–hEGF gene cloning and transference to an α-factor yeast expression vector.

Molecular Cloning[8]

A 100-fold molar excess of the 5′-phosphorylated gene was added to 100 ng (0.03 pmol) of *Eco*RI cut and alkaline phosphatase-treated pBR328 in a total of 20 μl of 1× KB, 1 mM DTT, 1 mM ATP. After 5 min at 37° and cooling to 14°, 1 μl of T_4 DNA ligase was added and the solution was left at 14° for 4 hr. The vector was precipitated by the addition of 5 μg of poly(A) in 20 μl of 0.2 M sodium acetate and 3 vol of 100% ethanol as described above. The dried pellet was taken up in 100 μl of distilled water, 7 μl of 1 M CaCl$_2$ and 100 μl of competent HB101 *Escherichia coli* cells was added. The transformation was left on ice for 45 min, heated to 42° for 90 sec, and transferred to a 10-ml culture tube containing 3 ml of L-broth, shaken for 45 min at 37°, and centrifuged to collect the cells. All but approximately 100 μl of the solution was removed and discarded. The cells were suspended in the remaining medium and spread onto an L-agar plate containing 0.1% ampicillin, air dried, and placed overnight in an incubator at 37°. A vector sample ligated without the synthetic gene was run as a control.

Twenty-four colonies were sampled and cultured overnight in 2 ml of L-broth containing 0.1% ampicillin. Plasmid DNA was isolated from 1 ml of the culture and screened for a 185-bp insert after cutting with EcoRI.

Large-scale plasmid preparations were conducted on two positive samples and the EcoRI inserts were gel isolated and electroeluted. Sequencing was performed by the dideoxy chain termination method after cloning into bacteriophage M13. The sequence obtained for both independent isolates was consistent with the synthesized fragments.

The pBR328 clone of hEGF was exhaustively digested with HgaI according to the manufacturer's recommendations and the 159-bp insert was gel isolated by 7% native polyacrylamide gel electrophoresis.

Details of the use of the HgaI cut hEGF gene in the construction of a yeast vector for expression and secretion of the mature peptide are described by Brake et al.[9] In brief, the plasmid pAB112, a pBR322 derivative containing the yeast α factor promoter, leader, processing signals and terminator was cut with HindIII and SalI to remove the α factor coding sequences. After gel isolation, the hEGF HgaI insert, adapted with four synthetic linkers to restore the Glu-Ala processing site and introduce two termination codons after the coding region was cloned into pAB112 to yield pαEGF-111 (Fig. 7). The EcoRI-cut α factor leader/hEGF hybrid was used to produce a yeast expression plasmid that was then employed in a transformation of yeast. After growing transformants to stationary phase in selective medium and pelleting the cells by centrifugation, 5–10 μg of hEGF was found per milliliter of culture as evidenced by a competitive receptor binding assay using ^{125}I-labeled mouse EGF and human foreskin fibroblasts.[27]

Further details of hEGF expression and characterization as well as modification of the α factor construct for optimal expression are described elsewhere.[9] The yeast-derived growth factor has been shown to promote corneal wound healing.[28]

Acknowledgment

A number of my colleagues at Chiron were involved in the synthesis, cloning, and expression of hEGF. Their names appear in the references cited and include A. Brake, D. Coit, U. Heberlein, L. Ku, F. Masiarz, G. Mullenbach, and R. Najarian. Also, I thank Ms. Dana Topping and Ms. Becky Tooman for preparation of the manuscript.

[27] S. Cohen and G. Carpenter, Proc. Natl. Acad. Sci. U.S.A. 72, 1317 (1975).
[28] J. R. Brightwell, S. L. Riddle, R. A. Eiferman, P. Valenzuela, P. J. Barr, J. P. Merryweather, and G. S. Schultz, Invest. Ophthalmol. Visual Sci. 26, 105 (1984).

[4] Purification of Polypeptide Growth Factors from Milk

By YUEN SHING, SUSAN DAVIDSON, and MICHAEL KLAGSBRUN

Milk contains polypeptide growth factors that are capable of stimulating DNA synthesis and cell division in cultured cells.[1-3] The growth factor activity in human milk can be detected throughout the lactation period. The major growth factor species in human milk has a molecular weight of 6000 and a pI of 4.5.[4] On the other hand, the growth factor activity in bovine milk can be detected only in colostrum, the milk produced in the first 3 days after the birth of a calf. The major growth factor species in bovine colostrum has a molecular weight of about 30,000 and a pI of 10. The chapter describes the procedures for the purification of these two growth factors.

Growth Factor Assay

Growth factor activity is assayed by measuring DNA synthesis in confluent monolayers of quiescent mouse BALB/c 3T3 cells.[5] These cells are grown at 37° in Dulbecco's modified Eagle's medium (Grand Island Biological Co., Grand Island, NY) supplemented with 4.5 glucose/liter, 10% calf serum (Colorado Serum Co., Denver, CO), 50 U penicillin/ml, and 50 μg streptomycin/ml. About 200 μl containing 10^4 cells is plated into 96-well microtiter plates (Costar, Cambridge, MA). The 3T3 cells are incubated 5–10 days without medium change in order to deplete the serum of growth-promoting factors. Test samples of up to 50 μl are added to the cells along with 10 μl of [^{3}H]thymidine (New England Nuclear, 6.7 Ci/mmol, final concentration of 1 μCi/0.25 ml/well). After an incubation of about 40 hr, the medium is removed and the cells are washed once in 0.15 M NaCl. Measurement of the incorporation of [^{3}H]thymidine into DNA is accomplished by the following successive steps: addition of methanol twice for periods of 5 min, four washes with water, addition of cold 5% trichloroacetic acid twice for periods of 10 min, and four washes with water. Cells are lysed in 200 μl of 0.3 N NaOH and counted in 2 ml of

[1] M. Klagsbrun, Proc. Natl. Acad. Sci. U.S.A 75, 5057 (1978).
[2] M. Klagsbrun and J. Neumann, J. Supramol. Struct. 11, 349 (1979).
[3] A. Sereni and R. Baserga, Cell Biol. Int. Rep. 5, 338 (1981).
[4] Y.W. Shing and M. Klagsbrun, Endocrinology 115, 273 (1984).
[5] M. Klagsbrun, R. Langer, R. Levenson, S. Smith, and C. Lillehei, Exp. Cell Res. 105, 99 (1977).

scintillation fluid (Beckman Ready Solv EP). In this assay, background incorporation is typically 200–1000 cpm and maximal incorporation as indication by addition of 50 μl of calf serum is 120,000–150,000 cpm. A unit of growth factor activity is defined as the amount of growth factor required to elicit half-maximal DNA synthesis.

Purification of Human Milk-Derived Growth Factor

The major growth factor species in human milk can be purified to homogeneity using the following procedures: (1) removal of fat and cellular residue by centrifugation, (2) acid precipitation at pH 4.3, (3) Sephadex G-100 gel filtration chromatography, (4) DEAE–Sephacel anionic-exchange chromatography, and (5) size exclusion HPLC on TSK-2000 columns.

Defatting and Acid Precipitation. Human milk can be obtained from private donors, hospitals, or milk banks. About 100 ml of whole human milk collected from several individuals in the first month of lactation is centrifuged at 10,000 g for 1 hr using a Sorvall RC-5 Superspeed centrifuge (Du Pont). Fat floating at the top of the centrifuge tube and the cells and debris sedimenting at the bottom of the centrifuge tube are discarded. The defatted, acellular fraction is adjusted to pH 4.3 by addition of 6 N HCl. After 1 hr at room temperature with stirring, the precipitate that is formed is removed by centrifugation at 10,000 g for 1 hr. The supernatant fraction ⌐acidified milk is dialyzed exhaustively against deionized distilled H_2O by ⌐ of Spectrapor tubing with molecular weight cutoff 3500 and lyophi-

ion Chromatography. Approximately 1.5 g of the lyophilized ' by acid precipitation of about 100 ml of milk is resus- ⌐quilibration buffer and is chromatographed at 4° on a ⌐ (5 × 90 cm) equilibrated with 0.1 M NaCl and ' 4.3. The elution is carried out with the same '/hr. Fractions of 18 ml are collected and ⌐eir ability to stimulate DNA synthesis. ⌐ growth factor activity are detected. · weight of abo⌐t 6000, accounts ⌐tions co⌐ ⌐aining this major fur⌐ ⌐r purification. The ights of about 30,000 ⌐tal activity respectively

⌐nple (50 mg of protein in ⌐ growth factor, prepared by

pooling the activity peak fractions of Sephadex G-100 column chromatography, is lyophilized, redissolved in 30 ml of water, and dialyzed exhaustively against 0.01 M sodium acetate, pH 5.6, by means of Spectrapor tubing with molecular weight cutoff 3500. After dialysis, the sample (50 mg/30 ml) is applied to a DEAE–Sephacel column (2.5 × 32 cm) equilibrated at 4° with the dialysis buffer. The column is washed with 100 ml of an NaCl gradient (0–0.1 M) in 0.01 M sodium acetate, pH 5.6, at a flow rate of 20 ml/hr. Growth factor activity is subsequently eluted from the column with 320 ml of a second NaCl gradient (0.1 to 0.8 M) in the same buffer. Fractions (8 ml) are collected and tested for the ability to stimulate DNA synthesis in 3T3 cells. Virtually all of the growth factor activity adheres to the anion exchanger. When the gradient of NaCl (0.1–0.8 M) is applied, two peaks of growth factor activity are eluted. The major one is eluted second at about 0.7 M NaCl and contains about 80% of the activity. The fractions containing this growth factor activity are collected, dialyzed exhaustively against deionized distilled H_2O by means of Spectrapor tubing with molecular cutoff 3500, and lyophilized for further purification. DEAE–Sephacel (Pharmacia, Sweden) is initially prepared for chromatography as follows: 200 ml of DEAE–Sephacel is stirred into 400 ml of 0.5 M sodium acetate, pH 5.6, and allowed to settle for 2 hr. The supernatant solution is decanted and replaced with 1 liter of 0.01 M sodium acetate, pH 5.6, four times. After chromatography, DEAE–Sephacel can be regenerated for reuse by washing 200 ml of used resin with 1 liter of 0.1 N HCl once, followed by 1 liter of distilled H_2O four times, 1 liter of 0.1 N NaOH twice, and 1 liter of distilled H_2O four times. The resin is then equilibrated with 400 ml of 0.5 M sodium acetate, pH 5.6, followed by 0.01 M sodium acetate, pH 5.6, four times. Conductivity and pH a monitored to make sure that the ion exchanger is properly equilibra DEAE–Sephacel is stored at 4° in the equilibration buffer suppleme with 0.4% sodium azide.

Size-Exclusion HPLC on TSK-2000 Columns. The lyophilized milk sample (3.4 mg) prepared by DEAE–Sephacel column chro phy is purified to homogeneity with a Beckman Model 334 HPL system (Beckman, Palo Alto, CA) using a TSK-2000 size exclus (7.5 × 60 mm, Varion, Walnut Creek, CA) equilibrated wit monium sulfate and 0.05 M potassium phosphate, pH 7. H out at room temperature at a flow rate of 1 ml/min. Samp applied to the column and fractions of 0.5 ml are c exhaustively against deionized distilled H_2O by means ing with molecular weight cutoff 3500, and tested stimulate DNA synthesis in 3T3 cells. The columns a following molecular weight standards: Blue Dextr

TABLE I
PURIFICATION OF HUMAN MILK-DERIVED GROWTH FACTOR

Step	Total protein (mg)	Total activity (Units)	Specific activity (U/mg)	Recovery (%)	Purification (-fold)
Human milk (100 ml)	1,400	21,800	15.6	100	1
Acid precipitation	1,100	18,000	16.4	83	1.1
Sephadex G-100	52	3,750	72	17	4.6
DEAE–Sephacel	3.4	960	282	4.4	18.1
HPLC, TSK-2000 (Tris–HCl buffer, pH 7)	0.0012	172	143,000	0.8	9,200

vine serum albumin (MW 67,000), ovalbumin (MW, 43,000), α-chymotrypsinogen (MW 25,700), and lysozyme (MW 14,300). The growth factor activity migrates as a sharp peak with a molecular weight of about 6000. This purified human milk-derived growth factor shows a single band with a molecular weight of about 6000[4] when analyzed by SDS–polyacrylamide gel electrophoresis[6] after visualization with the silver stain.[7] The purification scheme for human milk-derived growth factor is summarized in Table I. The growth factor is purified about 9000-fold. The recovery of growth factor activity is about 0.8%. Half-maximal stimulation (1 unit of activity) occurs at a concentration of about 25 ng/ml. The purified growth factor can be stored at $-20°$ in lyophilized form for at least 6 months without substantial loss of activity.

Large-Scale Purification of Bovine Colostrum-Derived Growth
    Factor

The major growth factor in bovine colostrum can be purified using the following procedures: (1) removal of fat and cellular residue by centrifugation; (2) Bio-Rex 70 cation exchange chromatography; (3) preparative isoelectric focusing; and (4) size-exclusion HPLC on TSK-3000 columns.

Defatting and Filtration. About 500 ml of whole bovine colostrum obtained from a dairy farm is defatted by centrifugation at 10,000 g for 1 hr. Fat floating at the top of the centrifuge tube and the cells and debris sedimenting at the bottom of the centrifuge tube are discarded. The defatted, acellular fraction (about 400 ml) is diluted four times with 0.01 M Tris–HCl, pH 7, centrifuged again at 10,000 g for 1 hr, and filtered through a 0.8 micron filter with vacuum to remove all residues.

[6] U. K. Laemmli, *Nature (London)* 227, 880 (1970).
[7] B. R. Oakley, D. R. Kirsch, and N. R. Morris, *Anal. Biochem.* 105, 361 (1980).

Cation Exchange Chromatography. About 2 liters (50 g of protein) of the filtrate prepared by the previous step is mixed thoroughly with 500 ml of Bio-Rex 70 (200–400 mesh, Bio-Rad) equilibrated at 4° with 0.1 M NaCl, 0.01 M Tris–HCl, pH 7. The Bio-Rex 70 is allowed to settle for 1 hr and the supernatant is discarded. A slurry of Bio-Rex 70 is packed into a column (5 × 25 cm), rinsed with 1 liter of equilibration buffer, and eluted with a gradient of 0.1 to 1.0 M NaCl at a flow rate of 40 ml/hr in the same buffer. Fractions (18 ml) are collected and monitored for growth factor activity. The active fractions elute at about 0.6 M NaCl and are collected for further purification. Bio-Rex 70 is initially prepared for chromatography as follows: 2 lb of Bio-Rex 70 is washed with 2000 ml of 0.5 M NaCl, 0.5 M Tris–HCl, pH 7, once and with 4000 ml of equilibration buffer (0.1 M NaCl, 0.01 M Tris–HCl, pH 7) four times. After chromatography, Bio-Rex 70 is regenerated for reuse by washing 2 lb of used resin with 4000 ml of 3 M NaCl, 0.01 M Tris–HCl, pH 7, twice followed by 4000 ml of 0.1 M NaCl, 0.01 M Tris–HCl, pH 7, several times. Conductivity and pH are monitored to make sure that the Bio-Rex 70 is properly equilibrated. Bio-Rex 70 is stored at 4° in the equilibration buffer supplemented with 0.4% sodium azide.

Preparative Isoelectric Focusing. A sample (50 mg of protein in 100 ml of buffer) of bovine colostrum growth factor, prepared by pooling the active fractions of Bio-Rex 70 ion exchange column chromatography, is dialyzed exhaustively against deionized distilled H_2O by means of Spectrapor tubing with molecular weight cutoff 8000 and lyophilized. The lyophilized sample is further purified on an LKB 8100 vertical electrofocusing column (LKB, Broma, Sweden) with a capacity of 110 ml. The column is prepared as follows: About 25 ml of a dense solution containing sucrose (60%, w/v) and 0.16 M H_3PO_4, pH 1.2, are poured into the bottom of the column. A 110-ml gradient of sucrose (5–50%, w/v) containing 6 M urea and carrier ampholytes (LKB), 1.8 ml of pH 3.5–10 (final concentration of 0.7%) and 2.4 ml of pH 9–11 (final concentration of 0.5%), is pumped onto the column above the dense solution. An 8-ml solution of 0.25 M NaOH is pumped on top of the gradient. Sample (50 mg of protein in 10 ml of water) is introduced into the gradient prior to gradient formation. The anode is placed in the lower dense electrode solution and the cathode is placed in the upper electrode solution. Isoelectric focusing is carried out with an LKB 2103 power supply at a constant voltage of 1100 V for about 40 hr at 9°. Fractions (2 ml) are collected, measured for pH, dialyzed against H_2O by means of Spectrapor tubing with molecular weight cutoff 8000, and tested for their ability to stimulate DNA synthesis in 3T3 cells. The fractions containing growth factor activity are focused into a sharp peak at about pH 10 and are pooled and lyophilized for further purification.

Size Exclusion HPLC on TSK-3000 Columns. The lyophilized bovine colostrum sample prepared by isoelectric focusing is purified on a Beckman Model 334 HPLC gradient system (Beckman, CA) using TSK size exclusion columns (Varion, CA). HPLC is carried out at room temperature at a flow rate of 1 ml/min. Samples of 100 μl (4 mg of protein) are applied to the column and fractions of 0.5 ml are collected, dialyzed exhaustively against deionized distilled H_2O by means of Spectrapor tubing with molecular weight cutoff 8000, and tested for their ability to stimulate DNA synthesis in 3T3 cells. Under standard conditions (TSK 2000, 7.5×60 mm, equilibrated with 0.6 M NaCl, 0.01 M Tris–HCl, pH 7), the growth factor activity elutes with a molecular weight of about 30,000. However, the peak of growth factor activity is contaminated with other proteins. To achieve purification, growth factor prepared by isoelectric focusing is chromatographed under dissociating conditions [TSK-3000, 7.5×50 mm, equilibrated with 6 M guanidine–HCl and 0.02 M 2-(N-morpholino)ethanesulfonic acid, pH 6.5]. Analysis by SDS–polyacrylamide gel electrophoresis[6] with silver stain[7] shows that growth factor purified in this manner has a native molecular weight of 30,000 and a monomer molecular weight of about 18,000–20,000. The purification scheme for bovine colostrum-derived growth factor is summarized in Table II. The growth factor is purified about 150,000-fold. The recovery of growth factor activity is about 1%. Half-maximal stimulation (1 U of activity) occurs at a concentration of about 15 ng/ml. The purified growth factor can be stored at $-20°$ without substantial loss of activity for at least 6 months.

TABLE II
PURIFICATION OF BOVINE COLOSTRUM-DERIVED GROWTH FACTOR

Step	Total protein (mg)	Total activity (Units)	Specific activity (U/mg)	Recovery (%)	Purification (-fold)
Bovine colostrum (500 ml)	30,000	60,000	2	100	1
Bio-Rex 70	50	8,000	162	13	80
Isoelectric focusing	4	2,000	500	3	250
HPLC, TSK-3000 (guanidine–HCl, pH 6.5)	0.002	600	300,000	1	150,000

TABLE III
BIOCHEMICAL CHARACTERIZATION OF MILK-DERIVED GROWTH FACTORS,
EGF, AND PDGF

Property	Human milk derived growth factor	EGF	Bovine colostrum-derived growth factor	PDGF
Molecular weight	6,000	6,000	30,000	30,000
Isoelectric point	4.5	4.5	10	10
Inactivation by 5 mM DTT	No	No	Yes	Yes
Inactivation by boiling (10 min)	ND[a]	ND[a]	No	No

[a] NA, Not determined.

Summary

There appear to be at least three growth factors for mouse BALB/c 3T3 cells in human milk. The purification of the predominant one is described in this chapter. Biochemical and immunological studies indicate that this growth factor is probably a form of human epidermal growth factor (EGF). Like EGF, the major human milk-derived growth factor has a molecular weight of about 6000, a pI of about 4.5, and is resistant to inactivation by dithiothreitol.[4] (See this volume, Harper et al. [1], for purification of human EGF.) In addition, Carpenter has shown that antibodies against human EGF will precipitate most of the growth factor activity for 3T3 cells found in human milk.[8] The EGF-like species of growth factor cannot be detected in bovine milk. Instead, the major growth factor in bovine colostrum appears to be biochemically similar to platelet-derived growth factor (PDGF). Like PDGF, the bovine colostrum-derived growth factor has a molecular weight of about 30,000, a pI of about 10, is totally inactivated by dithiothreitol but is stable to treatments with guanidine–HCl, urea, and heat. Biochemical characterizations of milk-derived growth factors, EGF, and PDGF are summarized in Table III. At present, very little is known about the physiological role of these growth factors in milk. The availability of these growth factors in homogeneous form will facilitate the studies in understanding their possible involvement in the growth process.

[8] G. Carpenter, *Science* **210**, 198 (1980).

[5] Biosynthesis of the Epidermal Growth Factor Receptor in Cultured Cells

By ANN MANGELSDORF SODERQUIST and GRAHAM CARPENTER

As one aspect of our studies into the role of epidermal growth factor (EGF) in the control of cell growth, we have examined the biosynthesis of the EGF receptor in cultured human cells.[1,2] We wanted to learn how this protein is synthesized, how the addition of carbohydrate contributes to receptor processing, and what the kinetics of its insertion into the plasma membrane are. The steps we have followed in an attempt to accomplish these goals are outlined in this chapter.

We have drawn on many techniques that have been developed independently to meet various needs, but that have been successfully applied in concert to the identification and characterization of cellular proteins of low abundance. Many of the techniques — metabolic labeling, polyacrylamide gel electrophoresis, iodinated ligand binding, to name a few — have been dealt with in detail in previous volumes of this series. In this chapter, however, we will concentrate on how we have used and sometimes modified these standard techniques to study the EGF receptor in cultured human cells.

We have worked with two cell types: an epidermoid carcinoma cell line, designated A431,[3] and human diploid fibroblasts. A431 cells have been used extensively for biochemical studies of the epidermal growth factor receptor because they possess an extremely large number of EGF receptors, 2×10^6 receptors/cell[4] (see also the chapter by Barnes, this volume [8]). Human fibroblasts (HF), on the other hand, have a more typical complement of EGF receptors, 1×10^5 receptors/cell.[5] Because of the relative paucity of receptors in these cells, they have not been studied as thoroughly as A431 cells. The methods described in this chapter are, in general, those used in A431 cells. Adaptations of these methods that have applied to fibroblasts will also be included.

[1] C. M. Stoscheck, A. M. Soderquist, and G. Carpenter, *Endocrinology* **116**, 528 (1985).
[2] A. M. Soderquist and G. Carpenter, *J. Biol. Chem.* **259**, 12586 (1984).
[3] D. J. Giard, S. A. Aaronson, G. J. Todaro, P. Arnstein, J. H. Kersey, H. Dosik, and P. Parks, *J. Natl. Cancer Inst. (U.S.)* **51**, 1417 (1973).
[4] H. Haigler, J. F. Ash, S. J. Singer, and S. Cohen, *Proc. Natl. Acad. Sci. U.S.A.* **75**, 3317 (1978).
[5] G. Carpenter, K. J. Lembach, M. M. Morrison, and S. Cohen, *J. Biol. Chem.* **250**, 4297 (1975).

General Methods for the Isolation and Identification of EGF Receptor Forms

The large numbers of EGF receptors on A431 cells have allowed purification of receptor from these cells with relative ease.[6] However, even in these cells, receptors comprise only 0.2% of the total cell protein.[7] The metabolism of receptors has, therefore, been followed by analysis of immunoprecipitated, metabolically labeled receptors separated on polyacrylamide gels.

Materials

A431 cells

HF: Human foreskin fibroblasts

Tissue cultureware: Disposable 35- and 60-mm dishes, 96-well plates

DM: Dulbecco's modified Eagle's medium, supplemented with 20 mM HEPES, pH 7.5, 2 g/liter sodium bicarbonate, 50 ng/ml Garamycin

Met⁻ MEM: Methionine-free modified Eagle's medium

CS: Bovine calf serum

^{35}S-Met: [^{35}S]methionine (New England Nuclear, > 800 Ci/mmol)

CMF-PBS: Calcium- and magnesium-free phosphate-buffered saline

TGP: 1% Triton X-100 (v/v), 10% glycerol (v/v), 1mM phenylmethyl sulfonyl fluoride (PMSF) in 20 mM HEPES, pH 7.4

RIPA: 1% Triton X-100, 1% sodium deoxycholate, 0.1% sodium dodecyl sulfate, 10 mM Tris (pH 7.4), 0.15 M NaCl, 1 mM ethylenediaminetetraacetic acid, 0.02% sodium azide. For solubilizing cells, we add aprotinin (Sigma) to a final concentration of 100 KIU/ml to the RIPA

Polypropylene microfuge tubes (1.5 ml)

Anti-EGF receptor rabbit serum, as previously described[8]

Preimmune rabbit serum

Staph A cells: 10% suspension of formalin-fixed *Staphylococcus aureus* cells (Sigma, Pansorbin). We wash the cells as suggested by Bethesda Research Laboratories for their Staph A preparation. The cells are pelleted; resuspended in an equal volume of 10% 2-mercaptoethanol, 3% SDS, in CMF-PBS; incubated at 95° for 30 min; washed three times and resuspended (bringing up to original volume) in buffer used for washing immunoprecipitates (RIPA, pH 8.5; see below)

RIPA, pH 8.5: RIPA made as above (without aprotinin), then adjusted to pH 8.5 with sodium hydroxide.

[6] S. Cohen, this series, Vol. 99, p. 379.
[7] C. M. Stoscheck and G. Carpenter, *J. Cell. Physiol.* **120,** 296 (1984).
[8] C. M. Stoscheck and G. Carpenter, *Arch. Biochem. Biophys.* **227,** 457 (1983).

[14]C-Labeled molecular weight standards (Bethesda Research Laboratories)

Vertical slab gel apparatus (Hoeffer Model 600)

Enlightening fluorographic enhancer (New England Nuclear)

Gel drier (Hoeffer)

Film (Kodak, X-OMat)

Intensifying screens (Du Pont, Cronex Hi-Plus)

Metabolic Labeling of Cells. For most studies of EGF receptor biosynthesis, we have metabolically labeled the receptor protein by incubating cells with [35S]methionine. Considering the number of receptors/cell, few A431 cells are needed. For many experiments, confluent wells of a 96-well culture dish (approximately 100,000 cells/well) are washed 2 × and refed with 100 μl fresh DM + 10% CS and labeled with 5 μCi of [35S]methionine (50 μCi/ml). Labeling for short periods of time (5 min–1 hr) is more efficient if cells are washed 2 × and refed with 100 μl of methionine-free Met⁻ MEM in the absence of serum. The cells, in this case, are preincubated in the Met⁻ medium for 30 min before the addition of label. If more labeled extract is needed, A431 cells in 35-mm dishes are washed and labeled in 1 ml of the appropriate medium. The dishes of cells are labeled with the same concentration of [35S]methionine, 50 μCi/ml.

Human fibroblasts are labeled basically as above, except that confluent 60-mm culture dishes of cells (approximately 800,000 cells/dish) are used. Dishes are washed and refed with 1.5 ml of either DM + 10% CS or Met⁻ MEM, depending on the intended labeling period, preincubated 30 min if methionine-free medium is used, and labeled with 50 μCi/ml [35S]methionine in either case.

Cell Solubilization. Efficient solubilization of the EGF receptor in A431 cells can be achieved using 1% Triton X-100. At the end of the labeling period, cells are washed 4× with cold CMF-PBS, then solubilized at room temperature with rocking for 20 min in TGP. When dishes of cells are used, they are solubilized at 2 × 10⁶ cells/ml, which is approximately equivalent to 2 mg of cellular protein/ml. Cells that are labeled in wells of 96-well plates are routinely solubilized in 200 μl of TGP to assure good recovery of the solubilized cells from the well. Triton does not solubilize nuclear membranes under either of these conditions. The resulting detergent extract can, therefore, be used immediately for immunoprecipitations, without prior centrifugation to remove debris and viscous DNA. Loss of receptors when small numbers of cells are used, is thus, kept to a minimum.

We have found that EGF receptors in human fibroblasts are not completely solubilized in 1% Triton alone. Complete extraction of receptors from these cells requires harsher treatment, and we routinely employ RIPA buffer to solubilize fibroblasts. A 60-mm dish of cells is washed 4 × with cold CMF-PBS and solubilized in 1 ml of RIPA, using the same conditions as used for A431 cells. Since RIPA solubilizes nuclear membranes, the released DNA must be pelleted or digested before the extracts can be used for immunoprecipitation. We centrifuge RIPA extracts at 100,000 g for 1 hr, then transfer the supernatants to fresh tubes for immunoprecipitations.

Immunoprecipitation. The most critical reagent in the study of EGF receptor biosynthesis is the antibody used to identify the various forms of the receptor. In principle, monoclonal antibodies to the EGF receptor might be expected to be highly specific for the receptor molecule (see also the chapter by Sato *et al.,* this volume [6]). In practice, however, the monoclonal antibodies that have been produced thus far are not as useful for specific immunoprecipitation of the EGF receptor as polyclonal antibodies have been. Several of the monoclonal antibodies are directed against the carbohydrate portion of the receptor.[9-11] Although they are highly specific for a particular carbohydrate sequence, the same antigenic determinants are present on other glycoproteins and glycolipids within the cells. As a result, unrelated proteins are present in the immunoprecipitates.

We have several rabbit antisera that have been raised against either affinity purified EGF receptor (native receptor) from A431 cells, or against purified receptor that has been subjected to SDS–polyacrylamide gel electrophoresis (denatured receptor). Each of these antisera has slightly different characteristics,[8] but all of them are highly specific for the EGF receptor. We have, therefore, found our crude sera to be entirely satisfactory for immunoprecipitations.

To isolate EGF receptors from A431 cells, we mix TGP-solubilized cell extract (2×10^6 cells/ml) with anti-receptor serum in a ratio of 10:1 (v/v) in a 1.5-ml polypropylene microfuge tube. The mixture is incubated for 15 min at room temperature. Since the wells of 96-well plates contain about 100,000 cells at confluency, 5 μl of antiserum is added to the extracts from each well. The appropriate ratio of antiserum to cell extract would, however, have to be determined empirically for other antisera.

We routinely use formalin-fixed *Staphylococcus aureus* cells to precipitate the antigen–antibody complexes, although we have also used protein A coupled to Sepharose in the past. The insolubilized protein A may bind fewer proteins from the cell extracts nonspecifically, but Sepharose beads

[9] P. Fredman, N. D. Richert, J. L. Magnani, M. C. Willingham, I. Pastan, and V. Ginsberg, *J. Biol. Chem.* **258**, 11206 (1983).
[10] P. J. Parker, S. Young, W. J. Gullick, E. L. V. Mayes, P. Bennett, and M. D. Waterfield, *J. Biol. Chem.* **259**, 9906 (1984).
[11] R. A. Childs, M. Gregoriou, P. Scudder, S. J. Thorpe, A. R. Rees, and T. Feize, *EMBO J.* **3**, 2227 (1984).

form a loose pellet even with centrifugation and are, thus, easily lost during repeated washes. After a 15-sec centrifugation in a microfuge, supernatants can be aspirated from Staph A cells without losses, and yet the cells can be resuspended with only moderately vigorous vortexing.

To precipitate the antigen–antibody complexes, a volume of the Staph A cell suspension is added that is 10 times the volume of serum. After incubation for 15 min at room temperature, the Staph A cells are pelleted and washed 4× with 1 ml RIPA buffer, pH 8.5, that has been warmed to 37°. We have found this regimen to be the most effective of several tried for reducing nonspecific binding of labeled proteins either to the rabbit serum or to the Staph A cells.[8]

For immunoprecipitation of receptors from extracts of A431 cells, the above method yields exceedingly clean results. If cells have been labeled with [35S]methionine at 50 μCi/ml for 24 hr in complete medium, or for 1 hr in methionine-free medium, then 50 μl of cell extract (or the contents of 1 well of a 96-well plate) contains sufficient labeled receptors to be seen as a dark band on a 24-hr fluorographic exposure of a dried polyacrylamide gel. The specificity of the receptor immunoprecipitation can be checked by incubating a duplicate sample of extract with the appropriate volume of preimmune rabbit serum followed by Staph A cells. Minor labeled bands visible in the immune precipitation are usually present in the non immune precipitation as well.

Since there are many times fewer receptors on human fibroblasts than A431 cells, more cells must be labeled in order to obtain sufficient radioactivity to visualize the receptor protein. On the other hand, proportionately much less antiserum is needed for complete immunoprecipitation of the fibroblast receptors. For each lane of a polyacrylamide gel, we immunoprecipitate receptors from an entire 60-mm dish of HF (1 ml of RIPA cell extract, representing 800,000 cells). Only 1 μl of our antiserum is needed for the immunoprecipitation. We extend the incubation time to 1 hr at room temperature, however, to assure interaction of the antiserum with the receptors. The antigen–antibody complexes are then precipitated with 10 μl formalin-fixed Staph A cells. The amount of radioactivity in fibroblast receptors immunoprecipitated in this fashion requires a 4- to 7-day fluorographic exposure of the gel.

There is a greater probem of fibroblast cellular proteins binding nonspecifically to Staph A cells than there is in immunoprecipitations from A431 cells. We find that a simple palliative is to prebind unlabeled fibroblast proteins to the Staph A cells. To do this, we solubilize unlabeled fibroblasts in RIPA at a concentration of approximately 800,000 cells/ml. After pelleting the DNA and debris, as described above, we mix the RIPA extract with a 10% suspension of Staph A cells in 1:1 ratio (v/v), incubate the mixture for 30 min at room temperature, pellet the fixed cells in a clinical centrifuge, and resuspend the pellet in RIPA to the original volume. Incubation of antigen–antibody complexes with the prebound Staph A cells is for the 15 min. The immune complex–Staph A pellets are then washed 4× with warm RIPA, pH 8.6.

Polyacrylamide Gel Electrophoresis. In order to characterize the various forms of the EGF receptor, immunoprecipitates from metabolically labeled cells are subjected to SDS–polyacrylamide gel electrophoresis, and the radiolabeled proteins are visualized by fluorography. The apparent molecular weights of the receptor forms can then be estimated by comparing their electrophoretic migration with that of radiolabeled proteins of known molecular weight.

To prepare immunoprecipitates for electrophoresis, washed Staph A pellets are aspirated as dry as possible without losing any cells. The cells are resuspended in 100 μl Laemmli sample buffer,[12] then incubated at 60° for 5 min to assure denaturation of the immunoprecipitated proteins. Alternatively, the samples can be heated in a boiling water bath for 2 min, but we find that microfuge tubes from various manufacturers tend to "pop" open under these conditions, resulting in substantial loss of sample. After heating, the samples are centrifuged for 1 min to pellet the Staph A cells tightly. The supernatants are then loaded on SDS–polyacrylamide gels, as is a 2-μl aliquot of ^{14}C-labeled molecular weight standards, which has also been heated in 100 μl Laemmli sample buffer.

We make discontinuous SDS–polyacrylamide gels as described by Laemmli.[12] Using a Model 600 Hoeffer gel apparatus, we make 7.5% gels that are 1.5 mm thick and have 2.5% stackers. The gels are run either overnight at 40 V, or for approximately 2 hr at 50 mA, whichever is more convenient. In either case, the gels are cooled with tap water.

When the dye front has reached the bottom of the gel, the gels are removed from the apparatus and the bottom, left-hand corner is clipped to mark the orientation of the gel. We treat all gels containing [35S]methionine-labeled proteins for 30 min with Enlightening before placing them on filter paper backing and drying them for 1.5 hr on a Hoeffer gel dryer set at 60°. The dried gels are exposed to Kodak X-OMat film at −70° in an X-ray film cassette containing a Du Pont Hi-Plus intensifying screen.

Methods for Studying the Kinetics of EGF Receptor Biosynthesis

In both A431 cells and human fibroblasts, the major protein identified by immunoprecipitation after a 24-hr labeling period with [35S]methionine migrates with an apparent molecular weight of approximately 170,000.[1,2]

---

[12] U. K. Laemmli, *Nature (London)* **227,** 680 (1970).

This protein is not immunoprecipitated from either cell type when cell extracts are incubated with preimmune rabbit serum. Nor is it identified in cells that have been labeled for 1 hr or less. Rather, a 160,000-Da EGF receptor precursor is present in both cell types labeled for short periods with [35S]mehionine. A 90,000-Da protein is also specifically immunoprecipitated from A431 cells labeled for short periods of time but is not found in human fibroblasts or several other cell types regardless of the labeling time. While both the 160,000- and 90,000-Da proteins are the major labeled proteins immunoprecipitated from cells labeled 1 hr or less they are both minor in cells labeled for 24 hr or more. The 90,000-Da protein is not a precursor of the membrane-bound EGF receptor.[2,13] Other methods for analyzing this receptor-related protein will be described below.

Pulse–Chase Labeling of the Receptor. The kinetics of conversion of the receptor precursor to its mature form can be studied by pulse labeling A431 cells with [35S]methionine, then chasing them for various times with fresh medium. The 160,000-Da receptor precursor is labeled when A431 cells are exposed to [35S]methionine for as little as 5 min.[1,2] However, we routinely pulse label the cells for 1 hr in methionine-free medium. Most of the receptor remains in its 160,000-Da form, but incorporates much more label. Several dishes, or wells of a 96-well plate, are preincubated in methionine-free medium and labeled with [35S]methionine for 1 hr at 37°. At the end of the pulse, one dish or well is washed, and the cells are solubilized as described above. The remainder of the dishes or wells are quickly washed 2× with DM + 10% CS at 37° to remove the unincorporated label, then refed with the appropriate volume of this medium. Incubation of these dishes or wells then continues in the complete medium for various lengths of time (1–4 hr, for example), to follow processing of the pulse-labeled receptor to its mature form. After each time interval, the cells are washed and solubilized as described above.

To avoid fluctuations in temperature during the period of the experiment, the dishes or 96-well plates are incubated in a small water bath set to 37°. The water level is adjusted so that it just covers a plastic test tube rack submerged in the water bath. The bottoms of the dishes or plates should be in contact with the water, but should rest on the rack, not float.

When working with dishes of cells, we usually label all of the dishes simultaneously (adding label to dishes at 20-sec intervals, for instance). The chase can be started in the same manner, and at the appropriate time each dish can be removed from the water bath for solubilization of the cells. It is more convenient when using multiwell plates to stagger the beginning of the labeling in each well so that all of the wells are ready to be

---

[13] A. M. Soderquist, C. M. Stoscheck, and G. Carpenter, in preparation.

solubilized at the same time. As long as the experiment is taking place over only a few hours, the difference in cell number between the first and last well labeled will be insignificant.

At the end of a pulse-chase experiment, receptors are immunoprecipitated from the detergent extracts, separated on gels, and visualized by fluorography, as described above. The rate of conversion of receptor precursor to mature form can then be determined by densitometric scans of the 160,000- and 170,000-Da receptor bands on the resulting film. Such scans reveal that the conversion in A431 cells occurs with half-time of 1.7 hr.[1] Also apparent in a pulse-chase of A431 cells is the disappearance of the 90,000-Da protein that coprecipitates with the receptor precursor in early time points. The disappearance of the 90,000-Da protein is coincident with that of the receptor precursor in these cells.

Immunoprecipitation of Cell Surface Receptors. Since ligand binding takes place on the cell surface, it is of interest to determine how long it takes newly synthesized EGF receptors to be transported to the cell surface.

Our anti-receptor sera recognize the extracellular domain of the receptor molecule.[8] As a result, we have been able to monitor the appearance of newly synthesized receptor molecules on the surface of A431 cells by combining a typical pulse-chase technique with immunoprecipitation of cell surface receptors. To minimize the amount of antiserum needed, these experiments are done on confluent wells of 96-well dishes. The pulse labeling and chase are performed as described above, so that all of the time points end simultaneously. At that point, instead of solubilizing the cells, the 96-well plate is transferred to ice and washed 2× and refed with 100 μl cold serum-free DM. Either preimmune or anti-receptor rabbit serum (5 μl) is added to each well and incubated with the cells for 1 hr at 4°. The unbound rabbit serum is removed from cells by four washes with 200 μl cold CMF/PBS, and the cells are solubilized by a 20-min incubation at room temperature with 200 μl TGP. The solubilized cells are transferred to microfuge tubes, and the immune complexes are precipitated by a 15-min incubation with 50 μl Staph A cells. The resulting Staph A pellets are washed and prepared for electrophoresis.

One potential problem with this technique would be exchange of the antiserum between cell surface and intracellular receptor molecules once the cells are solubilized. One can show that this is not happening (at least when cells are solubilized with TGP) by labeling cells for 4 hr with [^{35}S]methionine, then immunoprecipitating cell surface receptors as described. Other experiments have shown that there is incorporation of [^{35}S]methionine into the 90,000-Da protein and the 160,000-Da receptor precursor, as well as the mature, 170,000-Da receptor during a 4-hr labeling period. All three receptor species are detectable in immunoprecipitates from detergent extracts of cells labeled for 4 hr. However, when cells are labeled for the same time, but the anti-receptor serum is added to intact rather than solubilized cells, only the 170,000-Da receptor is immunoprecipitated. Although immunoprecipitation of cell surface receptors is specific, this technique is not quantitative. At least one of the possible reasons for this is readily apparent: antibodies added to intact cells growing in monolayer have access to only the upper surfaces of the cells. Nevertheless, this modified immunoprecipitation technique has been useful. We have shown that EGF receptors pulse labeled for 1 hr begin to appear on the plasma membrane of A431 cells within the first hour of the chase period.[1] Maximum antiserum detection of receptors on the cell surface occurs between 3 and 4 hr of chase.

Methods for Analyzing EGF Receptor Glycosylation

To study the glycosylation of the EGF receptor in A431 cells, we have metabolically labeled cells with either [^{35}S]methionine to follow the protein portion or ^{3}H-labeled monosaccharides to follow the carbohydrate portion of the receptor. The extent of receptor glycosylation has then been determined by treating cells with tunicamycin, an inhibitor of N-linked glycosylation, and examining the immunoprecipitable receptor species. We have corroborated these data by treating immunoprecipitated receptors with Endo H, an enzyme that removes high-mannose N-linked oligosaccharides from glycoproteins. To further characterize the N-linked chains of the receptor, peptides from its ^{3}H-labeled monosaccharide-labeled precursor and mature forms have been subjected to Con A chromatography.

Materials
A431 cells
Glc$^-$ MEM: Glucose-free modified Eagle's medium
[^{3}H]GlcNH$_2$: D-[1^{3}H]Glucosamine hydrochloride (Pathfinder Laboratories, 10–30 Ci/mmol)
[^{3}H]Man: D-[2-^{3}H]Mannose (Pathfinder Laboratories, 10–30 Ci/mmol)
TM: Tunicamycin (Calbiochem). A 10 mg/ml stock solution is made up in dimethyl sulfoxide and stored at 4°. It is diluted with sterile water to the appropriate working concentration just before use
Endo H: Endoglycosidase H (Miles). The enzyme is stored at −20° at a concentration of 1 U/ml in sterile filtered 50 mM sodium citrate, pH 6.0
Endo H buffer: 0.1 M sodium citrate (pH 6.0), 0.1% Triton X-100
2× Laemmli buffer: Double-strength Laemmli (12) buffer
Pronase: Type XIV, nonspecific protease from *Streptomyces griseus* (Sigma)

Pronase buffer: 0.1 M Tris (pH 8.0), 1mM calcium chloride, 10 mg/ml
  Pronase
Pasteur pipets
Glass wool
Con A – Sepharose: Concanavalin A – Sepharose (Pharmacia)
Con A elution buffer: 10 mM Tris (pH 8.0); 1 mM CaCl$_2$, 1 mM
  MgSO$_4$, 150 mM NaCl, 0.02% sodium azide
α-m-Glc: α-methylglucoside (Sigma)
α-m-Man: α-methylmannoside (Sigma)
Aqueous scintillation fluid

Metabolic Labeling of Receptor Oligosaccharides. The covalent addition of carbohydrate to the EGF receptor in A431 cells can be demonstrated by incorporation of [³H]glucosamine or [³H]mannose into the immunoprecipitable 170,000-Da receptor molecule. This is accomplished in the same manner as metabolic labeling of the receptor polypeptide. That is, cells are refed with fresh medium and incubated for either 24 or 1 hr at 37° in the presence of labeled monosaccharide. Typically we have labeled 35-mm dishes of A431 cells in 1 ml of medium. For short pulse labeling we have washed the cells twice and refed them with Glc⁻ MEM, then preincubated them for 30 min in this medium before adding the labeled monosaccharide. For longer labeling times, the cells have been washed 2× and refed with DM + 10% CS, then labeled immediately. In either case, 50 μCi of ³H-labeled monosaccharide is added. At the end of the labeling period, the cells are washed and solubilized as described for [³⁵S]methionine labeling. For immunoprecipitation, 100-μl aliquots of cell extract (representing 200,000 cells) are immunoprecipitated with 10 μl anti-receptor serum followed by 100 μl of Staph A cells and are processed for electrophoresis as previously described.

Treatment of the gels with Enlightening enhancer makes a dramatic difference in the time necessary to see [³H]glucosamine- or [³H]mannose-labeled receptors. Dark bands are visible on film exposed for 3 days at −70° with an intensifying screen, if the gel has been soaked 30 min in Enlightening solution. Without Enlightening solution, no bands are apparent even after a 7-day exposure under the same conditions.

Both [³H]glucosamine and [³H]mannose are incorporated into primarily the mature 170,000-Da receptor in A431 cells labeled for 24 hr. After a 1-hr incubation with either labeled sugar, the major bands immunoprecipitated correspond to the 160,000-Da receptor precursor and the 90,000-Da protein.

Tunicamycin Inhibition of N-Linked Glycosylation. The antibiotic, tunicamycin, inhibits the cotranslational addition of N-linked oligosaccharides to protein by interfering with the first step in synthesis of the core oligosaccharide.[14] The contribution of N-linked oligosaccharides to glycosylation of the EGF receptor can, thus, be assessed by immunoprecipitating receptors from cells treated with tunicamycin and comparing their apparent molecular weights with receptors from untreated cells.

A431 cells in 35-mm dishes or the wells of 96-well plates are washed 2× and refed with the appropriate volume of DM + 10% CS, as described above. Tunicamycin is then added to a final concentration of 2 μg/ml, a concentration that has been found to inhibit glycosylation in A431 cells by approximately 80%, while inhibiting protein synthesis by only 20%.[2] After a 30-min pretreatment with tunicamycin, either [^{35}S]methionine or ^{3}H-labeled monosaccharides (50 μCi/ml in either case) are added, and the cells are labeled for 24 hr. Solubilization of the cells, immunoprecipitation of receptors, electrophoresis, and fluorography are as described above.

The [^{35}S]methionine-labeled receptor band from tunicamycin-treated cells migrates at 130,000 Da.[2] This 40,000-Da difference compared with receptors from untreated cells suggests that several N-linked oligosaccharides are added to the EGF receptor in these cells. No labeled band is visible in immunoprecipitates from tunicamycin-treated cells labeled with either [^{3}H]GlcNH$_2$ or [^{3}H]Man; therefore, 130,000 should represent the molecular weight of the protein portion of the receptor.

Endoglycosidase H Digestion of Immunoprecipitated Receptors. Although the specificity of endo H is greater than often realized, in general this enzyme removes high mannose-type-N-linked oligosaccharides from glycoproteins. Thus, the Glc$_3$Man$_9$GlcNAc$_2$ oligosaccharide cores that are added cotranslationally to nascent polypeptides within the endoplasmic reticulum are sensitive to endo H removal. On the other hand, complex N-linked chains, in which the glucose residues and most of the mannose residues have been replaced by sialic acid, *N*-acetylglucosamine and galactose are resistant to endo H digestion.

Endo H digestion can be performed on radiolabeled, immunoprecipitated receptors by resuspending the washed Staph A pellets in 50 μl endo H buffer and incubating the pellets with 5 mU endo H for 18 hr at 37°. After digestion, 50 μl of 2 × Laemmli buffer is added, the sample is heated to 60° for 5 min, the Staph A cells are pelleted, and the supernatant is loaded on a gel.

Endo H digestion of the immunoprecipitated 160,000-Da receptor precursor results in a 30,000 shift in molecular weight to 130,000.[2] This is the same molecular weight as the aglycoreceptor synthesized in the presence of TM, and therefore suggests that all of the N-linked oligosaccharides of the receptor precursor are high mannose, as expected. Endo H removes

[14] J. S. Tkacz and J. O. Lampen, *Biochem. Biophys. Res. Commun.* **65**, 248 (1975).

only approximately 10,000 Da from the mature 170,000-Da receptor, indicating that most but not all of the N-linked chains on the mature receptor have been processed to complex type.

Concanavalin A-Sepharose Chromatography of Receptor Peptides. Differential elution of glycopeptides from Con A-Sepharose columns[15,16] can be used to further characterize the N-linked oligosaccharides of the EGF receptor.[17] Aglycopeptides do not interact with the Con A and thus pass through the column. Glycopeptides can be separated into three categories based on the affinity of their binding to Con A. Complex-type N-linked oligosaccharides that exhibit tri- and tetraantennary branching are not bound to Con A and can be collected in the flow-through of the column. Biantennary complex chains are bound by Con A but can be eluted using relatively mild means. High-mannose chains are tightly bound by Con A and require high concentrations of α-methylmannoside for elution.

To prepare receptor glycopeptides, EGF receptors are first immunoprecipitated from A431 cells labeled for 1 or 24 hr with either [^{3}H]glucosamine or [^{3}H]mannose, as described above. If receptors are immunoprecipitated from three to four 100-μl aliquots of labeled cell extract, there will be several thousand cpm of receptor glycopeptides to load on the Con A column. The immunoprecipitates are subjected to SDS–polyacrylmide electrophoresis and the radiolabeled bands are visualized by fluorography. Using the fluorograph as a guide, the labeled bands are cut out of the dried gel, rehydrated in 1 ml Pronase buffer, then incubated at 60° for 3 hr. After digestion, 3 ml of water is added to the digested bands, and they are boiled for 10 min. The liquid is transferred to a fresh tube, and the gel bands are boiled for 10 min in another 3 ml of water. The combined washes are lyophilized to dryness, then resuspended in 1 ml water.

Con A-Sepharose is brought to room temperature before columns are poured. One-milliliter columns are poured in disposable Pasteur pipets plugged with glass wool and are washed with at least two column volumes of Con A elution buffer. The 1-ml sample is applied to the column, allowed to run in, and followed by 1 ml of elution buffer. These 2 ml are collected in tube 1. The column is then washed with four more 2-ml aliquots of elution buffer, collected as fractions 2–5. These first five fractions will contain receptor aglycopeptides, as well as glycopeptides having complex N-linked oligosaccharides that are not bound by Con A. To elute peptides containing biantennary complex oligosaccharides, the column is then

[15] R. D. Cummings and S. Kornfeld, *J. Biol. Chem.* **257**, 11235 (1982).
[16] R. Merkel and R. D. Cummings, this series, Vol. 138, p. 232.
[17] R. D. Cummings, A. M. Soderquist, and G. Carpenter, *J. Biol. Chem.*, **260**, 11944 (1985).

washed with 10 ml of elution buffer containing 10 mM α-methylglucoside. Five 2-ml fractions are collected. Peptides containing high-mannose oligosaccharides are eluted with 10 ml of elution buffer containing 100 mM α-methylmannoside, which has been heated to 60° before addition to the column. Five 2-ml fractions are collected. Aliquots (for instance, 200 μl of each 2-ml fraction) of all of the column fractions are then counted in aqueous scintillation fluid. The percentage of radioactivity collected in the flow-through and under each of the two elution conditions is calculated. The flow-through is designated peak I; the α-m-Glc wash, peak II; and the α-m-Man wash, peak III.

If peptides from the mature EGF receptor are applied to a Con A column, 74% of the radioactivity from [³H]glucosamine-labeled glycopeptides is collected in peak I, 15% in peak II, and 11% in peak III, indicating that the N-linked oligosaccharides of the mature receptor are mostly of the complex type.[17] The profile from [³H]mannose-labeled mature receptor indicates that 44% of the radioactivity is in peak I, 12% is in peak II, and 43% is in peak III. If peptides from either [³H]mannose- or [³H]glucosamine-labeled receptor precursors are subjected to Con A chromatography, however, all of the radioactivity requires α-methylmannoside for elution, confirming that all of the N-linked oligosaccharides of the receptor precursor are high mannose.

Methods for Identifying the Secreted Receptor Form from A431 Cells

Anti-EGF receptor serum specifically immunoprecipitates a 110,000-Da protein from the medium of A431 labeled with [³⁵S]methionine for several hours.[13] In A431 cell-conditioned medium that has been ultracentrifuged to remove any membrane fragments or whole cells, the 110,000-Da protein is the only labeled protein immunoprecipitated. The disappearance of the 90,000-Da receptor-related protein from the cells during pulse-chase experiments raises the possibility that the cellual 90,000-Da protein may be the precursor of the extracellular 110,000-Da receptor-related protein.

Materials
^{125}I-Labeled EGF and unlabeled EGF[18]
HEPES–BSA: A 10× stock solution of 200 mM HEPES, pH 7.4, 0.5% bovine serum albumin
A431 medium: Conditioned medium collected from A431 cells

---

[18] G. Carpenter, this series, Vol. 109, p. 101.

Biosynthesis of the Extracellular Receptor-Related Protein. The possibility that the 90,000-Da protein is the precursor of the 110,000-Da protein can be examined by performing a typical pulse-chase experiment in which the medium, as well as the solubilized cells from each time point, is immunoprecipitated. If the volume of TGP for cell solubilization is the same as the volume of medium in which the cells were grown, equal aliquots of solubilized cell extract and medium represent the same number of cells. We typically plan to have 35-mm dishes of A431 cells at a density of approximately 2×10^6 cells and in 1 ml of medium at the end of the pulse-chase experiment. The medium is removed, and the cells can then be solubilized at the normal ratio of 2×10^6 cells/ml. After collection, the medium is centrifuged at 100,000 rpm for 1 hr to remove any whole cells or cellular debris. Immunoprecipitations are performed on 50-μl aliquots of cell extract and the corresponding medium for each time point. The amounts of serum and Staph A cells, and the incubation times for each, are the same as those described above.

We have found that the increase in 110,000-Da protein in the medium over time is inversely proportional to the decrease in 90,000-Da protein in the cells. In fact, densitometric scans show that the amount of 110,000-Da protein in the medium after a 4-hr chase is nearly equivalent to the amount of 90,000-Da protein present in the 1-hr pulse.[13]

EGF Binding Assay for Secreted Receptor Form. Immunoprecipitation of the 110,000-Da protein with anti-EGF receptor only demonstrates that it is antigenically related to the membrane-bound EGF receptor. Functional relatedness can be demonstrated by measuring the ligand-binding capacity of the 110,000-Da protein in the medium.

The ligand-binding protocol we have used is essentially that used by Stoscheck and Carpenter.[8] A431 cell-conditioned medium is collected from 35-mm dishes of cells and ultracentrifuged at 100,000 g for 1 hr. Triplicate 20-μl aliquots of medium are incubated for 30 min at room temperature in 100-μl reaction mixtures containing 40 ng ^{125}I-labeled EGF (approximately 50,000 cpm/ng) in 20 mM HEPES, pH 7.4, 0.05% BSA. A 25-fold excess (1 μg) of unlabeled EGF is added to 1 of the triplicate samples. To isolate the 110,000-Da protein specifically, the assay mixtures are then incubated for 60 min at room temperature with 4 μl of anti-EGF receptor serum, and the antigen–antibody complexes are precipitated by the addition of 40 μl Staph A cells. After a 15-min incubation, the Staph A cell pellets are washed 4X with 20 mM HEPES, pH 7.4, 0.2% Triton X-100, and are resuspended in 1 ml of this buffer for counting in a gamma counter. Specific binding of ^{125}I-labeled EGF to each sample of medium can be calculated by subtracting nonspecific counts (in the presence of excess unlabeled EGF) from total counts (in the absence of unlabeled EGF).

That the secreted 110,000-Da protein from A431 cells is functionally related to the EGF receptor is shown by the specificity of ^{125}I-labeled EGF binding to it. As indicated above, our anti-receptor serum recognizes only the 110,000-Da protein in medium from [^{35}S]methionine-labeled A431 cells. If normal rabbit serum is substituted for anti-receptor serum in the above assay, very few counts are precipitated. If anti-receptor serum is used, but the A431 cell-conditioned medium is replaced with fresh DM + 10% CS, few counts result, as well. However, anti-receptor serum precipitates significant numbers of counts from A431 cell-conditioned medium that has been incubated with ^{125}I-labeled EGF.[13]

Acknowledgments

This work was supported by Research Grant CA24071 from the National Cancer Institute and BC294 from the American Cancer Society. A.M.S. was supported in part by National Cancer Institute Training Grant CA 09313. G.C. is the recipient of an Established Investigator Award from the American Heart Association.

[6] Derivation and Assay of Biological Effects of Monoclonal Antibodies to Epidermal Growth Factor Receptors

By J. DENRY SATO, ANH D. LE, and TOMOYUKI KAWAMOTO

In recent years a considerable number of monoclonal antibodies have been raised against epidermal growth factor (EGF) receptors[1-10] (Table I[8-14]). In each case EGF receptors of A431 human epidermoid carcinoma

[1] A. B. Schreiber, I. Lax, Y. Yarden, Z. Eshhar, and J. Schlessinger, *Proc. Natl. Acad. Sci. U.S.A.* **78,** 7535 (1981).

[2] M. D. Waterfield, E. L. V. Mayes, P. Stroobant, P. L. P. Bennet, S. Young, P. N. Goodfellow, G. S. Banting, and B. Ozanne, *J. Cell. Biochem.* **20,** 115 (1982).

[3] A. B. Schreiber, T. A. Libermann, I. Lax, Y. Yarden, and J. Schlessinger, *J. Biol. Chem.* **258,** 846 (1982).

[4] J. Schlessinger, I. Lax, Y. Yarden, H. Kanety, and T. A. Libermann, *in* "Receptors and Recognition" (M. F. Greaves, ed.), Vol. 17, pp. 279–303. Chapman and Hall, London, 1984.

cells[15] were used as the immunogen. These cells are a particularly good source of EGF receptors as they express an unusually large number of receptor molecules (1 to 3×10^6/cell).[15,16] Thus, whole A431 cells and A431 plasma membrane vesicles[9] in addition to partially purified EGF receptors[5,6] have been successfully used to raise EGF receptor monoclonal antibodies.

The receptor antibodies that have been described in the literature are classified in Table I with respect to their abilities to inhibit EGF binding to A431 cells and to their specificities for carbohydrate antigenic determinants. Although Table I is incomplete, several interesting properties of these antibodies are apparent. A large majority of the antibodies bind to oligosaccharides; an additional 10 monoclonal antibodies directed against A431 EGF receptors[4] have also been found to react with carbohydrate determinants.[11] All of the monoclonal antibodies with carbohydrate specificities recognize blood group antigen-related determinants.[8–14] This result may not be surprising in view of the fact that the extracellular portion of the human EGF receptor has 12 potential asparagine-linked glycosylation sites[17] of which 11 are modified in A431 cells.[18] However, as a conse-

---

[5] T. Kawamoto, J. D. Sato, A. Le, J. Polikoff, G. H. Sato, and J. Mendelsohn, *Proc. Natl. Acad. Sci. U,S.A.* **80**, 1337 (1983).

[6] J. D. Sato, T. Kawamoto, A. D. Le, J. Mendelsohn, J. Polikoff, and G. H. Sato, *Mol. Biol. Med.* **1**, 511 (1983).

[7] N. D. Richert, M. C. Willingham, and I. Pastan, *J. Biol. Chem.* **258**, 8902 (1983).

[8] M. Gregoriou and A. R. Rees, *EMBO J.* **3**, 929 (1984).

[9] J. E. Kudlow, M. J. Khosravi, M. S. Kobrin, and W. W. Mak, *J. Biol. Chem.* **259**, 11895 (1984).

[10] P. J. Parker, S. Young, W. J. Gullick, E. L. V. Mayes, P. Bennett, and M. D. Waterfield, *J. Biol. Chem.* **259**, 9906 (1984).

[11] H. C. Gooi, E. F. Hounsell, I. Lax, R. M. Kris, T. A. Libermann, J. Schlessinger, J. D. Sato, T. Kawamoto, J. Mendelsohn, and T. Feizi, *Biosci. Rep.* **5**, 83 (1985).

[12] H. C. Gooi, J. Schlessinger, I. Lax, Y. Yarden, T. A. Libermann, and T. Feizi, *Biosci. Rep.* **3**, 1045 (1983).

[13] P. Fredman, N. D. Richert, J. L. Magnani, M. C. Willingham, I. Pastan, and V. Ginsburg, *J. Biol. Chem.* **258**, 11206 (1983).

[14] H. C. Gooi, J. K. Picard, E. F. Hounsell, M. Gregoriou, A. R. Rees, and T. Feizi, *Mol. Immunol.,* **22**, 689 (1985).

[15] R. N. Fabricant, J. E. De Larco, and G. Todaro, *Proc. Natl. Acad. Sci. U.S.A.* **74**, 565 (1977).

[16] H. T. Haigler, J. F. Ash, S. J. Singer, and S. Cohen, *Proc. Natl. Acad. Sci. U.S.A.* **75**, 3317 (1978).

[17] A. Ullrich, L. Coussens, J. S. Hayflick, T. J. Dull, A. Gray, A. W. Tam, J. Lee, Y. Yarden, T. A. Libermann, J. Schlessinger, J. Downward, E. L. V. Mayes, N. Whittle, M. D. Waterfield, and P. H. Seeburg, *Nature (London)* **309**, 418 (1984).

[18] E. L. V. Mayes and M. D. Waterfield, *EMBO J.* **3**, 531 (1984).

TABLE I
EGF RECEPTOR MONOCLONAL ANTIBODIES

Antibody	Ig class	Inhibits EGF binding	Binds to carbohydrate
2G2 IgM	IgM	+ (1)[a]	ND[b]
EGFR1	IgG$_{2b}$	− (2)	− (11)
TL5 IgG	IgG$_{2a}$	− (3)	+ (12)
29.1 IgG	IgG$_1$	− (4)	+ (11)
528 IgG	IgG$_{2a}$	+ (5, 6)	− (11)
579 IgG	IgG$_{2a}$	+ (5, 6)	− (11)
225 IgG	IgG$_1$	+ (5, 6)	− (11)
455 IgG	IgG$_1$	− (5, 6)	+ (11)
101	IgG$_3$	− (7)	+ (13)
EGR/G49	IgG$_3$	− (8)	+ (8, 14)
B1D8	IgG$_{2a}$	+ (9)	− (9)
1A	IgG$_3$	ND	+ (10)
2A	IgM	ND	+ (10)
3A	IgG$_3$	ND	+ (10)
4A	IgG$_3$	ND	+ (10)
8A	IgM	ND	+ (10)
9A	IgG$_3$	ND	+ (10)
10A	IgG$_3$	ND	+ (10)
11A	IgG$_3$	ND	+ (10)
13A	IgG$_3$	ND	+ (10)
14A	IgG$_3$	ND	+ (10)

[a] Numbers in parentheses refer to footnotes.
[b] ND, No data available.

quence, the carbohydrate nature of the determinants for at least some of these antibodies restricts their reactivity to EGF receptors of A431 cells.[7,8,10] Furthermore, antibodies with carbohydrate specificities may be expected to react with other glycoproteins, glycolipids, and polysaccharides,[11,19] which would bear upon their usefulness as diagnostic and analytical reagents. A second feature of the monoclonal antibodies in Table I is that the antibodies which inhibit the binding of EGF to A431 cells apparently do not react with carbohydrate determinants, and, conversely, those antibodies with carbohydrate specificities that have been tested do not block EGF binding. Thus, these two properties may be mutually exclusive for EGF receptor antibodies. The reactivity patterns of these antibodies suggest that oligosaccharides are not involved in the binding of EGF to its receptor. They further indicate that the methods used to screen hybrid-

[19] T. Feizi, *Nature (London)* **314**, 53 (1985).

omas for EGF receptor antibodies can be selected to bias the binding specificities as well as the potential uses of those antibodies.

Last, only three of the antibodies in Table I are IgM molecules with the remainder being IgG molecules. This distribution is probably a reflection of the immunization protocols used to raise the antibodies, but it is of interest because of all the EGF receptor antibodies described only 2G2 IgM has been reported to mimic the biological activity of EGF.[1] This antibody, which inhibits EGF binding to target cells, induced morphological changes in A431 cells,[1] stimulated the activity of the EGF receptor protein kinase,[1] induced receptor clustering and internalization,[3] stimulated DNA synthesis by human fibroblasts *in vitro*,[1] and stimulated prolactin synthesis by GH$_3$ rat pituitary cells.[20] These properties of 2G2 IgM have led to the hypothesis that the information necessary to elicit the delayed cellular responses to EGF resides in the receptor molecule.[1,3] The production and characterization of additional IgM monoclonal antibodies to the EGF receptor may be very useful in understanding how ligand–receptor interactions trigger complex cellular responses.

Maintenance of Cell Lines

A431 human epidermoid carcinoma cells,[15] HeLa S cells, and early passage human foreskin fibroblasts were maintained in DME-F12 medium [1:1 mixture by volume of Dulbecco's modified Eagle's medium (Gibco) and Ham's F12 medium (Gibco)] supplemented with 5% newborn calf serum (NCS) (Irvine Scientific).[6] The medium included 15 mM HEPES (pH 7.6), 2.2 g/liter NaHCO$_3$, 40 mg/liter penicillin, 8 mg/liter ampicillin, and 90 mg/liter streptomycin. The cells were incubated at 37° in a humidified atmosphere of 5% CO$_2$.

NS-1-Ag4-1[21] (NS-1) and P3-X63-Ag8.653[22] (X63.653) mouse myeloma cells were maintained at 37° in RDF + 5F (factor) medium[6] supplemented with 1% fetal calf serum (FCS) (batch T48006; Reheis). This medium consisted of a 1:1 mixture (by volume) of RPMI 1640 (Gibco) and DME-F12 medium (RDF nutrient medium), 10 μg/ml bovine insulin (Sigma), 10 μg/ml human transferrin (Sigma), 10 μM 2-aminoethanol (Sigma), 1 × 10^{-9} M sodium selenite (Sigma or Difco), and 10 μM 2-mercaptoethanol (Sigma). RDF nutrient medium included 2 mM L-glutamine, 0.01% sodium pyruvate, 15 mM HEPES, 2.2 g/liter NaHCO$_3$, 40 mg/liter

[20] J. Hapgood, T. A. Libermann, I. Lax, Y. Yarden, A. B. Schreiber, Z. Naor, and J. Schlessinger, *Proc. Natl. Acad. Sci. U.S.A.* **80**, 6451 (1983).

[21] G. Kohler and C. Milstein, *Eur. J. Immunol.* **6**, 511 (1976).

[22] J. F. Kearney, A. Radbruch, B. Liesegang, and K. Rajewsky, *J. Immunol.* **123**, 1548 (1979).

penicillin, 8 mg/liter ampicillin, and 90 mg/liter streptomycin. NS-1-503 myeloma cells,[23] cloned from NS-1-Ag4-1, were passaged in RDF + 5F medium to which was added 1 mg/ml fatty acid-free BSA (Miles) complexed with oleic acid (Sigma) in a 1:2 molar ratio. BSA–oleic acid was prepared by adding dropwise 20 μl of oleic acid solution (20 mg/ml in 100% ethanol) per 1 ml of sterile fatty acid-free BSA (50 mg/ml in PBS or RDF) with constant shaking at room temperature.[23] Newly formed hybridomas were plated in the medium used to maintain their myeloma parent cells. While being established as cell lines, hybridomas were adapted to grow in KSLM[23,24] serum-free medium or in LDL (low density lipoprotein)-free KSLM medium.[24] KSLM medium is RDF + 5F medium supplemented with 0.5 mg/ml BSA–oleic acid and 2 μg/ml human LDL.[23,24]

Production of EGF Receptor Monoclonal Antibodies

For an excellent discussion of the techniques used to produce murine monoclonal antibodies the reader is referred to Ref. 25. EGF receptor monoclonal antibodies were raised by immunizing BALB/c mice intravenously three or four times at fortnightly intervals with approximately 10 μg of A431 EGF receptors. Receptors from A431 membranes solubilized with Triton X-100 were prepared by chromatography on EGF Affi-Gel as described by Cohen et al.,[26] and they were estimated to be about 30% pure by SDS–polyacrylamide gel electrophoresis[27] and silver staining.[28] One month after the last immunization the mice were given an intraperitoneal boost of 1×10^6 to 5×10^6 paraformaldehyde-fixed A431 cells. Four days later 2 to 5×10^7 myeloma cells were fused with a 10-fold excess of splenocytes from immunized mice as described by Galfré et al.[29]; we used 1 ml of 37% polyethylene glycol 1540 (Baker) in RDF medium as the fusing agent. Fusion products were resuspended in the medium used to maintain the parent myeloma cell line, HAT (hypoxanthine, aminopterin, thymidine)[30] was added, and the cells were plated at densities of 2 to 5×10^5 splenocytes per microtest well in 100 μl of medium. Subsequent

[23] T. Kawamoto, J. D. Sato, A. Le, D. B. McClure, and G. H. Sato, Anal. Biochem. 130, 445 (1983).
[24] T. Kawamoto, J. D. Sato, D. B. McClure, and G. H. Sato, this series, Vol. 121, p. 266.
[25] P. Parham, this series, Vol. 92, p. 110.
[26] S. Cohen, G. Carpenter, and L. King, Jr., J. Biol. Chem. 255, 4834 (1980).
[27] U. K. Laemmli, Nature (London) 277, 680 (1970).
[28] B. R. Oakley, D. R. Kirsch, and N. R. Morris, Anal. Biochem. 105, 361 (1980).
[29] G. Galfré, S. C. Howe, C. Milstein, G. W. Butcher, and J. C. Howard, Nature (London) 266, 550 (1977).
[30] J. W. Littlefield, Science 145, 709 (1964).

fusion experiments have suggested that a single intraperitoneal immunization of BALB/c mice with whole A431 cells may generate more EGF
receptor antibodies than the immunization protocol described here (A. D.
Le and H. Masui, unpublished results).

Screening of Hybridomas and Characterization of EGF Receptor Antibodies

Two weeks after fusion hybridoma medium was assayed for EGF
receptor monoclonal antibodies. In the primary assay medium conditioned by hybridomas was used to inhibit the binding of ^{125}I-labeled EGF
to paraformaldehyde-fixed A431 cells. Receptor-grade EGF (Collaborative
Research) was labeled with Na^{125}I (Amersham) to a specific activity of
5×10^4 cpm/ng by the chloramine-T method of Hunter and Greenwood.[31]
Confluent cultures of A431 cells in 10-cm plates were washed with PBS
and fixed with 0.2% paraformaldehyde in PBS for 10 min at room temperature. The cells were harvested and resuspended at a density of 1×10^6
cells/ml in PBS containing 0.2% BSA (fraction V; Sigma). A431 cells
(2×10^4) were spotted onto glass fiber filters in punctured microtest wells
(V and P Enterprises, San Diego, CA; or Isolab, Inc., Akron, OH), and the
cells were washed twice under vacuum with 0.25% gelatin in PBS, or
DME-F12 medium, containing 1% BSA or 5% NCS. ^{125}I-Labeled EGF in
50 μl of hybridoma medium was incubated with the A431 cells at 37°;
control wells contained ^{125}I-labeled EGF alone or ^{125}I-labeled EGF and a
100-fold molar excess of unlabeled EGF in 0.2% BSA. After 2 to 3 hr the
cells were washed at least four times under vacuum with 0.25% gelatin and
5% NCS in either PBS or DME-F12, and the filters were air dried. The
binding of ^{125}I-labeled EGF to A431 cells in the presence of hybridoma
supernatants was expressed as a percentage of maximum specific cell-
bound radioactivity.

Seventy-two (3.8%) hybridoma supernatants of 1886 wells from 4 fusion experiments were initially found to inhibit ^{125}I-labeled EGF binding to
A431 cells to various degrees.[6] Only three wells eventually yielded hybridomas secreting antibodies that inhibited EGF binding: these hybridomas
have been designated 579, 528, and 225.[6] The remaining 69 wells proved
to be negative upon retesting the original hybridoma cultures or cloned
derivatives. As shown in Fig. 1, 579 IgG, 528 IgG, and 225 IgG maximally
inhibited the binding of ^{125}I-labeled EGF to A431 cells by 95–99%. None
of these three antibodies has been found to recognize carbohydrate deter-

[31] W. M. Hunter and F. C. Greenwood, *Nature (London)* **194**, 495 (1962).

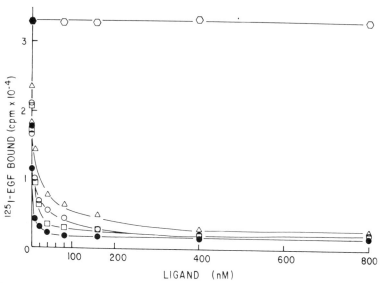

FIG. 1. Inhibition of ^{125}I-labeled EGF binding to A431 cells. Paraformaldehyde-fixed A431 cells (2×10^4) were incubated in duplicate for 2.5 hr at 37° in 50 μl of DME-F12 medium with 100 pg ^{125}I-labeled EGF (1.5×10^6 cpm/ng) and increasing concentrations of 528 IgG (○), 225 IgG (□), 579 IgG (△), 455 IgG (○), or EGF (●). (From Ref. 6.)

minants,[11] and their binding activities are not restricted solely to A431 cells, but these antibodies do appear to be specific for human EGF receptors.[6]

On several occasions hybridoma supernatants were first tested for their ability to bind rabbit anti-mouse ^{125}I-labeled IgG (RAM; Miles) to paraformaldehyde-fixed A431 cells; 2×10^4 cells in PBS with 0.2% BSA were adsorbed to glass fiber filters in punctured microtest wells and were washed under vacuum with 0.25% gelatin and 5% NCS in PBS or DME-F12 medium. The cells were incubated with 50 μl of hybridoma medium for 2 hr at 37°, the cells were washed, and they were incubated for a further 2 hr with 10 μl of ^{125}I-labeled RAM. The cells were washed four times with gelatin/NCS buffer, and cell-bound radioactivity was measured with a gamma counter. Hybridomas 579 and 455[6] were cloned from wells which gave 6.5-fold and 12-fold, respectively, more binding of ^{125}I-labeled RAM to A431 cells than control wells. As mentioned above, 579 IgG blocked the binding of EGF to A431 cells; 455 IgG, on the other hand, had no effect on EGF binding but was able to immunoprecipitate A431 EGF receptors.[6] Furthermore, 455 IgG has been shown to recognize a blood group A-related carbohydrate determinant.[11]

Immunoprecipitation of EGF Receptors

To confirm the EGF receptor specificities of the antibodies which inhibited EGF binding to A431 cells the antibodies were used to precipitate EGF receptors from solubilized A431 membranes. Antibodies were purified from conditioned medium or ascites fluid with protein A-agarose[32] following anion-exchange chromatography (see below). Purified IgG was covalently attached to Affi-Gel 10 (Bio-Rad) (0.5 to 2.0 mg/ml gel) according to the manufacturer's instructions. A431 membranes were prepared by hypotonic shock from confluent cell cultures as described by Thom *et al.*[33] Membrane protein (50 μg) solubilized in 0.25% Triton X-100/10 mM HEPES (pH 7.4)/5% glycerol was incubated with 50 μl of antibody-Affi-Gel for 16 hr at 4°. Bound proteins were recovered by centrifugation in a microfuge (Eppendorf), and the immunoprecipitates were washed five times with 1 ml of 0.5% Triton X-100/0.1% SDS/20 mM HEPES (pH 7.4)/10% glycerol. Adsorbed proteins were eluted from the Affi-Gel at 100° in SDS sample buffer.[27] The immunoprecipitates and nonprecipitated proteins were analyzed by SDS-polyacrylamide gel electrophoresis. Figure 2 is a silver-stained gel of antibody-fractionated A431 membrane proteins. All three antibodies that inhibited EGF binding precipitated M_r 165,000 and M_r 145,000 polypeptides which are characteristic of the intact EGF receptor and a degradation product generated by a calcium-dependent protease.[26,34-36] In addition these two polypeptides were precipitated by 455 IgG, the antibody which had no effect on EGF binding.

Binding of Labeled Antibodies to Cell Membranes

To characterize the EGF receptor monoclonal antibodies with respect to binding affinities and numbers of binding sites ^{125}I-labeled purified antibodies were reacted with A431 cells, HeLa S cells, and human foreskin fibroblasts in saturation binding assays. The assays were done in duplicate on glass fiber filters in punctured microtest wells. The antibodies were iodinated by the chloramine-T method[31] to specific activities of 2×10^4 to 2×10^5 cpm/ng. Increasing concentrations of ^{125}I-labeled antibodies were incubated in 50 μl of DME-F12 medium with 1×10^4 paraformaldehyde-fixed A431 cells or 5×10^5 fixed HeLa S cells or human fibroblasts which

[32] P. L. Ey, S. J. Prowse, and C. R. Jenkin, *Immunochemistry* **15,** 429 (1978).
[33] D. Thom, A. J. Powell, C. W. Lloyd, and D. A. Rees, *Biochem. J.* **168,** 187 (1977).
[34] S. Cohen, H. Ushiro, C. Stoscheck, and M. Chinkers, *J. Biol. Chem.* **257,** 1523 (1982).
[35] D. Cassel and L. Glaser, *J. Biol. Chem.* **257,** 9845 (1982).
[36] R. W. Yeaton, M. T. Lipari, and C. F. Fox, *J. Biol. Chem.* **258,** 9254 (1983).

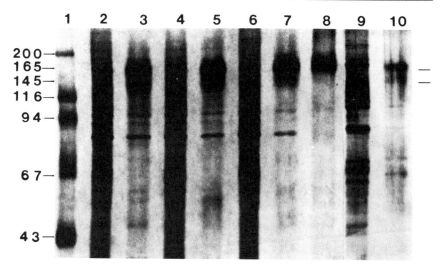

FIG. 2. Immunoprecipitation of A431 EGF receptors. A431 membrane proteins solubilized with Triton X-100 were incubated at 4° with monoclonal antibodies covalently bound to Affi-Gel. Precipitated and nonprecipitated proteins were electrophoresed in a 7.5% SDS-polyacrylamide gel and were stained with silver: molecular weight markers (×10⁻³) (1); 528 IgG, supernatant (2), and precipitate (3); 225 IgG, supernatant (4), and precipitate (5); 579 IgG, supernatant (6), and precipitate (7); 528 IgG, precipitate (8); 455 IgG, supernatant (9), and precipitate (10). (From Ref. 6.)

had been washed with gelatin/NCS buffer. Control reactions included a 100-fold molar excess of unlabeled ligand. After 2 to 3 hr at 37° the cells were washed under vacuum at least four times with 0.25% gelatin and 5% NCS in DME-F12 medium and were air dried. Cell-bound radioactivity

TABLE II

BINDING AFFINITIES AND NUMBERS OF BINDING SITES OF EGF RECEPTOR MONOCLONAL ANTIBODIES[a]

Monoclonal antibody	A431		HeLa S		Human fibroblasts	
	K_d ($\times 10^9 M$)	Sites ($\times 10^{-6}$/cell)	K_d ($\times 10^9 M$)	Sites ($\times 10^{-4}$/cell)	K_d ($\times 10^9 M$)	Sites ($\times 10^{-4}$/cell)
528 IgG	1.5–2.5	1.2–2.2	1.6	1.5	1.5	1.5
225 IgG	0.6–1.3	0.8–2.2	1.1	1.5	1.0	1.9
579 IgG	1.2–1.8	1.1–1.4	1.2	1.8	0.7	1.4
455 IgG	20	1.8	ND	ND	ND	ND
EGF	2–5	1.5–3	2.1	1.7	2.1	2.6

[a] Summary of Scatchard analyses of saturation binding of ¹²⁵I-labeled ligands to A431 cells, HeLa cells, or human fibroblasts. ND, Not determined. From Ref. 6.

was measured with a gamma counter and corrected for nonspecific binding that occurred in the presence of unlabeled antibody. The binding data were analyzed by the method of Scatchard[37] and are summarized in Table II. 528 IgG, 579 IgG, and 225 IgG bound to 1 to 2×10^6 sites on A431 cells and 1 to 2×10^4 sites on HeLa S and human fibroblasts with dissociation constants of 1 to 2×10^{-9} M; 455 IgG bound to a similar number of A431 sites but with a 10- to 20-fold lower affinity than the other antibodies ($K_d = 20 \times 10^{-9}$ M). Thus, the three antibodies that competitively inhibited EGF binding bound to EGF receptors with affinities similar to that of EGF.[6]

Purification of Monoclonal Antibodies

The methods described here are for the purification of IgG antibodies. For a method of purifying IgM antibodies the reader is referred to Ref. 25.

Method 1 (Protein A – Agarose)[32]

Medium conditioned by hybridomas or ascites fluid produced in BALB/c mice was centrifuged at 800 g to remove cells. The supernatants were applied to a column of DEAE–cellulose (Pharmacia) at 4° and the column washed with medium or PBS. The flow-through fraction was adjusted to pH 8.0 and applied to a 5-ml column of protein A–agarose (available from Boehringer-Mannheim). Antibody was eluted from the column with 1 M acetic acid (pH 3.0)/100 mM glycine/20 mM NaCl. The eluate was dialyzed at 4° several times against 2 liters of PBS followed by DME-F12 medium. Antibody concentrations were measured by absorbance at 280 nm (1 mg/ml = 1.4 OD$_{280}$ with a 1-cm light path). Antibody solutions were sterilized by filtration through 0.2-μm filters (Millipore) and stored at −20°.

Method 2 (DEAE – Affi-Gel Blue)[38]

DEAE–Affi-Gel Blue is available from Bio-Rad (Richmond, CA). Up to 6 liters of serum-free KSLM medium[23,24] conditioned by hybridomas was filtered (Whatman No. 1) to remove particulate matter. Solid ammonium sulfate was slowly stirred into the filtrate at room temperature to a final concentration of 50%, and the filtrate was stirred for a further 2 hr. The precipitate was recovered by centrifugation at 10,000 g for 30 to 60 min, and it was dissolved in 50 ml of 20 mM Tris (pH 8.0)/28 mM

---

[37] G. Scatchard, *Ann. N.Y. Acad. Sci.* **55**, 660 (1949).
[38] C. Bruck, D. Portetelle, C. Glineur, and A. Bollen, *J. Immunol. Methods* **53**, 313 (1982).

NaCl/0.02% NaN$_3$. The dissolved precipitate was dialyzed against several changes of the resuspension buffer at 4°. The dialysate was applied to a DEAE–Affi-Gel Blue column in 20 mM Tris (pH 8.0)/28 mM NaCl/0.02% NaN$_3$, and the column was eluted with the same buffer. After use the column was purged with 20 mM Tris (pH 8.0)/1.4 M NaCl/0.02% NaN$_3$ followed by 2 M guanidine hydrochloride.

Column fractions were monitored by absorbance at 280 nm, and the first protein peak was collected. If antibodies are being purified from ascites fluid the column should be eluted with a continuous salt gradient, and fractions should be monitored with a serological assay. An 80% saturated solution of ammonium sulfate was stirred into the pooled fractions to a final concentration of 40%. The precipitate was collected by centrifugation at 10,000 g, redissolved in 10 to 20 ml of PBS, and dialyzed against several changes of PBS at 4°. The antibody concentration of the dialysate was measured by absorbance, and the antibody solution was sterilized by filtration. The purity of each antibody preparation was assessed by SDS–polyacrylamide gel electrophoresis and silver staining. Approximately 50 mg of purified IgG was obtained from 6 liters of conditioned medium.

Biological Effects of EGF Receptor Antibodies

As an intensively studied peptide growth factor, EGF induces several well-defined immediate and delayed responses in receptor-bearing target cells.[39,40] The immediate responses include morphological changes, increased nutrient uptake, increased ion fluxes, EGF receptor clustering and internalization, and stimulation of the receptor tyrosine-specific protein kinase activity. The delayed responses of DNA synthesis and cytokinesis require cells to be exposed to EGF for at least 8 hr.[39] Some cell lines such as A431 epidermoid carcinoma cells,[41,42] GH$_4$ rat pituitary cells,[43] and T222 human lung epidermoid carcinoma cells[44] respond to EGF with inhibited growth rather than increased proliferation. This knowledge of the effects of EGF on responsive cells has made it possible to test EGF receptor monoclonal antibodies for agonistic and antagonistic biological activities.

[39] G. Carpenter and S. Cohen, *Annu. Rev. Biochem.* **48**, 193 (1979).

[40] G. Carpenter, *Mol. Cell. Endocrinol.* **31**, 1 (1983).

[41] D. Barnes, *J. Cell Biol.* **93**, 1 (1982).

[42] G. N. Gill and C. S. Lazar, *Nature (London)* **293**, 305 (1981).

[43] A. Schonbrunn, M. Krasnoff, J. M. Westendorf, and A. H. Tashjian, Jr., *J. Cell Biol.* **85**, 786 (1980).

[44] H. Masui, T. Kawamoto, J. D. Sato, B. Wolf, G. H. Sato, and J. Mendelsohn, *Cancer Res.* **44**, 1002 (1984).

Effects on Cell Growth in Vitro

We first tested our monoclonal antibodies[5,6] for intrinsic growth-promoting activity for A431 cells and human fibroblasts, cells which exhibit
opposite growth responses to nanomolar concentrations of EGF. A431
cells were plated in 24-well plates (Costar) at an initial density of 1.5×10^4
cells per well in 1 ml of serum-free DME-F12 medium and increasing
concentrations of EGF (tissue culture grade; Collaborative Research) or
purified monoclonal IgG (Fig. 3). After 4.5 days at 37° the cells in duplicate cultures were trypsinized and counted with a model Z_f Coulter
counter. A 9.5-fold increase in cell number occurred in the absence of EGF
or IgG, and 10 nM EGF inhibited cell growth by about 80%. 528 IgG,
579 IgG, and 225 IgG, which inhibited EGF binding, also individually
inhibited A431 growth. Maximal inhibition occurred at approximately
10 nM IgG with each antibody and was comparable to that caused by
EGF; this immunoglobulin concentration was near saturation for each

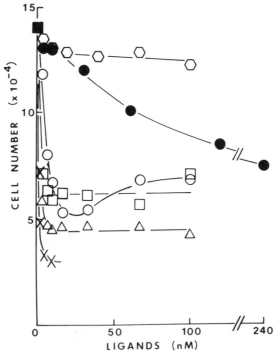

FIG. 3. Effects of monoclonal antibodies on A431 cell growth. A431 cells (1.5×10^4/well)
were incubated for 4.5 days at 37° in DME-F12 medium with 528 IgG (○), 528 Fab (●),
225 IgG (△), 579 IgG (□), 455 IgG (⬡), or EGF (✕). Each point represents the mean cell
count of duplicate wells. (From Ref. 6.)

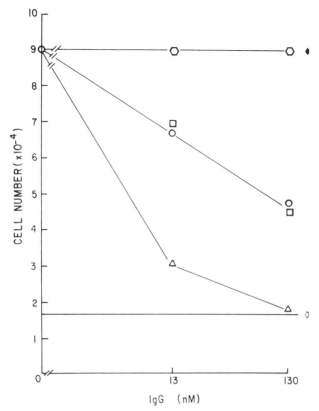

FIG. 4. Effects of monoclonal antibodies on EGF-induced growth of human fibroblasts. Human fibroblasts (1×10^4/well) were incubated for 6 days at 37° in serum-free medium with 528 IgG (O), 225 IgG (△), 579 IgG (□), or 455 IgG (○) in the presence or absence of 3 nM EGF. Cell growth in the presence of monoclonal antibodies only or 3 nM EGF alone is indicated by the open and closed arrowheads, respectively. Each point represents the mean cell count of duplicate wells. (From Ref. 6.)

antibody. But 455 IgG, which binds to a carbohydrate determinant remote from the EGF binding site of the EGF receptor, neither stimulated nor inhibited A431 growth.

   Human foreskin fibroblasts were plated in 24-well plates (Costar) at a density of 1×10^4 cells per well in 1 ml of serum-free DME-F12 medium. The medium included 2 μg/ml insulin (Sigma), 2 μg/ml transferrin (Sigma), 4 μg/ml cold insoluble globulin (plasma fibronectin), and 1 μg/ml hydrocortisone (Sigma). Each purified monoclonal IgG was added to concentrations of 0, 13, and 130 nM in the presence and absence of 3 nM EGF (Collaborative Research). After 6 days at 37° cells in duplicate wells were trypsinized and counted (Fig. 4). In 3 nM EGF the number of fibroblasts increased 9-fold whereas none of the antibodies alone stimulated cell growth. The three antibodies which are competitors of EGF inhibited EGF-stimulated fibroblast growth: 130 nM 528 IgG or 579 IgG reduced cell growth by 50%, and 130 nM 225 IgG completely abolished the stimulatory effect of EGF; 455 IgG had no effect on fibroblast growth in either the presence or absence of EGF.

Of the other EGF receptor monoclonal antibodies described in the literature (Table I) five have been tested for their effects on cell proliferation. Only EGR/G49[8] was tested on A431 cells and was found to fully inhibit cell growth at a concentration of 100 nM. This antibody appears not to bind to EGF receptors of other human or rodent cell lines and was not tested on cells that both bound EGR/G49 and were stimulated by EGF. As mentioned previously, 20 nM 2G2 IgM[1] alone stimulated human fibroblast proliferation, and 10 nM 2G2 IgM stimulated thymidine incorporation by human fibroblasts and 3T3 mouse fibroblasts 40% as well as 2 nM EGF. Neither 2G2 Fab nor TL5 IgG were mitogenic for human fibroblasts unless cross-linked with at least 50 μg/ml goat anti-mouse IgG[3]; under these conditions [^{3}H]thymidine incorporation was enhanced 2- to 3-fold. EGFR1[2] did not block EGF binding and was not mitogenic for human fibroblasts. B1D8[9] alone at concentrations up to 100 μg/ml neither stimulated nor inhibited human fibroblast proliferation in 2% FCS, but at 1.3 μM it inhibited EGF-stimulated cell growth by 70%, and at 100 μg/ml B1D8 substantially reduced EGF-induced [^{3}H]thymidine incorporation. These results indicate that most of the EGF receptor monoclonal antibodies tested were without intrinsic mitogenic activity, and those that blocked EGF binding acted as EGF antagonists in fibroblast growth assays; 528 IgG, 579 IgG, 225 IgG and EGR/G49 mimicked EGF in that they inhibited A431 proliferation by binding to either oligosaccharide (EGR/49) or peptide determinants on A431 EGF receptor molecules.

We have also investigated the combined effects of EGF and EGF receptor antibodies on A431 cell growth. As our antibodies were not mitogenic in themselves, and three of the antibodies inhibited A431 growth, we expected growth inhibition to result from adding individual antibodies and EGF to A431 cultures. In contrast to this expectation EGF at normally inhibitory concentrations was mitogenic for A431 cells in the presence of a saturating concentration of purified 528 IgG[5]; cell growth was stimulated relative to cultures that received 528 IgG only, but it never exceeded the degree of proliferation in control cultures which received neither antibody nor EGF. In the presence of 3 nM EGF in serum-free DME-F12 medium 528 IgG at saturating concentrations stimulated A431 growth; 225 IgG at concentrations up to 5 nM stimulated cell growth but higher concentrations of this antibody inhibited growth.[6] These results, with antibodies that competitively inhibited EGF binding to A431 cells, in conjunction with the observation that picomolar concentrations of EGF stimulated rather than inhibited A431 growth,[5] raised the possibility that A431 cells possessed two populations of receptors with different affinities for EGF with the high-affinity receptors involved in growth stimulation and the low-affinity receptors involved in growth inhibition. By using ^{125}I-labeled EGF, labeled to a high specific activity, in a saturation binding assay in the presence of 100 nM 528 IgG, a minor population of high-affinity EGF receptors was detected on A431 cells.[5] These receptors comprised approximately 0.2% of the total number of EGF receptors, and they bound EGF with an estimated dissociation constant of 7×10^{-11} M.

Although there is no direct evidence that this high-affinity population of receptors plays a role in the stimulation of A431 growth by EGF, the concentration of EGF needed to half-saturate these receptors is within the range of EGF concentrations that stimulate growth.[5] It is possible that functionally distinct populations of EGF receptors are synthesized from different mRNAs as several species of EGF receptor mRNAs have been detected in A431 cells.[17,45,46] Alternatively, different classes of A431 EGF receptors could arise during the posttranslational modification of a single species of EGF receptor polypeptide. As a number of monoclonal antibodies that recognize oligosaccharide determinants on A431 EGF receptors do not react with receptors from other human cell lines,[7,8,10] in principle different classes of EGF receptors could result from differential glycosylation.

Effects on Receptor Internalization

EGF receptor clustering and internalization[47-52] are relatively rapid events which are induced by EGF in both cells that are stimulated (human fibroblasts) and inhibited (A431 cells) by EGF. Receptor internalization

[45] C. R. Lin, W. S. Chen, W. Kruiger, L. S. Stolarsky, W. Weber, R. M. Evans, I. M. Verma, G. N. Gill, and M. G. Rosenfeld, *Science* **224**, 843 (1984).

[46] Y.-H. Xu, S. Ishii, A. J. L. Clark, M. Sullivan, R. K. Wilson, D. P. Ma, B. A. Roe, G. T. Merlino, and I. Pastan, *Nature (London)* **309**, 806 (1984).

[47] J. Schlessinger, Y. Shechter, M. C. Willingham, and I. Pastan, *Proc. Natl. Acad. Sci. U.S.A.* **75**, 2659 (1978).

[48] H. Haigler, T. J. A. McKanna, and S. Cohen, *J. Cell Biol.* **81**, 382 (1979).

[49] G. Carpenter and S. Cohen, *J. Cell Biol.* **71**, 159 (1976).

[50] A. Le, J. Mendelsohn, G. H. Sato, H. Sunada, and C. MacLeod, *in* "Growth and Differentiation of Cells in Defined Environment" (H. Murakami, I. Yamane, D. W. Barnes, J. P. Mather, I. Hayashi, and G. H. Sato, eds.), pp. 303–306. Springer-Verlag, New York, 1985.

[51] J. Schlessinger, A. B. Schreiber, T. A. Libermann, I. Lax, A. Avivi, and Y. Yarden, *in* "Cell Membranes: Methods and Reviews" (E. Elson, W. Frazier, and L. Glaser, eds.), Vol. 1, pp. 117–149. Plenum Press, New York, 1983.

[52] J. Schlessinger, A. B. Schreiber, A. Levi, I. Lax, T. A. Libermann, and Y. Yarden, *CRC Crit. Rev. Biochem.* **14**, 93 (1982).

and degradation result in a reduction of as much as 80% in the number of surface receptors available to EGF. A number of EGF receptor monoclonal antibodies have also been found to induce EGF receptor internalization; these include 2G2 IgM and cross-linked 2G2 Fab,[3] TL5 IgG,[3] EGFR1,[2] and EGR/G49.[8] B1D8 does not cause receptor internalization.[9] We originally reported that none of our EGF receptor antibodies induced receptor internalization.[5,6] In those experiments trace amounts of ^{125}I-labeled antibodies were incubated with confluent cultures of A431 cells. At intervals the cells were washed with ice-cold DME-F12 medium and surface-bound antibody was eluted with 0.2 N acetic acid in 0.5 M NaCl (pH 2.3). Under these conditions there was no reduction in the maximum level of eluted surface-bound antibody whereas with 3.3 nM ^{125}I-labeled EGF the number of surface receptors decreased by 70% within 3 hr at 37°. In subsequent experiments[50] saturating concentrations of ^{125}I-labeled antibodies were incubated with A431 cells at 37°. Acid-resistant internalized radioactivity increased to a maximum level in 1 hr with 20 nM ^{125}I-labeled 225 IgG and decreased by about 60% over the following 8 hr at 37°.[50] Receptor internalization was also observed with 20 nM 528 IgG and 20 nM 579 IgG, but internalization was not induced by 200 nM 455 IgG (A. D. Le et al., unpublished results). Also, 225 IgG has been found to cause receptor down regulation; a maximum reduction of approximately 70% in EGF binding sites was induced by 225 IgG at 20 to 200 nM.[50] That receptor internalization requires saturating concentrations of antibody but only subsaturating concentrations of EGF suggests that these events should not be considered equivalent and may in fact occur by different mechanisms. Furthermore, as a number of EGF receptor antibodies are internalized and yet are not mitogenic, receptor internalization per se cannot be a sufficient signal for mitogenesis.

Effects on EGF Receptor Protein Kinase Activity

Of the EGF receptor monoclonal antibodies tested for the ability to stimulate the receptor kinase only 2G2 IgM[1,3] and 2G2 Fab fragments[3] had stimulatory activity. 528 IgG,[6] 579 IgG,[6] 225 IgG,[6] and B1D8,[9] which block EGF binding, inhibited EGF-enhanced kinase activity but were not themselves active; 455 IgG,[6] TL5 IgG,[3] and EGR/G49,[8] which do not block EGF binding, neither stimulated receptor kinase activity nor inhibited EGF-induced kinase activity. Whether or not EGF receptor antibodies that do not interfere with EGF binding can enhance the stimulation by EGF of the EGF receptor kinase has not been explored.

We used a modification of the method of Ushiro and Cohen[53] to assay our antibodies for effects on the activity of the EGF receptor protein

[53] H. Ushiro and S. Cohen, *J. Biol. Chem.* **255**, 8363 (1980).

kinase.[6] Membranes were prepared by hypotonic shock[33] from 1×10^8 A431 cells and were resuspended in 500 μl of 20 mM HEPES (pH 7.4)/1 mM MgCl$_2$. Membrane samples (20 μl) were incubated at 0° with or without 20 nM EGF (Collaborative Research) in the presence or absence of 1 μM purified antibody in 200 μl of HEPES/MgCl$_2$ buffer. After 10 min 12 μCi of [^{32}P]ATP (Amersham) was added to each reaction, and the mixtures were incubated for a further 10 min at 0°. BSA (fraction V; 2 mg/ml in PBS) was added to 400 μg/ml and the reactions were stopped with 150 μl of ice-cold 50% TCA in 10 mM pyrophosphate. The membranes were collected by centrifugation in a microfuge and were washed three times with 10% TCA. The pellets were solubilized in 0.3 N NaOH and counted by liquid scintillation. This *in vitro* kinase assay can also be done with membranes solubilized with nonionic detergent such as Triton X-100[53,54] and can include sodium vanadate to inhibit endogenous phosphatase activity.[55,56] Finally, artificial substrates can be used to assay EGF receptor kinase activity; histones,[53,57] tubulin,[34] antibodies to the pp60^{src} transforming protein of Rous sarcoma virus,[58,59] and synthetic peptides corresponding to a phosphate acceptor site of pp60^{src}[60] have been used as exogenous substrates for the EGF receptor kinase. To examine receptor kinase substrates other than the receptor itself intact cells prelabeled to a steady state with ortho [^{32}P] phosphate have proved most useful.[61]

Biological Effects in Vivo

Thus far only three EGF receptor monoclonal antibodies have been tested for their effects on tumor cell growth *in vivo*.[44] In this study 528 IgG[6] and 225 IgG,[6] which inhibit EGF binding to human cells, and 455 IgG,[6] which is not an inhibitor of EGF binding, were examined for effects on tumor formation by A431 epidermoid carcinoma cells[15] in BALB/c nude mice. Also, 528 IgG was tested on tumor formation by T222 human lung epidermoid carcinoma cells,[44] Li-7 human hepatoma cells,[62] and HeLa cells. Tumors were initiated with subcutaneous injections of 1×10^7 A431 cells or 5×10^7 HeLa cells in medium or minced A431, T222, and Li-7

[54] S. Cohen, G. Carpenter, and L. E. King, Jr., *J. Biol. Chem.* **255**, 4834 (1980).
[55] G. Swarup, S. Cohen, and D. L. Garbers, *Biochem. Biophys. Res. Commun.* **107**, 1104 (1982).
[56] J. Leis and N. O. Kaplan, *Proc. Natl. Acad. Sci. U.S.A.* **79**, 6507 (1982).
[57] G. Carpenter, L. E. King, Jr., and S. Cohen, *J. Biol. Chem.* **254**, 4884 (1979).
[58] M. Chinkers and S. Cohen, *Nature (London)* **290**, 516 (1981).
[59] J. E. Kudlow, J. Buss, and G. N. Gill, *Nature (London)* **290**, 519 (1981).
[60] L. Pike, B. Gallis, J. E. Casnellie, P. Bornstein, and E. G. Krebs, *Proc. Natl. Acad. Sci. U.S.A.* **79**, 1443 (1982).
[61] T. Hunter and J. A. Cooper, *Cell* **24**, 741 (1981).
[62] S. Hirohashi, Y. Shimosato, T. Kameya, T. Koide, T. Mukojima, Y. Taguchi, and K. Kageyama, *Cancer Res.* **39**, 1819 (1979).

tumors. The antibodies were used as ascites fluid or were purified from ascites fluid by 50% ammonium sulfate precipitation and DEAE–cellulose chromatography, and they were administered intraperitoneally. An initial experiment with 528 ascites fluid and 225 ascites fluid demonstrated that these antibodies, when given thrice weekly for 3 weeks from the time of tumor initiation, completely prevented the formation of A431 tumors. No tumors developed in the 2 months following the end of the antibody treatment. Thirty micrograms EGF (Collaborative Research) administered on the same schedule as the antibodies had no effect on A431 tumor growth. In a second experiment 528 ascites fluid administered twice weekly starting 7 days after tumor initiation markedly slowed but did not prevent the growth of A431 tumors. With purified antibodies twice weekly doses of 0.2 mg 528 IgG were sufficient to prevent A431 tumor formation by cell suspensions and to reduce the size of tumors formed by minced A431 tumors, but 2-mg doses were necessary to prevent A431 tumors formed by minced tumor tissue. In similar experiments 0.2-mg doses of 225 IgG significantly reduced tumor size, and 2-mg doses prevented A431 tumor formation; the growth of A431 tumors was also slowed by 0.2-mg doses of 455 IgG and prevented by 2-mg doses of 455 IgG. Twice weekly injections of 0.2 mg 528 IgG additionally prevented T222 tumor formation when administered from the time of tumor initiation, but 528 IgG at doses up to 2 mg had no effect on the growth of Li-7 or HeLa tumors.

The effects of these antibodies on A431 tumor growth *in vivo* are of interest in view of the results obtained with A431 cells *in vitro*.[6] Both 528 IgG and 225 IgG severely inhibited A431 growth *in vitro*, but they were not cytotoxic; 455 IgG had no apparent effect on the proliferation of A431 cells *in vitro*. That all three antibodies were able to kill A431 cells *in vivo* strongly implies that an antibody-dependent cellular immune response was involved in the destruction of A431 cells in nude mice. As nude mice lack mature T cells, monocytes, macrophages, or killer cells, but not cytotoxic T cells, may have been involved in this cell-mediated immunity.

Several studies on the effects of IgG isotypes on antibody-dependent cell-mediated cytotoxicity have concluded that cell-mediated tumor killing *in vitro* and *in vivo* was best directed by IgG_{2a} mouse monoclonal antibodies,[63–69] and macrophages[63–66] or nonadherent killer cells were

[63] H. Koprowski, Z. Steplewski, D. Herlyn, and M. Herlyn, *Proc. Natl. Acad. Sci. U.S.A.* **75**, 3405 (1978).

[64] D. M. Herlyn, Z. Steplewski, M. F. Herlyn, and H. Koprowski, *Cancer Res.* **40**, 717 (1980).

[65] D. Herlyn and H. Koprowski, *Proc. Natl. Acad. Sci. U.S.A.* **79**, 4761 (1982).

[66] Z. Steplewski, M. D. Lubeck, and H. Koprowski, *Science* **221**, 865 (1983).

[67] I. Hellström, K. E. Hellström, and M.-Y. Yeh, *Int. J. Cancer* **27**, 281 (1981).

[68] K. Imai, M. A. Pellegrino, B. S. Wilson, and S. Ferrone, *Cell. Immunol.* **72**, 239 (1982).

[69] T. J. Kipps, P. Parham, J. Punt, and L. A. Herzenberg, *J. Exp. Med.* **161**, 1 (1985).

shown to act as effector cells. In light of those results it is somewhat surprising that 225 IgG and 455 IgG, which are IgG_1 antibodies,[6] killed A431 cells, albeit less effectively than 528 IgG, in nude mice. However, several other examples of IgG_1 antibodies directing cell-mediated target cell lysis have been reported.[70-72] Whatever the mechanism by which our antibodies kill A431 cell *in vivo,* it is probably not dependent on high densities of EGF receptors, as T222 cells, with lower numbers of receptors, were also killed by 528 IgG *in vivo.*

The results obtained by Masui *et al.*[44] and by Trowbridge and Domingo,[73] who showed that the human transferrin receptor monoclonal antibody B3/25 prevented M21 melanoma tumors in nude mice, provide evidence that unmodified monoclonal antibodies to cell surface receptors for peptide growth factors or essential transport proteins may be useful in killing some tumor cells *in vivo.* This cytotoxic effect is most likely mediated by host effector cells. In addition, this antibody-dependent cell killing appears not to be indiscriminant in that not all cells expressing the appropriate receptors are killed *in vivo.* The basis for this differential cytotoxicity is unknown. Another unresolved issue is the significance of the deprivation of growth factors and essential nutrients caused by receptor antibodies in the killing of tumor cells *in vivo.* It is possible that cells which internalize receptor–antibody complexes will largely escape cell-mediated lysis but will succumb to prolonged deprivation of required growth factors and nutrients. Clearly more research is required into the mechanism(s) by which receptor antibodies direct cell killing *in vivo* and into the possible toxic effects of receptor antibodies on normal cells, but the results obtained thus far suggest that receptor monoclonal antibodies may prove to be useful therapeutic agents.

Acknowledgments

This chapter is dedicated to the memories of Drs. Sidney P. Colowick and Nathan O. Kaplan.

We thank our colleagues Drs. J. Polikoff, H. Masui, J. Mendelsohn, and G. H. Sato for their contributions to the original research described here. J.D.S. and T.K. thank Dr. R. Knowles (Memorial Sloan-Kettering Cancer Center, New York, NY) for discussions on hybridoma fusions. J.D.S. thanks Dr. K. Itakura (Beckman Research Institute of the City of Hope, Duarte, CA) for his interest and support. The original antibodies described were produced and characterized in G. H. Sato's laboratory at the University of California, San Diego Cancer Center. This research was supported by N.I.H. grants to G. H. Sato. J.D.S. was the recipient of an N.I.H. postdoctoral fellowship.

[70] A. H. Greenberg and P. M. Lydyard, *J. Immunol.* **123,** 861 (1979).
[71] K. Imai, A.-K. Ng, M. C. Glassy, and S. Ferrone, *Scand. J. Immunol.* **14,** 369 (1981).
[72] P. Ralph and I. Nakoinz, *J. Immunol.* **131,** 1028 (1983).
[73] I. S. Trowbridge and D. L. Domingo, *Nature (London)* **294,** 171 (1981).

[7] Purification of Functionally Active Epidermal Growth Factor Receptor Protein Using a Competitive Antagonist Monoclonal Antibody and Competitive Elution with Epidermal Growth Factor

By GORDON N. GILL and WOLFGANG WEBER

Methods

Introduction

The epidermal growth factor (EGF) receptor is a 1186-amino acid transmembrane phosphorylated glycoprotein which mediates the biological effects of EGF.[1] The binding site for EGF, located in the extracellular 621-amino acid portion of the molecule, is separated by a 23-amino acid hydrophobic membrane spanning section from the 542-amino acid intracellular region of the molecule which contains protein-tyrosine kinase activity catalyzing the transfer of the γ-phosphate of ATP to tyrosine residues in substrate proteins.[2] The EGF receptor is a protooncogene with its intracellular domain corresponding to *erbB*, the transforming protein of avain erythroblastosis virus.[3] Amplification of the EGF receptor gene and enhanced expression of EGF receptor protein occurs in several human malignancies.[2,4-6] The function of the EGF receptor is regulated not only by binding of ligands EGF and transforming growth factor (TGF) α[7] but also by phosphorylation—both intramolecular self-phosphorylation at tyrosines 1068, 1148, and 1173[8] and phosphorylation at threonine 654 by the tumor promoter receptor, protein kinase C.[9,10] Phosphorylation at other sites on the EGF receptor may also prove to be important.[11]

[1] T. Hunter and J. A. Cooper, *Annu. Rev. Biochem.* **54**, 897 (1985).
[2] A. Ullrich, L. Coussens, J. S. Hayflick, T. J. Dull, G. Gray, A. W. Tam, J. Lee, Y. Yarden, T. A. Libermann, J. Schlessinger, J. Downward, E. L. V. Mayes, N. Whittle, M. D. Waterfield, and P. H. Seeburg, *Nature (London)* **309**, 418 (1984).
[3] L. Frykberg, S. Palmieri, H. Beug, T. Graf, M. J. Hayman, and B. Vennstrom, *Cell* **32**, 227 (1983).
[4] C. R. Lin, W. S. Chen, W. Kruijer, L. S. Stolarsky, W. Weber, R. M. Evans, I. M. Verma, G. N. Gill, and M. G. Rosenfeld, *Science* **224**, 843 (1984).
[5] G. T. Merlino, Y.-H. Xu, S. Ishii, A. J. L. Clark, K. Semba, K. Toyoshima, T. Yamamoto, and I. Pastan, *Science* **224**, 417 (1984).
[6] T. Libermann, H. Nusbaum, N. Razon, R. Kris, I. Lax, H. Soreq, N. Whittle, M. D. Waterfield, A. Ullrich, and J. Schlessinger, *Nature (London)* **313**, 144 (1985).
[7] G. J. Todaro, C. Fryling, and J. E. DeLarco, *Proc. Natl. Acad. Sci. U.S.A.* **77**, 5258 (1980).
[8] J. Downward, P. Parker, and M. D. Waterfield, *Nature (London)* **311**, 483 (1984).

To define regulation and metabolism of the EGF receptor and to study its expression and role in controlling proliferation of normal and malignant cells, purification of the EGF receptor protein is frequently important. Several methods of purification have been developed. Hock et al.[12] described sequential ion exchange, Cibacron blue–Sepharose, wheat germ agglutinin–Sepharose, and gel filtration chromatographies. As with other multistep purification schemes, yields were low. Cohen et al.[13] developed an EGF Affi-gel affinity medium which has been useful in purifying active EGF receptor/kinase, but yields with this medium are low. Recently, immunoaffinity chromatographies with monoclonal anti-EGF receptor antibodies have been described. Parker et al. described adsorption to an immobolized monoclonal antibody directed against carbohydrate side chains with competitive elution of active protein with N-acetylglucosamine.[14] We have developed a single-step immunoaffinity chromatography procedure using a monoclonal anti-EGF receptor antibody which is a competitive inhibitor of EGF binding.[15,16] Desorption with urea gives quantitative recovery of denatured protein; competitive elution with EGF yields homogenous active protein.[17] This procedure provides a convenient single-step, high-yield purification of EGF receptor protein.

Materials

528 IgG: This anti-EGF receptor monoclonal IgG class 2a was developed by Kawamoto et al.[15] using EGF receptor from human epidermoid carcinoma A431 cells. The antibody (and antibodies 225 and 579) is a competitive inhibitor of EGF binding and an antagonist of EGF-stimulated protein–tyrosine kinase activity.[16] It is directed against protein, not carbohydrate epitopes. (Antibody is obtained from Drs. John Mendelsohn and Hideo Masui, Sloan Kettering/Memorial Hospital Cancer Center, New York, NY.)

[9] C. Cochet, G. N. Gill, J. Meisenhelder, J. A. Cooper, and T. Hunter, *J. Biol. Chem.* **259**, 2553 (1984).
[10] T. Hunter, N. Ling, and J. Cooper, *Nature (London)* **311**, 480 (1984).
[11] T. Hunter and J. Cooper, *Cell* **24**, 741 (1981).
[12] R. A. Hock, E. Nexo, and M. D. Hollenberg, *J. Biol. Chem.* **255**, 10737 (1980).
[13] S. Cohen, G. Carpenter, and L. King, Jr., *J. Biol. Chem.* **255**, 4834 (1980).
[14] P. J. Parker, S. Young, W. J. Gullick, E. L. V. Mayes, P. Bennett, and M. D. Waterfield, *J. Biol. Chem.* **259**, 9906 (1984).
[15] T. Kawamoto, J. D. Sato, A. Le, J. Polikoff, G. H. Sato, and J. Mendelsohn, *Proc. Natl. Acad. Sci. U.S.A.* **80**, 1337 (1983).
[16] G. N. Gill, T. Kawamoto, C. Cochet, A. Le, J. D. Sato, H. Masui, C. McLeod, and J. Mendelsohn, *J. Biol. Chem.* **259**, 7755 (1984).
[17] W. Weber, P. J. Bertics, and G. N. Gill, *J. Biol. Chem.* **259**, 14631 (1984).

Immune absorbent 528 IgG agarose A-15m is prepared by the cyanogen bromide coupling method.[18,19] The 528 IgG at 1.5 mg/ml is incubated by gentle shaking with an equal volume of CNBr-activated agarose A-15m for 2 hr at room temperature. The agarose is then filtered, resuspended in 2 vol of 0.25 M ethanolamine and shaken for 1 hr to remove any unreacted sites. The gel is then washed extensively with 0.1 M Tris–HCl, pH 7.4, and equilibrated with buffer C for immune chromatography. Coupling of antibody is complete as judged by removal of protein from the antibody solution. The chromatography medium is stable for more than 2 years when stored in 5 mM Tris–HCl, pH 7.4, 145 mM NaCl, and 0.1% sodium azide. The 528 IgG agarose has an antibody concentration of about 10 μM. One milliliter of affinity matrix can bind about 900 μg, i.e., about 7 nmol of EGF receptor of protein core 131 kDa.[2] The 528 IgG can be coupled to Sepharose 4B with equivalent results.

EGF: This is prepared from adult male mice submaxillary glands obtained from Pel Freeze.[20] Yields are approximately 20 mg/200 glands. EGF is available from Collaborative Research. Sepharose 4B is obtained from Pharmacia. Agarose A-15m is obtained from Bio-Rad. Aprotinin is from Sigma.

Cells: Human epidermoid carcinoma A431 cells, clone 29R,[21] are grown to confluence in Dulbecco's modified Eagle's medium containing 10% calf serum. These cells, which contain amplified EGF receptor genes, are a rich source of EGF receptor protein. The purification method can be applied to other human cells, to A431 cell medium,[22] and to A431 cell tumors carried in nude mice. The purification using 528 IgG is limited to human EGF receptors because this antibody does not react with rodent EGF receptors.

Buffers
A: 0.1 M Sodium bicarbonate, pH 8.3
    0.5 M NaCl
B: 10 mM HEPES, pH 7.4
    4 mM Benzamidine
    3 mM EDTA
    1mM Dithiothreitol
    1% Triton X-100

[18] P. Cuatrecasas, *J. Biol. Chem.* **245**, 3059 (1980).
[19] J. Porath, this series, Vol. 24, p. 13.
[20] C. R. Savage, Jr., and S. Cohen, *J. Biol. Chem.* **247**, 7609 (1982).
[21] T. Kawamoto, J. Mendelsohn, A. Le, G. H. Sato, C. S. Lazar, and G. N. Gill, *J. Biol. Chem.* **259**, 7761 (1984).
[22] W. Weber, G. N. Gill, and J. Spiess, *Science* **224**, 294 (1984).

10% Glycerol
0.5 Trypsin inhibitor U/ml of aprotinin
C: 20 mM HEPES, pH 7.4
1mM EDTA
6 mM 2-Mercaptoethanol
0.05% Triton X-100
10% Glycerol
130 mM NaCl
D: Buffer C plus 0.15 trypsin inhibitor U/ml of aprotinin
2 mg/ml Bovine serum albumin
E: Phosphate-buffered saline (137 mM NaCl, 2.6 mM KCl, 0.5 mM
MgCl$_2$, 0.68 mM CaCl$_2$, 15.7 mM Na$_2$HPO$_4$, 1.4 mM NaH$_2$PO$_4$)
1 mg/ml Bovine serum albumin
0.1% Sodium azide

Protein Tyrosine Kinase Assay

Reaction mixtures contain 20 mM HEPES, pH 7.4, 2 mM MnCl$_2$, 5 mM MgCl$_2$, 50 μM Na$_3$VO$_4$, 2 mM angiotensin II, 5 μM [γ-^{32}P]ATP (about 3 Ci/mmol), and enzyme in a final reaction volume of 50 μl. Kinase reactions are initiated by addition of ATP, incubated for 3 min at 30°, and stopped by the addition of trichloroacetic acid to a final concentration of 5%. Bovine serum albumin (100 μg) is added to facilitate precipitation of the EGF receptors and reactions are centrifuged for 10 min at 10,000 g. [32]P incorporated into peptide substrate is determined by diluting an aliquot of the supernatant with an equal volume of water and spotting the solution onto 2 × 3 cm phosphocellulose paper squares. These are washed three times with 75 mM phosphoric acid, dried, and counted for adsorbed radioactivity. Adsorption to the paper is quantitative.[17]

Procedure

Confluent dishes of cells are washed twice with cold phosphate-buffered saline (PBS) containing 1 mM EDTA and 1 mM EGTA. Cells are removed from the tissue culture dishes by scraping with a rubber policeman into ice-cold PBS containing EDTA and EGTA (1 mM each). Cells are transferred into 50-ml plastic centrifuge tubes, and collected by centrifuging at 1000 g for 5 min. The cell pellet is resuspended in 2 vol of buffer B and homogenized with five strokes of a glass:glass homogenizer. The homogenate is centrifuged in Corex tubes at 10,000 rpm for 10 min at 4°. The pellet is reextracted with 2 vol of buffer C, centrifuged to remove insoluble material, and the supernatants are combined. All procedures are carried out with ice-cold solutions. Chelators, EDTA and EGTA, are necessary to block the action of a calcium-sensitive neutral protease which removes about 20 kDa of carboxyl terminal sequence from the EGF receptor.[23]

The clarified cell extract is filtered through gauze to remove lipid and mixed with 528 IgG A-15m. The amount of 528 IgG A-15m required can be estimated from knowledge of the amount of ^{125}I-labeled EGF binding to cells, assuming a stoichiometry of one EGF binding site per EGF receptor protein.[17] The number of EGF receptors varies widely in different cell types from 10^4 to 10^7 molecules per cell. An excess of affinity chromatography medium is convenient and often used.

The protein:affinity chromatography mixture in a plastic tube is rotated in a cold room for 60 min. Alternatively, the cell extract is passed through a column containing 528 IgG A-15m at a flow rate of one column volume per hour. Longer mixing by rotation or recycling of the cell extract through the column can be carried out without causing degradation of the EGF receptor. Either procedure results in complete absorption of EGF receptor to the affinity matrix.

The affinity matrix is then washed to remove nonspecific protein. Batchwise washing consists of five washes with buffer C, two washes with buffer C containing I M NaCl, two washes with buffer C containing 1 M urea, and two washes with buffer C. Each wash is for 5 min; the affinity matrix is collected by centrifugation at 1500 g for 5 min. Batch washing is somewhat more complete, but the affinity matrix may be conveniently washed in the column with 10 column volumes of buffer C, 10 vol of buffer C with 1 M NaCl, 5 vol of buffer C with 1 M urea, and 5 vol of buffer C.

To obtain enzymatically active EGF receptor/kinase, the drained affinity chromatography medium is incubated with 0.1 mM EGF in buffer C for 1 hr at room temperature followed by an additional 1-hr incubation in the cold room. The solution is then collected either by centrifugation at 1500 g for 5 min and careful aspiration from above the agarose beads or by draining the solution from a column. Under these conditions about 80% of bound receptors are removed. The protein is homogeneous by the criteria of sodium dodecyl sulfate–polyacrylamide gel electrophoresis and amino terminal sequence.[17,22] Specific activity is about 15 mol ^{32}P incorporated into 1 mM angiotensin II min^{-1} mol^{-1} receptor protein. Enzyme stored in aliquots at $-70°$ is stable for up to 2 years.

[23] R. W. Yeaton, M. T. Lipari, and C. F. Fox, J. Biol. Chem. 258, 9254 (1983).

TABLE I
IMMUNOAFFINITY PURIFICATION OF EGF RECEPTOR
PROTEIN

	CONCENTRATION	
STEP	(μg)	(%)
Estimated starting material[a]	650	100
First elution with EGF	350	54
Second elution with EGF[b]	150	23
Elution with 6 M urea[c]	100	15

[a] Starting material was clone 29R A431 cells containing about 3 pmol EGF binding sites per 10^6 cells.[21] From 1.65×10^9 cells about 650 μg of EGF receptor protein with a protein core molecular weight of $131K^2$ can be estimated.

[b] A second elution with 0.1 mM EGF was carried out. Alternatively, recycling of first elution buffer may be used; 500 μg of active EGF receptor protein was obtained in the combined EGF elutions.

[c] After elution with EGF, all remaining protein was desorbed with 6 M urea. An additional 100 μg of EGF receptor was recovered.

Comments

This procedure, based on competitive elution with EGF ligand of EGF receptor protein specifically absorbed to an immobolized competitive antagonist monoclonal antibody, allows purification to homogeneity of active enzyme. A typical purification is shown in Table I. With this procedure active enzyme is obtained in high yield in a single step. The major limitation is that enzyme is eluted with EGF and is not suitable for kinetic or mechanistic analysis of EGF activation. It is useful for other kinetic analyses such as effects of self-phosphorylation.[24]

Absorption to immobilized 528 IgG is essentially complete. Careful washing of 528 IgG A-15m containing bound EGF receptor is essential to obtain homogeneous protein. With proper washing only EGF receptor is retained on the affinity matrix. This can be readily demonstrated by elution with 6 M urea in buffer C. This results in complete desorption of denatured EGF receptor, which is homogeneous by several criteria. The 528 IgG·A15 matrix can also be used to efficiently purify the EGF recep-

[24] P. J. Bertics and G. N. Gill, *J. Biol. Chem*, **260**, 14642 (1985).

tor-related protein secreted by A431 cells.[22] This protein synthesized on a specific mRNA corresponds to the amino terminal cell surface EGF binding segment of the EGF receptor protein.[2]

Acknowledgments

We would like to thank Paul J. Bertics, Gary J. Heiserman, and Gordon M. Walton for helpful discussions. This work was supported by grants from the National Institute of Arthritis, Diabetes, and Digestive and Kidney Diseases (AM13149) and the American Cancer Society (BC-209).

[8] *In Vitro* Growth of A431 Human Epidermoid Carcinoma

By David Barnes

The A431 human epidermoid carcinoma cell line is one of a series of human tumor-derived lines established by Giard *et al.*[1] Little characterization of the line was attempted until Fabricant *et al.*[2] reported that the cells expressed an unusually high number of cell surface receptors for epidermal growth factor (EGF). This property of A431 cells allowed biochemical approaches to the study of hormone–receptor interactions and biological effects that had not been possible previously.[3-5] These cells also have been used for studies of transferrin and lipoprotein receptors, extracellular matrix biosynthesis, and tumor promoter effects.[6-9]

EGF at very low concentrations stimulates the growth of A431 cells *in vitro*,[10] but higher concentrations of this growth factor, in the range of 1 to

[1] D. A. Giard, S. A. Aaronson, G. J. Todaro, P. Arnstein, J. H. Kersey, H. Dosik, and W. P. Parks, *J. Natl. Cancer Inst. (U.S.)* **51**, 1417 (1973).

[2] R. N. Fabricant, J. E. DeLarco, and G. J. Todaro, *Proc. Natl. Acad. Sci. U.S.A.* **74**, 565 (1977).

[3] G. Carpenter, L. King, and S. Cohen, *Nature (London)* **276**, 409 (1978).

[4] H. Haigler, J. F. Ash, S. J. Singer, and S. Cohen, *Proc. Natl. Acad. Sci. U.S.A.* **75**, 3317 (1978).

[5] M. Chinkers, J. A. McKanna, and S. Cohen, *J. Cell Biol.* **83**, 260 (1979).

[6] C. R. Hopkins and I. S. Trowbridge, *J. Cell Biol.* **97**, 508 (1983).

[7] R. G. H. Anderson, M. S. Brown, and J. L. Goldstein, *J. Cell Biol.* **88**, 441 (1981).

[8] D. S. Salomon, L. A. Liotta, M. Panneerselvam, V. P. Terranova, A. Sahi, and P. Fehnel, *Cell Cult. Methods Mol. Cell Biol.* **1**, 295 (1984).

[9] A. Sahai, K. B. Smith, M. Panneerselvam, and D. S. Salomon, *Biochem. Biophys. Res. Commun.* **109**, 1206 (1982).

[10] T. Kamamoto, D. J. Sato, A. Le, J. Polikoff, G. Sato, and J. Mendelson, *Proc. Natl. Acad. Sci. U.S.A.* **80**, 1337 (1983).

100 ng/ml, are inhibitory for A431 cell growth.[11,12] EGF at 1 to 10 ng/ml is growth stimulatory for most cell types in culture. In this chapter we describe methods for the growth of A431 cells for the study of EGF effects in both serum-containing and serum-free media, and also describe methods developed by Buss *et al.*[12] for the isolation of A431 variants exhibiting altered growth responses to EGF.

Source of Cells

A431 cells are available from the American Type Culture Collection (Rockville, MD) and from a number of laboratories active in peptide hormone research. Evidence exists that sublines from different laboratories exhibit somewhat different phenotypes,[13] and investigators should be aware of potential problems in this regard when attempting to relate their data to work of others. In addition, some lines are heavily contaminated with mycoplasma, and precautions should be taken that such contamination does not influence experimental results or lead to contamination of other cultures present in the laboratory.

Growth of A431 Cells in Serum-Containing Medium

A431 cells were derived in Dulbecco's modified Eagle's medium (DME) containing high glucose (4.5 g/liter) supplemented with 10% heat-inactivated (56°, 30 min) fetal calf serum, 200 U/ml penicillin, and 200 μg/ml streptomycin. Heat inactivation of the serum is often omitted in routine culture of the cells. When supplemented with serum, DME is superior to Ham's F12 for the growth of the cells, but is slightly inferior to a one-to-one mixture of DME and F12 (DME:F12).[14] Other basal nutrient media formulations have not been examined extensively for the ability to support A431 cell growth with 10% serum supplementation. Powdered formulations of these media are available from a number of commercial cell culture supply houses, and were supplemented with 15 mM 4-(2-hydroxyethyl)-1-piperazineethanesulfonic acid (HEPES), pH 7.4, and 1.2 g/liter sodium bicarbonate in the experiments cited.[14]

Cells are routinely grown in a 5% CO_2–95% air atmosphere at 37°. A431 cells grow well in DME or F12:DME supplemented with as little as 1% serum,[14] although higher concentrations are almost invariably used for

[11] D. Barnes, *J. Cell Biol.* **93**, 1 (1982).

[12] J. E. Buss, J. E. Kudlow, C. S. Lazar, and G. N. Gill, *Proc. Natl. Acad. Sci. U.S.A.* **79**, 2574 (1982).

[13] D. Gospodarowicz, G. M. Liu, and K. Gonzales, *Cancer Res.* **42**, 3704 (1982).

[14] D. Barnes, *Adv. Exp. Med. Biol.* **172**, 49 (1984).

routine growth. Doubling time is about 30 hr. The cells in serum-containing media can be propagated from clonal density, and also can be grown in semisolid agar (see chapter by Rizzino [32], this volume). Trypsinization procedures are routine (see volume LVIII, this series), and can be accomplished with a solution containing 0.1% crude trypsin and 1 mM ethylenediaminetetraacetic acid (EDTA) in phosphate-buffered saline.

EGF can be added directly to serum-containing medium with demonstrable inhibition of cell growth at concentrations as low as 1 ng/ml.[11,12] Although proliferation of the cells is prevented in the presence of EGF, the growth factor is not toxic, and proliferation will resume with a delay of about 72 hr if EGF is removed from the medium.[11] EGF for these experiments is added as a small aliquot from a concentrated stock.

Growth of A431 Cells in Serum-Free Medium

Most cell types in culture can be grown under serum-free conditions in a rich basal nutrient medium to which is added an optimized combination of hormones, binding proteins, attachment factors, and supplementary nutrients.[15] A431 cells can be grown in F12:DME containing HEPES and bicarbonate as described above and in which serum is replaced by commercially available bovine insulin, human transferrin, human fibronectin, and ethanolamine.[11] Antibiotics used are penicillin (200 U/ml), streptomycin (200 μg/ml) and ampicillin (25 μg/ml). Antibiotics and HEPES are stored frozen as concentrated stocks (100×) and added to the medium after dissolving the powdered formulation in a volume of water near that of the final volume.

Insulin is prepared at 1 mg/ml in 20 mM HCl; transferrin is made as a 5 mg/ml stock in phosphate-buffered saline. Both insulin and transferrin stocks can be filter sterilized after preparation. EGF and fibronectin can be obtained as sterile, lyophilized powders from several commercial sources. These should be reconstituted with sterile solutions, and not filtered. Ethanolamine is prepared as a 50 mM stock in water and membrane sterilized. Final concentrations added to the basal nutrient medium are as follows: insulin, 10 μg/ml; transferrin, 10 μg/ml; fibronectin, 5 μg/ml; ethanolamine 0.5 mM.

Stock solutions of all four supplements may be stored in the refrigerator and may also be stored long term in the freezer in aliquots. Multiple freeze–thaws of these solutions should be avoided. Water for the preparation of concentrated stocks and the medium is purified by passage through

---

[15] D. Barnes and G. Sato, *Cell* **22,** 649 (1980).

a Milli-Q (Millipore) water purification system immediately prior to use. F12:DME may be stored frozen for a period of several weeks without serious degradation of nutritional components. All solutions are stored in reusable plastic containers. Polypropylene tubes are used for concentrated stocks and 250-ml polystyrene flasks are used for storing media. All pipets and culture vessels are disposable plastic. Toxic residual detergent remaining on glassware makes the washing and reuse of pipets or glass containers risky for serum-free procedures. The tendency of some of the medium components to stick to glass surfaces also makes the use of glass disposable pipets undesirable.

Trypsinization and subculturing can be accomplished with the trypsin/EDTA solution described above, with the additional step of dilution of the cell suspension into an equal volume of F12:DME containing 0.1% soybean trypsin inhibitor prior to centrifugation of the cells and resuspension in fresh medium. Supplements should be added to the medium as small aliquots immediately before addition of cells to the culture vessels. Supplements cannot be added directly to the medium and stored for later use, as one might routinely do with a serum supplement. Insulin and fibronectin and possibly other components do not remain active under such conditions.

An alternative formulation for serum-free growth of A431 cells has been developed by Barka and van der Noen.[16] This procedure calls for DME as a basal medium and supplementation with fetuin 0.5 mg/ml), insulin (50 μg/ml), transferrin (10 μg/ml), biotin (1 μg/ml), and oleic acid (5 μg/ml) bound to fatty acid-free bovine serum albumin (1 mg/ml). The fetuin in this medium does not represent a well-defined supplement, and may be providing both attachment factors and peptide growth factors.[15]

Isolation of A431 Variants Exhibiting Altered Growth Responses to EGF

Variant A431 cells may be obtained by treating cultures with 1.25 μg/ml N-methyl-N'-nitro-N-nitrosoguanidine (MNNG) for 2 hr, followed by growth in DME supplemented with 10% calf serum and 100 ng/ml EGF.[12] The MNNG concentration used results in about 30% cell survival, and subsequent growth in the presence of EGF results in a frequency of variant appearance of about 2 per 10^5 cells, but only about 10% of these are hardy colonies. The degree to which these variants represent genuine mutational events versus existing resistant cells present in the A431 population prior to

[16] T. Barka and H. van der Noen, *Am. J. Anat.* **165**, 187 (1982).

mutagenesis is unclear.[12] Cells isolated in this way exhibit a reduced number of cell surface EGF receptors and reduced EGF-stimulated tyrosine kinase activity, and respond variously to EGF with regard to growth rate, morphology, cloning efficiency, and saturation density. Some of these clones are stimulated somewhat to proliferate by EGF at concentrations that are inhibitory to the parent line.[12]

Section II
Transforming Growth Factors

[9] Radioreceptor Assays for Transforming Growth Factors

By CHARLES A. FROLIK and JOSEPH E. DE LARCO

Transforming growth factors (TGF) are a group of polypeptides that can induce a transformed phenotype in the appropriate indicator cells — for example, the normal rat kidney fibroblast (NRK-49F) cell line.[1,2] Two different classes of TGF, TGFα and TGFβ, have been purified to homogeneity and have been partially or completely sequenced. TGFα [also called TGF type 1 or epidermal growth factor (EGF)-like TGF] has been obtained from the serum-free conditioned medium (SF-CM) of retrovirally transformed and tumor-derived cell lines.[3-6] The amino acid sequence of TGFα shows strong sequence homology with EGF.[4,5] Recently, relatively large quantities of TGFα have become available through both solid phase synthesis[7] (also see Tam [12], this volume) and molecular cloning.[8] These methods yield hundreds of milligrams of pure peptide that appear to have properties similar, if not identical, to those of the natural products isolated from the SF-CM of transformed cells. Type-β TGF (or TGF type 2) has been purified from human platelets[9] (also see Assoian [14], this volume), human placenta,[10] and bovine kidney[11] as well as from virally transformed and tumor-derived cell lines.[12,13] The N-terminal portion of this peptide

---

[1] J. E. De Larco and G. J. Todaro, *Proc. Natl. Acad. Sci. U.S.A.* **75,** 4001 (1978).

[2] A. B. Roberts, L. C. Lamb, D. L. Newton, M. B. Sporn, J. E. De Larco, and G. J. Todaro, *Proc. Natl. Acad. Sci. U.S.A.* **77,** 3494 (1980).

[3] H. Marquardt and G. J. Todaro, *J. Biol. Chem.* **257,** 5220 (1982).

[4] H. Marquardt, M. W. Hunkapiller, L. E. Hood, D. R. Twardzik, J. E. De Larco, J. R. Stephenson, and G. J. Todaro, *Proc. Natl. Acad. Sci. U.S.A.* **80,** 4684 (1983).

[5] H. Marquardt, M. W. Hunkapiller, L. E. Hood, and G. J. Todaro, *Science* **223,** 1079 (1984).

[6] J. Massagué, *J. Biol. Chem.* **258,** 13606 (1983).

[7] J. P. Tam, H. Marquardt, D. Rosberger, N. H. Heath, and G. J. Todaro, *Nature (London)* **309,** 376 (1984).

[8] R. Derynck, A. B. Roberts, M. E. Winkler, E. Y. Chen, and D. V. Goeddel, *Cell* **38,** 287 (1984).

[9] R. K. Assoian, A. Komoriya, C. A. Meyers, D. M. Miller, and M. B. Sporn, *J. Biol. Chem.* **258,** 7155 (1983).

[10] C. A. Frolik, L. L. Dart, C. A. Meyers, D. M. Smith, and M. B. Sporn, *Proc. Natl. Acad. Sci. U.S.A.* **80,** 3676 (1983).

[11] A. B. Roberts, M. A. Anzano, C. A. Meyers, J. Wideman, R. Blacher, Y.-C. E. Pan, S. Stein, S. R. Lehrman, J. M. Smith, L. C. Lamb, and M. B. Sporn, *Biochemistry* **22,** 5692 (1983).

[12] J. Massagué, *J. Biol. Chem.* **259,** 9756 (1984).

does not share sequence homology with any other known protein.[11,12] Similar to other growth factors, the TGFs have been shown to initially interact with the cell through membrane receptors[1,14-19] (also see chapters by Massagué [10, 13, 17] and Wakefield [16], this volume). It is the purpose of this chapter to describe the methods developed for the detection and characterization of these receptors.

Radioreceptor Assays for TGFα

Binding Assay for TGFα. The binding assay for TGFα is the same as that employed for EGF. Because the EGF assay has been thoroughly reviewed in a previous volume[20] the method will not be described here. Originally the TGFα utilized as standard in the assay was isolated from SF-CM. However, with the knowledge of the amino acid sequence of TGFα from two different sources,[4,5] two groups have produced milligram quantities of TGFα using either a peptide synthesis route to produce rodent TGFα[7] or the amino acid sequence to clone TGFα from a human source.[8] The synthetic products from either method appear to have biological properties and binding characteristics that are similiar to those of the native TGFα's isolated from the SF-CM of transformed cells. These peptides are readily radioiodinated using conventional techniques and will be of use in the future in radioreceptor assays as well as radioimmunoassays.

Purification of TGFα by Receptor Binding. Initial observations demonstrated that murine sarcoma virus (MSV)-transformed cells lacked available receptors for binding exogenously added radiolabeled EGF.[15] The assumption was made that these cells were producing and releasing an EGF-like peptide (TGFα) that bound to the EGF receptor. This receptor occupancy prevented the binding of exogenous radiolabeled EGF and, in turn, acted as a mitogen on these cells. Based on this model, SF-CM from an MSV-transformed 3T3 cell line was examined for the presence of a peptide that could compete with ^{125}I-labeled EGF in a radioreceptor assay.

[13] L. L. Dart, D. M. Smith, C. A. Meyers, M. B. Sporn, and C. A. Frolik, *Biochemistry,* **24,** 5925 (1985).

[14] J. E. De Larco and G. J. Todaro, *J. Cell. Physiol.* **102,** 267 (1980).

[15] G. J. Todaro, J. E. De Larco, and S. Cohen, *Nature (London)* **264,** 26 (1976).

[16] J. Massagué, *J. Biol. Chem.* **258,** 13614 (1983).

[17] C. A. Frolik, L. M. Wakefield, D. M. Smith, and M. B. Sporn, *J. Biol. Chem.* **259,** 10995 (1984).

[18] R. F. Tucker, E. L. Branum, G. D. Shipley, R. J. Ryan, and H. L. Moses, *Proc. Natl. Acad. Sci. U.S.A.* **81,** 6757 (1984).

[19] J. Massagué and B. Like, *J. Biol. Chem.* **260,** 2636 (1985).

[20] G. Carpenter, this series, Vol. 109, p. 101.

It was assumed that such a peptide would be able to interact with the EGF receptor system on indicator cell lines as well as on the original transformed cell line. The SF-CM from these cells did indeed contain mitogenic factors that were able to compete with radiolabeled EGF in a radioreceptor assay.[1] These activities, however, were unable to compete with EGF in a radioimmunoassay. This suggested that the species of interest were functionally similar to EGF but antigenically distinct. The concentrations of these factors (TGFα's) were sufficient to act as mitogens but they were too low for purification by conventional methods.

Therefore, attempts were made to purify one of the radiolabeled TGFα's from a crude mixture of radiolabeled peptides in order to further characterize it. To do this, the EGF receptors of the A431 cell line (2×10^6 receptors per cell) were used as an affinity reagent. These cells could be fixed to tissue culture dishes using 5% formaldehyde in phosphate-buffered saline. After the cells were fixed to the plastic surface their EGF receptors were still able to bind and release EGF and EGF-like peptides.[21] This "reagent" was able to preferentially bind EGF or EGF-like peptides from a complex mixture with the bound material being released by acidification. SF-CM from MSV-transformed 3T3 cells was chromatographed on a Bio-Gel P-60 column and an aliquot of the eluent containing EGF competing activity (approximately 0.1% TGFα) was radioiodinated using a modification of the Bolton–Hunter procedure.[22] The radiolabeled material was diluted with binding buffer[15] and bound to a fixed cell sheet of A431 cells. At the end of the binding period the unbound material was removed, the cell sheet washed twice with binding buffer and twice with phosphate-buffered saline to remove unbound radioactive material, and the bound material was eluted with three washes of 0.1% acetic acid. The eluted material was concentrated by lyophilization and redissolved in binding buffer. The cycled material could be retested for its ability to bind to either fixed A431 cells or to the EGF receptor system from other sources. Specific binding was calculated by subtracting the nonspecific binding [determined by adding a large excess of unlabeled EGF (50 μg/ml) to the radiolabeled binding mixture] from the total binding.

Table I lists the ratios of the specific binding to the nonspecific binding as well as the percentage of the input counts that bound specifically to the EGF receptors for the uncycled and cycled preparations of TGFα. After two cycles of binding and elution, there was a dramatic increase in both the ratio of specific to nonspecific binding and the percentage of input counts

[21] J. E. De Larco, R. Reynolds, K. Carlberg, C. Engle, and G. J. Todaro, *J. Biol. Chem.* **255**, 3685 (1980).
[22] J. E. De Larco, Y. A. Preston, and G. J. Todaro, *J. Cell. Physiol.* **109**, 143 (1981).

TABLE I
PURIFICATION OF ^{125}I-LABELED TGFα USING BINDING AND ELUTION
FROM AN EGF RECEPTOR-RICH CELL

Sample[a]	Specific binding/ nonspecific binding	Input counts bound specifically (%)
Uncycled	0.18	0.12
After first cycle	5.6	2.8
After second cycle	19	24

[a] Sample came from the EGF competing region of a BioGel P-60 column effluent to which the SF-CM from MSV-transformed 3T3 cells had been applied.[21]

that bound specifically. In the case of the twice cycled material, approximately 24% of the input counts bound specifically to the A431 cells; whereas only 0.12% of the uncycled input counts bound specifically under the same conditions. This recycled TGFα did not bind to a clone of mouse 3T3 cells which lacked the EGF receptor. The cycled radiolabeled TGFα had therefore been separated from most of the other radiolabeled peptides found in the original preparation. This "purified" or cycled radiolabeled peptide allowed the characterization of the ligand with respect to its physicochemical properties and its binding kinetics. It was shown that the twice-cycled radiolabeled TGFα bound to EGF receptors with characteristics similar to those of mouse submaxillary gland EGF.[21]

Characterization of TGFα Binding. Antibodies to the EGF receptor not only block the effects of EGF but also block the effects of TGFα. The mitogenic and morphological effects induced by both of these ligands were inhibited by antibodies developed against purified EGF receptors.[23] This suggests that activation of the EGF receptor is necessary for cells to respond to TGFα.

TGFα has been purified to homogeneity from SF-CM of several transformed cell lines.[3-6] These purified peptide growth factors can readily be iodinated using modifications of the Bolton–Hunter procedure.[22] The radiolabeled peptides have been used for kinetic studies and for the cross-linking of the radiolabeled ligand to its receptor system. The kinetic experiments indicate that most of the binding parameters are similar for EGF and TGFα. The one difference that stands out is the pH optimum for the binding of these ligands.[16] The EGF binding curve is broader and starts at a

[23] G. Carpenter, C. M. Stoscheck, Y. A. Preston, and J. E. De Larco, *Proc. Natl. Acad. Sci. U.S.A.* **80,** 5627 (1983).

lower pH than that for TGFα. This difference can be accounted for by the differences in the p*I* values of the two proteins—EGF having a p*I* of approximately 4.4 whereas TGFα has a p*I* of 6.8.

Purified radiolabeled TGFα has also been used to identify the membrane receptor that binds TGFα. To do this, the radiolabeled ligand was covalently attached to its receptor system on membrane preparations using divalent cross-linking reagents. The affinity labeled membrane preparations from either human placenta or A431 cells were characterized using sodium dodecyl sulfate–polyacrylamide gel electrophoresis and autoradiography.[16,24] Both ^{125}I-labeled EGF and TGFα yielded a predominant 170-kDa species in placenta and two species of approximately 160 and 145 kDa in A431 membrane preparations. The labeling of these species was inhibited to the same extent when an excess of either unlabeled native TGFα or unlabeled native EGF was added during the incubation of the membrane components with the labeled ligand. These binding and cross-linking experiments along with the EGF receptor antibody data indicate the TGFα's are binding to and using the EGF receptor system for the mediation of their biological effects.

Radioreceptor Assays for TGFβ

Radiolabeling of TGFβ. TGFβ is most easily purified from human platelets[9] although TGFβ's purified from other sources appear to be identical to platelet-derived TGFβ in all respects so far studied. A number of methods have been developed for the iodination of purified human TGFβ These include a modified chloramine-T method,[17] a Bolton–Hunter procedure,[18] and an immobilized lactoperoxidase glucose oxidase method.[19] All three procedures produced ^{125}I-labeled TGFβ with a specific activity of 40 to 175 μCi/μg. In our laboratory, we have found the modified chloramine-T method to most reproducibly yield biologically active ^{125}I-labeled TGFβ. This procedure utilizes 5 to 10 μg of purified TGFβ in 10 μl of solution. The exact composition of the solution used does not appear to be critical and can be 4 m*M* HCl or the solvent used in the last purification step. Lyophilization of TGFβ to dryness is not recommended as only 30 to 50% of the starting material will redissolve due to the tenacity with which TGFβ will stick to glass even after siliconization. To this TGFβ solution is added 10 μl of 1.5 *M* sodium phosphate, pH 7.5, and 10 μl of Na^{125}I (1 mCi). The reaction is initiated by addition of 5 μl of chloramine-T solution (100 μg/ml).[22] A second 5-μl aliquot is added after 2 min at room

[24] J. Massagué, M. P. Czech, K. Iwata, J. De Larco, and G. J. Todaro, *Proc. Natl. Acad. Sci. U.S.A.* **79**, 6822 (1982).

temperature followed 1.5 min later by a third 5-μl aliquot. One minute after the last addition of chloramine-T solution the reaction is terminated by adding 20 μl of 50 mM N-acetyltyrosine, 200 μl of 60 mM potassium iodide, and 200 μl of ultrapure urea (1.2 g/ml of 1 M acetic acid). After approximately 5 min, the bound ^{125}I is separated from the free ^{125}I by passing the reaction mixture over a Sephadex G-25 column (0.7 × 20 cm) equilibrated and eluted with 4 mM HCl, 75 mM sodium chloride, 0.1% bovine serum albumin. The void volume fractions were stored at $-20°$ in 4 mM HCl, 1% bovine serum albumin at a concentration of 60 to 80 μCi/ml. When stored for 5 weeks under these conditions, the ^{125}I-TGFβ still retained 93% of its initial activity in being able to bind to the cell membrane receptor.

The physical and biological integrity of the iodinated growth factor should be determined using a combination of several methods. Precipitation with 10 to 15% trichloroacetic acid gives an indication as to how much of the ^{125}I present in the preparation is protein bound. Values in the 95 to 99% range are usually considered acceptable. Because the biological activity of TGFβ depends on the presence of the dimeric form of the growth factor, the ^{125}I-labeled TGFβ should be run on a sodium dodecyl sulfate–polyacrylamide gel using the discontinuous buffer system of Laemmli[25] to ascertain that the iodinated product migrates at approximately 25 kDa. Typically 92 to 94% of the applied ^{125}I-labeled TGFβ can be expected to be found in this region. Finally the biological activity of the ^{125}I-labeled TGFβ can be determined by measuring the induction of growth of normal rat kidney fibroblasts in soft agar[2,14] and by determining the percentage of iodinated protein that can specifically bind to the fibroblast cell. To do the latter experiment, ^{125}I-labeled TGFβ is incubated with a fibroblast monolayer for 2 hr at 4°. The cell-bound ^{125}I is measured as described below, the medium pooled, and aliquots removed for determination of the trichloroacetic acid-precipitable ^{125}I left in the medium. The medium is then reincubated with fresh cells. The procedure is repeated for a total of five times. Nonspecific binding is determined by using an excess of unlabeled TGFβ in one well for each incubation. The results are plotted as ^{125}I-labeled TGFβ bound (percentage of initial) versus trichloroacetic acid-precipitable ^{125}I (percentage of initial). With a fresh preparation of ^{125}I-labeled TGFβ 95 to 98% should be able to bind specifically to the normal rat kidney fibroblast.

^{125}I-Labeled TGFβ Binding Assay. No cell lines have yet been found that contain a high concentration of TGFβ receptors. Therefore, investigators have tended to use those cells which have been shown to be responsive to TGFβ. These include the NRK (clone 49F) fibroblast (16,000–19,000

---

[25] U. K. Laemmli, *Nature (London)* **227**, 680 (1970).

binding sites per cell with a dissociation constant of 25 to 78 pM[17,19]) and the mouse embryo-derived AKR-2B (clone 84A) cell (10,500 binding sites per cell with a dissociation constant of 33 pM[18]).

When using the NRK fibroblast cell line, cells are routinely seeded in 24-well plates at 8×10^4 cells/ml per well in Dulbecco's modified Eagle's medium containing 10% calf serum and 50 μg/ml gentamicin. After an overnight incubation at 37° in a humidified 5% CO_2 in air atmosphere, the cells are washed twice with serum-free Eagle's medium and then incubated as before in 1 ml of Eagle's medium containing 1% calf serum. This second incubation in a low-serum medium was found to be useful for generating more reproducible data by minimizing the interference caused by the endogenous TGFβ present in calf serum. After 18 to 24 hr, the monolayers are washed two times with binding buffer (Dulbecco's modified Eagle's medium containing 0.1% bovine serum albumin and 25 mM HEPES buffer) followed by the addition of 200 μl of binding buffer containing ^{125}I-labeled TGFβ with or without various concentrations of competitors. For a standard competition assay, the final concentration of ^{125}I-labeled TGFβ was 50 to 100 pM (0.25 to 0.5 ng/well, approximately 50,000 cpm) with increasing concentrations of unlabeled TGFβ (0.1 to 500 ng/ml) or competitor being added. Cells are incubated with gentle agitation for 3 hr at 4° in a 5% CO_2 atmosphere. It was noted that when the reaction is incubated at room temperature a small amount of trichloroacetic acid-soluble radioactivity appeared in the medium indicating internalization and metabolism of the ^{125}I-labeled TGFβ—an undesirable event when determining equilibrium binding. After the incubation, cells are washed four times with ice-cold Hanks' solution containing 0.1% bovine serum albumin to remove unbound growth factor. The washed cells are solubilized by a 20-min incubation at room temperature with 0.6 ml of Triton X-100 solution (20 mM HEPES, 1% Triton X-100, 10% glycerol, 0.01% bovine serum albumin, pH 7.4[26]). Aliquots (500 μl) are then counted on a gamma spectrometer. Treatment of the monolayers with 1 ml collagenase (1 mg/ml water) for 5 min at room temperature has also been reported to be an effective way to remove cells from the plate for counting.[18] Solubilization with strong acid or alkali caused the release of plastic-bound ^{125}I-labeled TGFβ resulting in a dramatic increase in the nonspecific counts found in solution. The amount of nonspecific ^{125}I-labeled TGFβ bound to the plastic dish decreased with an increase in the confluency of the cells. The results of a typical competition binding assay using unlabeled human platelet TGFβ as the inhibitor are shown in Fig. 1.

Characterization of the TGFβ Receptor. The TGFβ receptor has been independently characterized from three different laboratories using two

[26] C. H. Heldin, B. Westermark, and A. Wasteson, *Proc. Natl. Acad. Sci. U.S.A.* **78**, 3664 (1981).

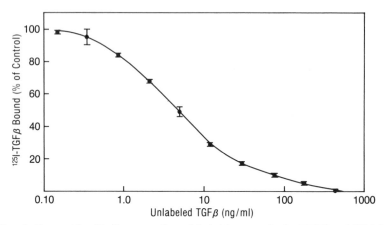

FIG. 1. Competitive binding assay for unlabeled human platelet TGFβ and ¹²⁵I-labeled TGFβ on NRK cells.

different cell lines (NRK clone 49F and AKR-2B clone 84A) and the results have all shown the presence of a high-affinity (25–78 pM) specific receptor at low concentrations (10,500 to 19,000 binding sites per cell).[17-19] One difference observed between the receptors on the AKR-2B cell and those on the NRK cell was that the former receptors required Mg^{2+} ion (5 mM) for maximal binding of TGFβ.[18] In addition to this high-affinity receptor on the NRK fibroblast cell membrane, a low-affinity binding site (dissociation constant of 4 nM and 180,000 sites per cell) has been reported.[17] It is still not clear whether this second binding site represents a nonspecific adherence of TGFβ to membrane components unrelated to a physiological response or whether it plays a part in a second, as yet undefined, role of TGFβ in the cell. In this regard affinity labeling experiments employing TGFβ have demonstrated that in addition to a high-affinity receptor with a molecular weight of approximately 280,000, there is a second protein of 70 to 90 kDa that is able to bind TGFβ with low affinity.[19]

[10] Purification of Type-α Transforming Growth Factor from Transformed Cells

By Joan Massagué

Type-α transforming growth factor (TGFα) is a polypeptide whose expression is selectively induced upon cellular transformation by a variety of oncogenes. TGFα was first found in 3T3 mouse fibroblasts transformed by Moloney murine sarcoma virus which carries an activated *mos* oncogene.[1] Production of TGFα has been subsequently documented in fibroblasts transformed by retroviruses containing activated *fes*, Ki-*ras*, *abl*, and other oncongenes, as well as in various cell lines of human tumor origin.[2] The discovery of TGFα in culture fluids from retrovirally transformed cells was prompted in part by the observation that these media could induce anchorage-independent growth, the "transformed phenotype," in certain normal rodent fibroblasts.[1] Later, it became apparent that the induction of the transformed phenotype was due to the action of a second factor, TGFβ, acting on target cells when they were fully stimulated mitogenically by TGFα (or in its place EGF) plus platelet-derived growth factor, insulin-like growth factor, and possibly other factors present in serum.[3-6]

Mature TGFα is a 50-amino acid, single-chain polypeptide with significant (35-40%) amino acid sequence identity with mouse and human epidermal growth factors (EGF).[4,7-10] In particular, the six cysteine residues which determine the positions of the three disulfide bridges in EGF have been preserved in TGFα The results from molecular cloning of rat and human TGFα suggest that this peptide is part of a larger, 160-amino acid precursor with an amino-terminal hydrophobic sequence which may

[1] J. E. De Larco and G. J. Todaro, *Proc. Natl. Acad. Sci. U.S.A.* **75**, 4001 (1978).
[2] G. J. Todaro, C. Fryling, and J. E. De Larco, *Proc. Natl. Acad. Sci. U.S.A.* **77**, 5258 (1980).
[3] A. B. Roberts, M. A. Auzano, L. C. Lamb, J. M. Smith, C. A. Frolik, H. Marquardt, G. J. Todaro, and M. B. Sporn, *Nature (London)* **295**, 417 (1982).
[4] J. Massagué, *J. Biol. Chem.* **258**, 13606 (1983).
[5] R. K. Assoian, G. R. Grotendorst, D. M. Miller, and M. B. Sporn, *Nature (London)* **309**, 804 (1984).
[6] J. Massagué, B. Kelly, and C. Mottola, *J. Biol. Chem.* **260**, 4551 (1985).
[7] H. Marquardt, M. W. Hunkapiller, L. E. Hood, D. R. Twardzik, J. E. De Larco, and G. J. Todaro, *Proc. Natl. Acad. Sci U.S.A.* **80**, 4684(1983).
[8] H. Marquardt, M. W. Hunkapiller, L. E. Hood, and G. J. Todaro, *Science* **223**, 1079 (1984).
[9] R. Derynck, A. B. Roberts, M. E. Winkler, E. Y. Chen, and D. V. Goeddel, *Cell* **38**, 287 (1984).
[10] D. C. Lee, T. M. Rose, N. R. Webb, and G. J. Todaro, *Nature (London)* **313**, 489 (1984).

METHODS IN ENZYMOLOGY, VOL. 146

act as a signal for translocation of TGFα across the membrane.[9,10] A second stretch of hydrophobic amino acids located carboxy terminal with respect to the mature TGFα sequence suggests that the TGFα precursor may reside in the cell as an integral membrane protein. TGFα and EGF interact equipotently with a common receptor type in intact cells and membrane preparations,[8,11] and the biological potency of TGFα and EGF both *in vivo* and *in vitro* is very similar.[4,12,13]

TGFα was first purified to homogeneity by Marquardt and Todaro from culture fluids from the human melanoma A2058 cell line.[14] The original purification method with minor modifications has been successfully used to purify rat TGFα and mouse TGFα from retrovirally transformed cells.[4,7] This method as performed in our laboratory is the subject of the present chapter.

Radioreceptor Assay for TGFα

The most rapid, simple, and quantitative assay to detect TGFα during purification is the radioreceptor assay. Given the ability of TGFα to compete with ^{125}I-labeled EGF for binding to a common receptor type, and the high abundance of EGF/TGFα receptors in A431 human carcinoma cells,[15] membranes from this cell type and trace concentrations of ^{125}I-labeled EGF can be used as a convenient radioreceptor assay for TGFα.[1,11] Membrane preparations from A431 cells are preferred to freshly grown A431 cell monolayers or formalin-fixed monolayers[1] as a source of receptor for the assay because of their convenience and the possibility of using the same membrane preparation over several months to 1 year if stored under proper conditions.

  Reagents
  EGF and ^{125}I-labeled EGF: EGF is purified from mouse submaxillary glands by the method of Savage and Cohen.[16] EGF is labeled with ^{125}I to a specific activity of 150–200 Ci/g as previously described[11]
  A431 membrane preparation: Confluent monolayers of A431 cells grown in Dulbecco's modified Eagle's medium supplemented with 10% calf serum are washed twice with isotonic saline solution and

[11] J. Massagué, *J. Biol. Chem.* **258**, 13613 (1983).
[12] J. M. Smith, M. B. Sporn, A. B. Roberts, R. Derynck, M. E. Winkler, and H. Gregory, *Nature (London)* **315**, 515 (1985).
[13] J. P. Tam, *Science* **229**, 673 (1985).
[14] H. Marquardt and G. J. Todaro, *J. Biol. Chem.* **257**, 5220 (1982).
[15] R. N. Fabricant, J. E. De Larco, and G. J. Todaro, *Nature (London)* **292**, 259 (1981).
[16] C. R. Savage and S. Cohen, *J. Biol. Chem.* **247**, 7609 (1972).

then scraped into a solution containing 0.25 M sucrose, 10 mM Tris, 1 mM EDTA, pH 7.0 1 mM phenylmethylsulfonyl fluoride. Cells are homogenized in this solution by 10 strokes of a tight-fitting Dounce homogenizer. The homogenate is centrifuged at 3,000 g for 10 min. The resulting pellet is subjected to the same cycle of homogenization and centrifugation twice more. The supernatants from these three spins are combined and centrifuged at 100,000 g for 60 min on top of a cushion of 37% (w/v) sucrose in 10 mM Tris, 1 mM EDTA, pH 7.0, in a swinging bucket rotor. The membranous material sedimenting at the cushion interface is carefully removed, diluted with 4 vol of a solution containing 10 mM Tris, 1 mM EDTA, pH 7.0, and centrifuged at 30,000 g for 30 min. The resulting pellet is resuspended at a concentration of 5 mg of membrane protein per milliliter in the latter solution, and stored in small aliquots at $-70°$

Binding buffer: 128 mM NaCl, 5 mM KCl, 1.2 mM CaCl$_2$, 1.2 mM MgSO$_4$, 50 mM HEPES, pH 7.7, 10 mg/ml of bovine serum albumin

Procedure

Samples to be assayed are lyophilized to dryness in polypropylene tubes and then reconstituted with 250 μl of binding buffer containing 0.2 nM ^{125}I-labeled EGF ($4-6 \times 10^4$ cpm), and 1 μg of A431 membrane protein (or an amount of membrane protein which will bind 10–15% of the input radioactivity in the assay in the absence of competing activity). A calibration curve is prepared by subjecting known amounts of EGF to the assay. Incubation proceeds for 40 min at room temperature. It is terminated by dilution of the samples with 4.0 ml of ice-cold binding buffer and filtration under negative pressure through 20-mm cellulose ester filter disks (0.2-μm pore, Millipore) that have been soaked for at least 30 min in binding buffer before the assay. Filters are washed twice with 4.0-ml aliquots of ice-cold binding buffer (with 1 mg/ml bovine serum albumin). The radioactivity trapped in the filters is measured in a gamma counter. Nonspecific binding of ^{125}I-labeled EGF to the membranes is determined in the presence of 100 nM native EGF; it is usually <10% of the total binding in the absence of competing activity.

Results can be expressed as EGF nanogram equivalents, one EGF nanogram equivalent being the amount of competing activity which causes the same amount of competition in the radioreceptor assay as 1 ng of mouse EGF. In our experience, 0.5 ng of mouse EGF or rat TGFα causes about 50% displacement of the tracer ^{125}I-labeled EGF specifically bound to the membranes in this assay.

Cell Culture and Collection of Conditioned Medium

Purification of TGFα from cells which produce it is most efficiently achieved using as starting material serum-free medium conditioned by highly dense cells cultures. Growth of cells to achieve the desired density is done with standard culture medium, usually Dulbecco's modified Eagle's medium with high concentration of glucose (Gibco 430-2100) and supplemented with 5–10% calf serum. For collection of serum-free conditioned medium an enriched formulation such as Waymouth's medium (Gibco 430-1400) is adequate. To remove traces of serum proteins and growth factors cell cultures should be washed at least twice by incubation with serum-free medium over a 4- to 6-hr period before collection of conditioned medium begins.

All TGFα-producing cell types described to date produce very small amounts of this factor (<1 μg/liter). Therefore, large-scale culture is needed to obtain microgram amounts of purified TGFα. Cell culture in roller bottles has been successfully used for this purpose. Cells are grown in roller bottles under an atmosphere of about 5% CO_2 in air until they reach the desired density. The yield of serum-free conditioned medium from a culture depends on conditions such as the seeding density, the volume of medium during growth, washes and collections, and the speed of rotation of the bottles. Optimal conditions to maximize the yield of conditioned medium vary with each cell type and should be determined in preliminary experiments in each case. Transformed cells usually have a tendency to detach from the walls of the culture vessel and be lost during media changes. This problem can be aggravated by the shearing force of the medium if excessively large volumes of it or a too high rotation speed are used. In general, cultures grow well at 0.2 rpm with 100 ml of medium/850 cm^2 roller bottle. Washes can be done with 50 ml of serum-free medium per bottle, and collections of conditioned medium with 75–150 ml/bottle.

For maximal efficiency in large-scale culture the filling of bottles with medium can be done by direct flow from the sterilizing apparatus. For this purpose, a peristaltic pump-driven sterile filtration system (Millipore) is better than a nitrogen pressure system because it allows easy monitoring and stopping of the flow of medium. Gassing with a gently stream of CO_2 through a sterile plugged pipet is done immediately before closing each bottle.

Collection of conditioned medium is done by decantation every 24–72 hr depending on the cell type, and is done before the medium becomes too acidic (pH about 6.5). Refilling the bottles after each collection is done as described above but gassing with CO_2 is usually not necessary because the high metabolism of the culture rapidly generates enough CO_2 to maintain an appropriate pH.

Concentration and Preparation of the Acid-Soluble Extract

Phenylmethylsulfonyl fluoride (Sigma) is added from a concentrated (200 mM) stock solution in anhydrous ethanol to the freshly collected conditioned medium to a final 0.1 mM concentration. Before storage, the medium must be separated from detached cells and if the volume is too large it should also be concentrated. Clarification of the medium is easily done by regular or continuous-flow centrifugation, or by filtration through a Millipore Pellicon cassette system (PSVP cassette). We have used the latter system with minimal losses of TGFα activity. For concentration of the clarified medium, we have successfully used an Amicon DC10 hollow fiber concentrator with a H1OP5-20 cartridge (5000 kDa cut off). We have not tested the PCAC Millipore cassette (1000 kDa cut off) in the Pellicon system, but this cartridge is probably adequate to concentrate the medium without excessive losses of TGFα.

Once clarified and concentrated, the medium can be stored at −20° while consecutive collections are made during the following days. After collections are completed, they are all pooled and concentrated to a minimum volume. It is important to recirculate a small volume of isotonic saline solution through the concentrating apparatus to recover concentrated conditioned medium left in the dead volume.

The saline wash is pooled with the concentrated medium and dialyzed in Spectrapore 3 tubing (Spectrum Medical Industries) against 100 vol of 0.1 M acetic acid over a 3-day period in the cold. The material that precipitates during dialysis is removed by centrifugation at 10,000 g for 30 min. The clarified sample is then lyophilized.

Molecular Filtration Chromatography

The lyophilized conditioned medium is dissolved with 1 M acetic acid to a final concentration of 25 mg of protein/ml. Insoluble material is removed by centrifugation at 100,000 g for 45 min and/or filtration through a microporous filter assembly (Millex GV, Millipore). The sample is then loaded onto a column of BioGel P-60 (200–400 mesh, BioRad Laboratories) equilibrated and eluted with 1 M acetic acid at room temperature. We use a 5 × 90 cm of BioGel P60 eluted with an upward flow of 30 ml/hr to process 1–2 g of protein sample obtained from a 120-liter batch of medium conditioned by Snyder–Theilen feline sarcoma virus-transformed rat embryo cells (FeSV-Fre cells, obtained from Dr. G. J. Todaro, Oncogene, Seattle, WA.). Eluted protein is monitored spectrophotometrically at 280 nm. Mature (50-amino acid form) TGFα from transformed cells typically elutes in the 5- to 10-kDa range of the column. Larger forms of TGFα-like activity are usually observed in these chromatograms.[4,14] They are structurally related to TGFα as determined with TGFα-specific antibodies,[17,18] and correspond to the glycosylated extracellular domain of the TGFα precursor.[19]

Reversed-Phase High-Pressure Liquid Chromatography

Solvents

Water/TFA: 0.05% (v/v) trifluoroacetic acid (Pierce) in deionized H_2O of the highest quality available. Degas by sonication before use

Acetonitrile/TFA: 0.05% (v/v) trifluoracetic acid in acetonitrile (Burdick and Jackson)

1-Propanol/TFA: 0.035% (v/v) trifluoracetic acid in 1-propanol (Burdick and Jackson)

Columns

Columns of μBondapak C_{18} (Waters) in semipreparative and/or analytical size are used depending on the amount of sample to be processed

Instrument

We use a Waters HPLC system consisting of a UK injector with a 2.0-ml sample loop, two standard pumps, a standard absorbance detector, and a standard system controller

Sample Preparation

Pools of active fractions from the BioGel P60 step or previous HPLC steps are lyophilized to dryness in regular glass or polypropylene vessels. The dry material is dissolved in less than 2.0 ml (or the volume of the sample loop in the HPLC injector) of 0.06% (v/v) trichloracetic acid in H_2O. The reconstituted sample is filtered through a Millex GV filter assembly before HPLC. The filter is washed with small volumes of the same solvent, and the washes are pooled with the sample without exceeding a final 2.0-ml sample volume

First HPLC Step: Steep Acetonitrile Gradient

This and all other HPLC steps can be run at room temperature. Like in all subsequent steps the column is equilibrated with 0.05% TFA in water before loading the sample. After loading the sample, the column is eluted with a linear 0–45% acetonitrile gradient in 0.05% trifluoracetic acid at a flow rate of 1 ml/min over a 60-min period. Absorbance is monitored at 254 nm, or at 210 or 214 nm if higher sensitivity is required. One-milliliter

[17] P. S. Linsley, W. R. Hargreaves, D. R. Twardzik, and G. J. Todaro, *Proc. Natl. Acad. Sci. U.S.A.* **82**, 356 (1985).

[18] R.A. Ignotz, B. Kelly, R. Davis, and J. Massagué, *Proc. Natl. Acad. Sci. U.S.A.* **83**, 6307 (1986).

[19] J. Teixidó, R. Gilmore, D. C. Lee, and J. Massagué, *Nature (London)*, in press.

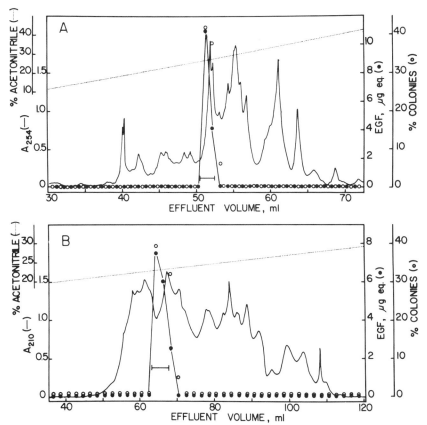

FIG. 1. Chromatography of TGFα on μBondapak C$_{18}$ eluted with acetonitrile. (A) Elution pattern of 27.8 mg of protein from BioGel P-60 TGFα pools on μBondapak C$_{18}$ support eluted with a 60-min, 0-45% steep linear acetonitrile gradient in 0.050% trifluoroacetic acid (dotted line). The lyophilized sample was reconstituted with 2.0 ml of 0.060% trifluoroacetic acid and chromatographed on a column (0.78 × 30 cm) as described in the text. Absorbance was monitored at 254 nm (—). Aliquots from the resulting 1.0-ml fractions were analyzed for anchorage-independent growth-promoting activity in NRK cells (○) and EGF-receptor binding activity (●). The horizontal bar indicates the fractions pooled for rechromatography. (B) Rechromatography of 6.89 mg of protein from the previous step on gradient-μBondapak C$_{18}$ eluted with a 5-min, 0-18% linear acetonitrile gradient followed by a 120-min, 18-28% linear acetonitrile gradient in 0.050% trifluoroacetic acid (dotted line). Absorbance was monitored at 210 nm (—). Aliquots from every other 1.0-ml fraction were assayed for anchorage-independent growth-promoting activity (○) and EGF-receptor binding activity (●). EGF-receptor binding activity is expressed as EGF microgram equivalents present in each fraction. The horizontal bar indicates fractions pooled for further purification of TGFα (From Ref. 4.)

samples are collected and aliquots are analyzed for EGF receptor competing activity. Rat TGFα elutes in this step as a sharp peak at 27% acetonitrile separated from the bulk of contaminating protein (Fig. 1).

Second HPLC Step: Shallow Acetonitrile Gradient

The pool of active fractions from the first HPLC step is chromatographed over the same column. Elution is with a 0–18% acetonitrile
gradient in 0.05% trifluoroacetic acid over a 5-min period followed by a
linear 18–28% acetonitrile gradient in 0.05% trifluoroacetic acid over a
120-min period. The flow rate is 1.0 ml/min and 1.0-ml fractions are
collected. Rat TGFα elutes in this step at 20–23% acetonitrile depending
on the age of the column. It elutes as a peak separated from most of the
contaminating protein (Fig. 1).

Third HPLC Step: Propanol Gradient

The pool of active fractions from the second HPLC step is rechromato-
graphed over an analytical μBondapak C$_{18}$ column. Elution is done with a
linear 0–15% 1-propanol gradient using the propanol/TFA mixture described above. The flow rate is 1.0 ml/min and 1.0-ml fractions are collected. Rat TGFα from FeSV-Fre cells elutes in homogeneous form as a

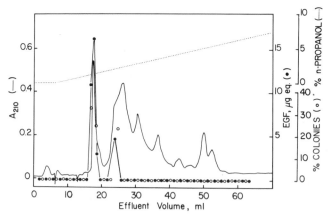

FIG. 2. Chromatography of TGFα on μBondapak C$_{18}$ eluted with 1-propanol. Rechromatography of 0.55 mg of protein from the shallow acetonitrile gradient μBondapak C$_{18}$ TGFα
pool. The sample reconstituted with 0.5 ml of 0.060% trifluoroacetic acid was chromatographed on a μBondapak C$_{18}$ column (0.39 × 30 cm) equilibrated with 0.050% trifluoroacetic
acid in water and eluted with a 90-min, 0–10% 1-propanol gradient in 0.35% trifluoroacetic
acid (dotten line). Absorbance was monitored at 210 nm (—). Aliquots from the indicated
1.0-ml fractions were assayed for anchorage-independent growth promoting activity (○) and
EGF-receptor binding activity (●). (From Ref. 4.)

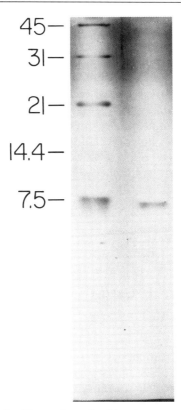

FIG. 3. Sodium dodecyl sulfate-polyacrylamide gel electrophoresis of purified TGFα. Protein (0.5 μg) from the major TGFα peak resolved in the μBondapak C_{18}/1-propanol step was heated for 1 min at 100° with 50 μl of sample buffer containing 50 mM dithiothreitol. This sample was electrophoresed on 15–18% linear polyacrylamide gradient gel and silver stained for protein detection. Arrow points at the major band in the TGFα-containing track. Approximately 0.5 μg of the following polypeptides were electrophoresed in the paralleled track as molecular weight markers (×10³): ovalbumin (M_r 45,000), carbonate dehydratase (M_r 31,000), soybean trypsin inhibitor (M_r 21,000), lysozyme (M_r 14,400), and insulin-like growth factor-II (M_r 7,500). (From Ref. 4.)

peak of 210 nm absorbance preceding all the contaminating proteins (Fig. 2). The elution of TGFα is at about 10% or lower 1-propanol concentration depending on the age of the column. After this and all other HPLC steps the column should be thoroughly washed with acetonitrile/TFA, and stored in this solvent.

Rat TGFα obtained in this step migrates as a single band in sodium dodecyl sulfate-polyacrylamide gels (Fig. 3). Although the molecular weight of rat TGFα is 5616,[8] it typically migrates as a 7.4K species in high-density polyacrylamide gels.[4,8]

Storage

Rat TGFα is stable for at least several weeks at 4° in the HPLC solvents after elution. Homogeneous preparations of TGFα can be lyophilized and stored at −70° in polypropylene tubes without any significant loss of receptor binding ability over at least 18 months.

Stock solutions of 0.25 mg/ml rat TGFα in 1 mM HCl prepared for routine use in experimental work retain full biological activity for at least 2 months at 4°. We have no information on whether the preservation of biological activity in TGFα preparations during storage is accompanied with complete chemical stability of the peptide.

[11] Separation of Melanoma Growth Stimulatory Activity and Human Type-α Transforming Growth Factor

By ANN RICHMOND, H. GREG THOMAS, and ROBERT G. B. ROY

Introduction

Melanoma growth stimulatory activity (MGSA) is a monolayer mitogen for cultured nevus cells and malignant melanoma cells. This growth factor is produced by the Hs0294 human melanoma cell line and is associated with the serum-independent proliferation of these cells in culture.[1,2] MGSA provides a major portion of the mitogenic stimulation for Hs0294 cells cultured in serum-free medium, since monoclonal antibodies developed to MGSA inhibit the serum-free growth of the Hs0294 cells by greater than 70% in a 9-day cell number assay.[3]

The biochemical properties of this growth factor have been partially characterized. MGSA is an acid- and heat-stable polypeptide which is sensitive to trypsin and dithiothreitol.[2] The activity differs from the transforming growth factors of the TGFα type which are also produced by human melanoma cells in that MGSA stimulates anchorage-dependent growth of melanoma and nevus cells, but does not enable normal rat kidney (NRK) indicator cells to exhibit anchorage-independent growth as

[1] A. Richmond, D. H. Lawson, and D. W. Nixon, *Cold Spring Harbor Conf. Cell Proliferation* **9**, 885 (1982).

[2] A. Richmond, D. H. Lawson, D. W. Nixon, J. S. Stevens, and R. K. Chawla, *Cancer Res.* **43**, 2106 (1983).

[3] D. H. Lawson, H. G. Thomas, R. G. B. Roy, D. S. Gordon, R. K. Chawla, D. W. Nixon, and A. Richmond, *J. Cell. Biochem.*, in press (1987).

METHODS IN ENZYMOLOGY, VOL. 146

does TGFα. Furthermore, epidermal growth factor (EGF) acts through the same receptor as TGFα, but has no effect on the [^{3}H]thymidine incorporation into Hs0294 cells.[4]

MGSA can be purified from acetic acid extracts of serum-free conditioned medium (CM) from the Hs0294 cell line by BioGel P-30 chromatography followed by RP-HPLC on a μBondapak C$_{18}$ column. When purified in this manner, the major peak of MGSA activity elutes at 35±3% acetonitrile just following the ^{125}I-labeled EGF competing TGFα peak which elutes at 30±4% acetonitrile.[5]

Methods of Analysis

MGSA Biological Assay for Growth Factor Activity

MGSA bioassays are performed by measuring stimulation in [^{3}H]thymidine incorporation in serum-depleted low-density cultures of Hs0294 human malignant melanoma cells according to a modification of the procedures of Iio and Sirbasku.[6] Eight thousand cells are plated in 28 × 61 mm glass scintillation vials (Wheaton #225288) in 2 ml of Ham's F10 medium supplemented with 10% FBS. Twenty-four hours later, cells are washed with 2 ml of PBS and placed on serum free F10 medium containing N-2-hydroxyethylpiperazine-N'-2-ethanesulfonic acid (HEPES) (30 mM) and ovalbumin (10 μg/ml). After a 24-hr incubation in serum-free medium, the medium is aspirated and aliquots of preparations of growth factor are added in a binding buffer composed of 2 ml of serum-free F10 with HEPES (30 mM) and ovalbumin (10 μg/ml). Eight hours later [^{3}H]thymidine (5 μCi) is added to each vial and the incubation is continued for another 16 hr. The reaction is stopped by addition of 2 ml of methanol:ethanol (3:1) for 10 min, then the methanol:ethanol is aspirated and the fixed cells are washed for 10 min with 4 ml of the same solution followed by a 10-min wash with 4 ml absolute methanol. The fixed cells are visually examined by phase microscopy to assess viability. Effects of toxic substances which might copurify with the growth factors can be identified in this manner. Ten milliliters of counting fluid (Scintiverse II) is added to each vial and radioactivity incorporated into DNA is counted in a Beckman liquid scintillation counter (LS-1800).

[4] A. Richmond, D. H. Lawson, D. W. Nixon, J. S. Stevens, and R. K. Chawla, *Cancer Res.* **43**, 2106 (1983).

[5] A. Richmond, D. H. Lawson, R. K. Chawla, and D. W. Nixon, *Cancer Res.* **45**, 6390 (1985).

[6] M. Iio and D. A. Sirbasku, *Cold Spring Harbor Conf. Cell Proliferation* **9**, 751 (1982).

A unit of MGSA is defined as the amount of growth factor necessary to stimulate an incorporation in [^{3}H]thymidine which is half-maximal to that produced by unfractioned Hs0294-conditioned medium. A large number of aliquots of whole Hs0294-conditioned medium are prepared and stored at −80° for use as a reference in all assays. This reference medium produces an average [^{3}H]thymidine incorporation which is ~200% of controls which receive nonconditioned medium.

Enzyme-Linked Immunoadsorbent Assay (ELISA) for MGSA. The monoclonal antibody, FB2AH7, was raised to a partially purified preparation of MGSA by the methods of Kohler and Milstein[7] with the cooperation of the Centers for Disease Control. Positive clones were selected on the basis of supernatant inhibition of [^{3}H]thymidine incorporation into DNA in serum-free cultures of Hs0294 cells and reduction in total cell number in serum-free cultures of Hs0294 cells during a 9-day assay as previously described.[3] Clone FB2AH7, an IgM$_K$, was able to markedly reduce thymidine incorporation and cell number in Hs0294 cells.[3] FB2AH7 antibodies are prepared from hybridoma cells which are maintained in Eagle's minimal essential medium supplemented with 350 mg glutamine, 1 ml insulin/transferrin/selenium (Collaborative Research, Inc., Waltham, MA), 40 mg gentamycin, 2.5 mg amphotericin B, 10% fetal bovine serum (FBS), 12 ml HEPES (1 M), and 12 ml of a 0.005 M 2-mercaptoethanol solution, pH 7.3. Hybridoma supernatant is clarified by centrifugation (1500 g for 5 min) then precipitated by dropwise addition of an equal volume of 4 M ammonium sulfate according to procedures of Garvey *et al.*[8] The precipitated antibody is subsequently dialyzed first for 5 days against 0.9% NaCl, then for 2 days against deionized H$_2$O, pH 7.8, prior to lyophilization and gel filtration chromatography on an ACA-34 Ultrogel column (0.05 M NaPO$_4$ buffer, pH 7.2). The antibody elutes in the excluded fractions. These antibody-containing fractions are pooled, dialyzed against deionized water (pH 7.8) for 2 days, then lyophilized. Antibody prepared from 1 liter of culture medium in this manner is dissolved in 5 ml of distilled water and the presence of antibody is verified at this concentration and after 1:10 and 1:100 dilution by double immunodiffusion using a rabbit antiserum to mouse immunoglobulins.

FB2AH7 has been demonstrated to bind reversed-phase high-pressure liquid chromatography (RP-HPLC)-purified MGSA in an ELISA developed in our laboratory.[9] The procedures of Hawkes *et al.* for dot-immuno-

[7] G. Kohler and C. Milstein, *Eur. J. Immunol.* **6**, 511 (1976).

[8] J. S. Garvey, N. E. Cremer, and D. H. Sussdorf, *in* "Methods in Immunology," p. 218. Addison-Wesley, Reading, MA, 1977.

[9] A. Richmond and G. Thomas, *J. Cell. Physiol.* **129**, 375 (1986).

binding assay for monoclonal antibodies[10] have been modified to yield a soluble enzyme–substrate product as produced in a standard ELISA, only in this case the antigen is bound to nitrocellulose paper. The assay will detect as little as 20 ng of antigen. Briefly, the antigen is placed on small circles of nitrocellulose paper that will fit into wells of a 96-well microtiter plate. Filters are allowed to dry thoroughly, then are washed for 5 min with Tris-buffered saline (TBS) (50 mM Tris · HCl, 200 mM NaCl, pH 7.4). To block nonspecific binding, to each well is added 200 μl of a 3% (w/v) solution of bovine serum albumin (BSA) in TBS. The microtiter plate is left at 4° overnight, then blocking solution is aspirated and 150 μl of a 1:25 dilution of the FB2AH7 antibody is added (1 vol antibody to 24 vol BSA blocking solution). After a 4-hr incubation at room temperature, the primary antibody is aspirated and the wells are washed 10 times with ice-cold TBS. At this time, the blocking step is repeated, then filters are incubated for 2 hr in 100 μl of alkaline phosphatase-conjugated goat anti-mouse immunoglobulin, diluted 1:1000 in blocking solution. Subsequently, the second antibody is aspirated, filters are washed 10× with TBS, then incubated with 100 μl of disodium p-nitrophenylphosphate (1 mg/ml) in diethanolamine buffer (49 mg MgCl · 6H$_2$O, 96 ml diethanolamine/liter H$_2$O, pH 9.8) at room temperature. Positive wells are easily distinguished by the bright yellow color. The reaction is stopped approximately 45 min later by addition of 50 μl of 3 M NaOH. After dissolution of the nitrocellulose paper, the plate is read on a Titertek Multiscan MC ELISA reader (Flow Laboratories) at 414–620 nm. Determinations of immunoreactive MGSA (IR-MGSA) are based upon standard curves utilizing dilutions of the FB2AH7 monoclonal antibody and correcting for equivalent moles of antigen bound per mole of antibody. Background binding is assessed by blocking nitrocellulose paper circles with 3% BSA and then incubating with FB2AH7 alkaline phosphatase-conjugated anti-mouse Ig and substrate as described above. The optical density of these controls is then subtracted from each sample value.

We have evaluated the binding of various agents to the FB2AH7 antibody to determine its specificity. No significant binding was seen with transferrin, bovine serum albumin, insulin, nerve growth factor, epidermal growth factor, multiplication stimulatory activity, or trasylol (aprotinin).

Transforming Growth Factor Assays

Transforming growth factors (TGFs) are monitored throughout the purification of MGSA by the NRK soft agar assay and the EGF radioreceptor assay. The details of each of these assays are given below.

[10] R. Hawkes, E. Niday, and J. Gordon, *Anal. Biochem.* **119**, 142 (1982).

Soft Agar Assay. Soft agar assays are performed according to the procedure of De Larco and Todaro[11] using nontransformed NRK-49F cells at a density of 2000 cells/35-mm Falcon plastic tissue culture dish. Aliquots of growth factors are added to the 1-ml underlayers [Dulbecco's modified Eagle's medium (DME), 10% calf serum, 0.5% Noble agar]. NRK-49F indicator cells are added in a 1-ml overlayer (DME, 10% calf serum, 0.3% Noble agar) at a density of 2×10^3 cells/ml. These cells are highly responsive to TGF's, which stimulate anchorage-independent growth resulting in the formation of colonies of NRK cells after incubation for approximately 3 weeks at 37° with an atmosphere of 95% air and 5% CO_2. Every 7 days the cultures are fed with 1 ml of DME/10% calf serum/0.5% agar. After 21 days the number of NRK colonies in each dish is determined by manual count on an inverted phase microscope. A colony is defined as a group of more than 10 cells. Each agar culture is counted independently by two individuals and the mean of the two counts is recorded as colonies present/cm^2, then multiplied by 9 to give total colonies per dish.

^{125}I-Labeled EGF Competitive Binding Assay

TGFα competes with ^{125}I-labeled EGF for binding sites in a radioreceptor assay (RRA). Consequently, the elution profile of TGFα can be readily followed with an EGF-RRA. Since NRK cells are the indicator cell in the TGFα soft agar assay, we have used NRK cells in the EGF-RRA.[5] NRK cells are plated into a 24-well Falcon plate at a density of 1.25×10^4 cells per well in DME containing 10% calf serum and 50 μg/ml gentamycin. Five days later, when the density is approximately 1×10^5 cells/well, the medium is aspirated and the cells are washed once with a binding buffer composed of serum-free DME containing 2 mg/ml ovalbumin, 30 mM HEPES, and 50 μg/ml gentamycin. The binding buffer is aspirated and duplicate wells are incubated with 1 ng/ml of ^{125}I-labeled EGF (228 μCi/ug) dissolved in binding buffer containing varying concentrations of unlabeled EGF (1 μg, 100, 10, 1, 0.1 and 0 ng/ml), or aliquots of RP-HPLC-purified TGF or MGSA. After incubating for 1 hr at 37° in a 95% air/5% CO_2 atmosphere, the medium is aspirated and wells are washed five times with 1 ml of ice-cold binding buffer. One milliliter of lysing buffer is added per well (0.5 g SDS, 0.121 g Tris, 0.037 g EDTA in 100 ml distilled H_2O). Lysed cells are aspirated and wells are washed with an additional 1 ml of lysing buffer. Radioactivity in the lysed cells is determined in a gamma counter (Micromedic Systems 4/200). Half-maximal binding is obtained with 1 ng of ^{125}I-labeling EGF, and maximal binding is achieved with 16 ng of ^{125}I-labeled EGF.[5] Nonspecific binding is determined in the pres-

[11] J. E. De Larco and G. J. Todaro, *Proc. Natl. Acad. Sci. U.S.A.* **75**, 4001 (1978).

ence of 1 μg/ml of unlabeled EGF. Data are recorded as counts per minute (CPM) bound minus nonspecific binding/10^6 cells or as percentage of maximal specific binding, and nanogram equivalents of EGF/fraction are based on these data.

Protein Determinations

Protein concentrations were determined by the methods of Bradford using the Bio-Rad dye-binding kit and bovine serum albumin as a standard.[12]

Purification

Source of MGSA. MGSA can be obtained from serum-free culture medium conditioned by confluent cultures of the Hs0294 human melanoma cell line as previously described.[2] This serum-free conditioned medium is collected sterilly, and the protease inhibitor trasylol is added (117 kallikrein inhibitory U/liter). The medium is centrifuged at 1500 g to remove any cells or cellular debris released into the culture medium, then ultracentrifuged (105,000 g) for 30 min to remove subcellular organelles. The resulting supernatant is lyophilized, extracted with 1 N acetic acid at room temperature, dialyzed exhaustively against 0.17 M acetic acid at 4°, lyophilized, and stored at −80°. Subsequently, MGSA can be partially purified from this extract by a combination of BioGel P-30 chromatography followed by RP-HPLC as previously published[5] or by a modified method which employs affinity chromatography using a monoclonal antibody to MGSA. For both methods the final step of purification is preparative polyacryamide gel electrophoresis (PAGE).

Purification: Method A

Step 1: BioGel P-30 Exclusion Chromatography. Acetic acid extracts of approximately 20 liters of lyophilized Hs0294-conditioned medium are dissolved in 30 ml of 1 N acetic acid and clarified by ultracentrifugation (105,000 g for 30 min). The resulting supernatant containing approximately 65 mg of protein is applied to a 2.5 × 90 cm BioGel P-30 column, equilibrated and eluted with 1 M acetic acid at a flow rate of approximately 9 ml/hr at room temperature as previously described.[2] Absorbance is monitored at 280 nm. Three-milliliter fractions are collected and 100-μl aliquots of each fraction are lyophilized and assayed for MGSA in the [³H]thymidine assay at 1:10 dilutions, or for TGFs in the soft agar assay at

---

[12] M. Bradford, *Anal. Biochem.* **72,** 248 (1976).

1:6 dilutions. When MGSA is assayed in this manner, the majority of the MGSA bioactivity elutes broadly and appears to reside in two major regions (13–25 and 5–10 kDa), just after the high (27,000) and low (11,000) M_r TGF activity (Fig. 1). At this stage of purification, inability to see MGSA bioactivity in the TGF peaks is a concentration-dependent phenomenon. When the BioGel fractions containing TGF activity but little or no MGSA are reassayed for MGSA at 10^{-2} or 10^{-3} dilutions, the results show that a significant portion of the MGSA bioactivity is also present in the TGF-containing fractions and thus the MGSA and TGF activities are not truly separated by BioGel P-30 chromatography. For subsequent purification, the high M_r MGSA/TGF fractions are combined [postvoid volume through high M_r TGF fractions (35–58)] and the low M_r MGSA/TGF fractions are combined [(fractions 59–90)]. The majority of the protein elutes in the void volume, such that from 65 mg of acid-soluble protein applied to the column, approximately 2 mg of protein elutes in the high M_r MGSA/TGF pool I and approximately 200 μg of protein elutes in the lower M_r MGSA/TGF pool II.

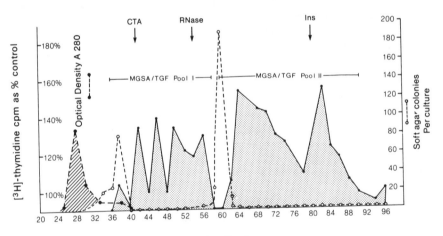

Fraction Number

FIG. 1. BioGel P-30 chromatography of the acetic acid-soluble concentrate from serum-free medium conditioned by the human malignant melanoma cell line Hs0294. Sixty milligrams of the lyophilized extract of Hs0294 CM was dissolved in 1 N acetic acid and chromatographed on a 90 × 2.5 cm BioGel P-30 column eluting with 1 N acetic acid. Three-milliliter fractions were collected, lyophilized, UV irradiated, and dissolved in binding buffer [F-10 plus ovalbumin (10 μg/ml) plus 30 mM 4-(2-hydroxyethyl)-1-piperazine-ethanesulfonic acid buffer]. Every other fraction was assayed for protein content (●---●), ability to stimulate [³H]thymidine incorporation in serum-depleted Hs0294 cells (●—●), and ability to stimulate anchorage-independent growth using Hs0294 cells or NRK cells in the soft-agar assay (O---O). Details of these assays are in the Methods of Analysis section. Reprinted with permission from Richmond et al.[2]

Step 2: RP-HPLC. After BioGel P-30 chromatography, MGSA can be further purified by RP-HPLC.[5] The fractions from the BioGel P-30 column which were pooled into high M_r MGSA/TGF pool I or low M_r MGSA/TGF pool II are lyophilized separately and dissolved in 1.2 ml of acetonitrile/water/TFA (6:94:0.05). Each sample is then injected into the μBondapak C_{18} cartridge column (8 × 100 mm) in a Z module (Waters Associates) and eluted with a 60-min linear gradient of acetonitrile/water/TFA (6:94:0.05 to 60:40:0.05, pH 2.2). Optical density is monitored at 206 nm. The flow rate is 1.5 ml/min and 1.5-ml fractions are collected. Aliquots (1/100) from each RP-HPLC fraction of TGF pool I and TGF pool II are lyophilized, UV irradiated, dissolved in serum-free Ham's F10 medium containing 10 μg/ml of ovalbumin, and further diluted 10^{-2} to 10^{-4} prior to assaying for MGSA using the standard [^{3}H]thymidine assay with Hs0294 cells. Diluting samples in this manner is crucial because the MGSA concentration in each fraction is unknown and the optimal concentration for a positive bioactivity assay is approximately 2 ng/ml. Higher concentrations of MGSA (10 ng/ml) are usually not mitogenic for the Hs0294 cells. Two other aliquots (each one-one hundredth of the sample) are lyophilized and assayed in the EGF-RRA and the soft agar assays to follow the elution profile of TGFs. The results from these assays reveal that the major MGSA peak elutes at approximately 35±3% acetonitrile and is separated from the major EGF-competing TGF activity peak, eluting at approximately 30±4% acetonitrile for both high and low M_r growth factor pools (Fig. 2A and B). Minor peaks of MGSA elute in the 12±5% and 25±5% acetonitrile range from both pools (Fig. 2C). The elution profile for MGSA and TGF is essentially the same for both TGF pools I and II, though there is a tendency for the TGF and MGSA peaks to elute a few fractions earlier with pool II and there are quantitative differences between pools in the proportion of activity eluting in each peak. Based upon repeated analyses of 20-liter batches of CM, the data show that approximately 72±10% of the MGSA activity is in TGF pool I. The proportion of activity from TGF pools I and II which elutes at 35±3% acetonitrile averages 64.5%, with the remainder of the activity eluting in the minor peaks (Table I).

Step 3: Reducing Sodium Dodecyl Sulfate (SDS) Gel Electrophoresis. The molecular weight of RP-HPLC-purified MGSA can also be analyzed by SDS–PAGE using the methods of Swank and Munkres.[13] This system is optimized for the separation of small peptides (2000–18,000). Using a higher relative proportion of cross linker to monomer and the inclusion of 8 M urea, separation of peptides with a molecular weight of 10,000 or less increases 2-fold. The resolving gel (0.75 mm) is 13.8% T and 9.1% C with a

---

[13] R. T. Swank and K. D. Munkres, *Anal. Biochem.* **39**, 462 (1971).

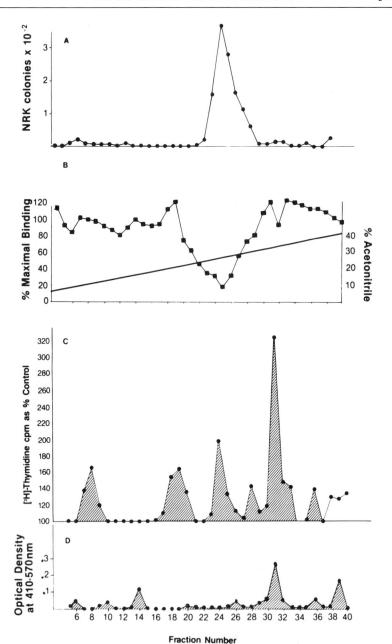

FIG. 2. RP-HPLC of TGF pool I on a μBondapak C$_{18}$ column. Fractions eluted from the BioGel P-30 column in MGSA/TGF pool I (Fig. 1) were combined, lyophilized, and dissolved in 2 ml of 6:94:0.05 acetonitrile:water:trifluoroacetic acid. The sample (1.9 mg pro-

TABLE I
PURIFICATION OF MGSA

Source of MGSA	MGSA U/liter[a]	Total activity (%)	Activity in BioGel pool (%)
Method A			
BioGel TGF pool I	35,000, total	76, total	
HPLC fraction 8	3,500	8	10
fraction 19	3,500	8	20
fraction 24	7,000	15	20
fractions 31 & 32	21,000	45	60
BioGel TGF pool II	11,200, total	24, total	
HPLC fraction 7	700	2	6
fraction 22	3,500	7	31
fraction 30–34	7,000	15	62

[a] One MGSA unit produces a stimulation in [³H]thymidine incorporation 50% of that produced by unprocessed Hs0294-conditioned medium. The average stimulation produced by unprocessed Hs0294-conditioned medium is 200% of untreated controls.

tein) was applied to a μBondapak C$_{18}$ column equilibrated with the same buffer. The gradient (—) consisted of 6–60% acetonitrile for 60 min at 25°, with a flow rate of 1.5 ml/min. Sixty fractions were collected with a volume of 1.5 ml each. Aliquots were used for ¹²⁵I-labeled EGF competition, soft agar assays, [³H]thymidine, and ELISA assays. (A) RP-HPLC elution pattern of TGFα activity based on stimulation of NRK cells to form colonies in soft agar. Aliquots (10 μl) of each fraction were lyophilized, UV irradiated, and added to each of two 1-ml underlayers of DME containing 10% calf serum and 0.5% agar. Two thousand NRK-49F cells are then placed in an overlayer of 0.7 ml of DME containing 10% calf serum and 0.3% agar. Cultures are fed weekly and after 21 days the total number of NRK colonies is determined as described in the Methods of Analysis section. (B) RP-HPLC elution pattern of EGF competing activity (■—■) in MGSA/TGF pool I. To determine EGF receptor competition, lyophilized aliquots of RP-HPLC fractions were dissolved in 1 ml binding buffer and incubated with confluent cultures of NRK cells (1.0 × 10⁵ cells/well in a 24-multiwell plate) containing 1 ng/ml of ¹²⁵I-labeled EGF. After a 60-min incubation at 37°, the percentage ¹²⁵I-labeled EGF bound is determined as described in the Methods of Analysis section. Reprinted with permission from Richmond et al.[5] (C) Elution pattern of MGSA bioactivity from RP-HPLC analysis of MGSA/TGF pool I based on stimulation in [³H]thymidine incorporation into DNA in serum-free cultures of Hs0294 cells. Aliquots (10 μl) of each RP-HPLC fraction were diluted with binding buffer to achieve a final dilution of 1:70,000 prior to assay in the standard MGSA bioassay as described in the Methods of Analysis section. Reprinted with permission from Richmond et al.[5] (●—●), [³H]Thymidine cpm expressed as a percentage of control. (D) Elution pattern of MGSA from RP-HPLC based on the FB2AH7 ELISA. Aliquots (10 μl) of RP-HPLC fractions were placed on nitrocellulose circles and dried. The ELISA for MGSA was performed as described in the Methods of Analysis section. The optical density at 414–620 nm indicates the elution profile of immunoreactive antigen (●—●).

stacking gel of 6.9% T and 9.1% C. Samples are prepared with a sample buffer of 0.01 M H_3PO_4, 8 M urea, 2.5% SDS, and 5% 2-mercaptoethanol equilibrated to a pH of 6.8 with Tris. The samples are then heated to 100° for 5 min and 25 μl is applied to each gel lane along with 25 μl of tracking dye solution. The electrophoresis buffer is 0.1 M H_3PO_4, made 1% SDS, equilibrated to a pH of 6.8 with Tris. Electrophoresis conditions are 35 mA constant current with cold tap water circulation. Electrophoresis time is set so the tracking dye just runs off the bottom of the gel. Molecular weight standards include myoglobin intact, 17,201; myoglobin I and II, 14,600; myoglobin I, 8240; myoglobin II, 6380; myoglobin III, 2560. Visualization of the electrophoresed proteins is achieved using the Bio-Rad silver stain kit. Complete separation of the major MGSA-containing fractions from the TGFs in MGSA/TGF pool I is evident when the bands in TGF-containing fractions at 12,000–13,000, 8000, and 5000 are no longer present and instead prominent bands appear at greater than 20,000, 20,000, 14,000–16,000 and 9000–10,000 coincident with a positive bioassay for MGSA[9] (Fig. 3).

Step 4: MGSA-ELISA Followed by Preparative Gel Electrophoresis. Using monoclonal antibody FB2AH7, a modified dot–blot procedure can be used to follow elution of the FB2AH7 antigen from the μBondapak C_{18} column. For both pool I and pool II the major immunoreactive peak elutes at 35% acetonitrile and is coincident with the major MGSA bioactivity peak for both the high and low M_r pools. Minor ELISA positive peaks also elute at 12±5% and 25±5% acetonitrile. Little if any immunoreactive protein is associated with the fractions containing the major portion of EGF-competing activity (Fig. 2D), further demonstrating that MGSA is separated from TGFα after RP-HPLC. ELISA results do not indicate major quantitative differences in the amount of FB2AH7 antigen between pools I and II, though the activity results suggest a more active form of the antigen may elute in pool I (Table I).[9]

Those RP-HPLC fractions which are positive for the FB2AH7 antigen in the ELISA are then analyzed by nonreducing SDS–PAGE according to the methods of Laemmli[14] as modified by Roberts et al.[15] Fractions eluting at 35% acetonitrile from pool II produce silver-staining bands at ~9–16 kDa.[14] The major bands in the MGSA-containing fractions eluting at 25±5% acetonitrile appear at <14 kDa, but faint bands at 24–26 and 14–16 kDa are also present (data not shown). MGSA bioactivity can be recovered from the nonreduced gels after acetic acid extraction, according to procedures of Roberts et al.[15] The bands are removed from the gels with

[14] U. K. Laemmli, *Nature (London)* **227**, 680 (1970).
[15] A. B. Roberts, C. A. Frolik, M. A. Anzano, R. K. Assoian, and M. B. Sporn, in "Methods in Molecular and Cell Biology" (D. Barnes, G. Sato, and D. Sirbasku, eds.), p. 181. Liss, New York, 1984.

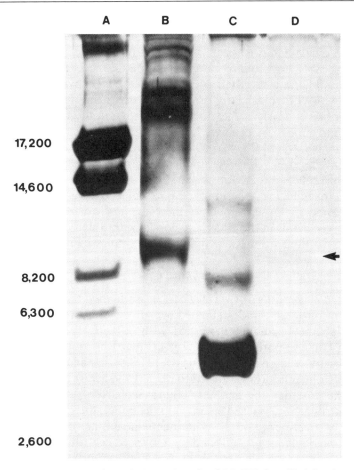

A   B   C   D

17,200

14,600

8,200

6,300

2,600

FIG. 3. SDS–polyacrylamide gel electrophoresis of RP-HPLC-purified fractions of ^{125}I-EGF-labeled competing TGF and MGSA. Lyophilized samples were dissolved in sample buffer containing 0.01 M H_3PO_4, 8 M urea, 2.5% SDS, and 5% 2-mercaptoethanol equilibrated to a pH of 6.8 with Tris. After heating to 100° for 5 min, samples were electrophoresed according to the Swank–Munkres procedure (12) in a 13.5% gel. The M_r standards shown at the arrows are 17,200, 14,600, 8200, 6200, and 2600, respectively. The second lane from the left (B) shows the electrophoretic profile of MGSA containing fraction 31 from pool I. The third lane from the left (C) shows the TGFα containing fraction 28 from pool I. The fourth lane (D) shows MGSA containing fraction 8, revealing a single band at 9300 (indicated by the arrow).

a scalpel, cut into small slices, and placed into pipet tips which have been sealed and plugged with glass wool. One milliliter of a solution of 1 M acetic acid made 3×10^{-6} M bovine serum albumin is added to each pipet tip. The acrylamide slices are then crushed with a Teflon rod. The protein is allowed to extract for 24 hr at room temperature. The pipet tips are cut and the eluate is collected. The crushed acrylamide is subsequently washed twice with 1 ml of the acetic acid–bovine serum albumin solution. The combined eluate and washes are frozen and lyophilized. Samples are then stored at $-80°$. Lyophilized samples are diluted 1:8, 1:80, or 1:800 in binding buffer and assayed for MGSA using the [³H]thymidine assay and the ELISA. MGSA bioactivity can be recovered from the 26, 24, 16, and < 14 kDa bands. Most of these polypeptides retain the FB2AH7 recognition site and are ELISA positive. Altogether these data suggest that MGSA exists in higher M_r (24,000–26,000) forms and lower M_r (~9000–16,000) forms.[9] Investigations are now underway to determine the nature of the relationship between these forms of MGSA.

When the reducing gels from the MGSA bioactive fractions from pool I are compared to those of pool II there are few qualitative differences. However, the bands in the 20-kDa region appear to stain more intensely in the active fractions from pool I than from pool II (see Fig. 4).

Purification Method B: Immunoaffinity Chromatography

Step 1: Preparation of an FB2AH7 Immunoaffinity Column. Four grams of cyanogen bromide-activated Sepharose 4B is swollen overnight in 0.01 M HCl at 4° and then washed with 0.01 M HCl. Approximately 10 mg of Ultrogel-purified FB2AH7 is dissolved in coupling buffer (8.4 g NaHCO₃ and 29.7 g NaCl/liter of H₂O, pH 8.3) and added to the swollen Sepharose. The mixture is adjusted to pH 8.0 and gently agitated overnight at 4°. The next day the antibody-bound Sepharose is collected by low-speed centrifugation, washed 3× in coupling buffer, then unbound sites are blocked by treatment with 1 M monoethanolamine, pH 8.0. The antibody-bound Sepharose is collected by low-speed centrifugation, incubated for 5 min with acetate buffer (58.4 g NaCl and 8.2 g NaC₃H₃O₂ per liter H₂O, pH 4), and then washed with TBS, pH 8.0. This procedure is repeated three times. The antibody-bound Sepharose is then washed with eluting buffer (0.5 M NaC₂H₃O₂, pH 4.5), followed by a final wash with TBS. The resulting column has a total volume of approximately 10 ml. The unbound FB2AH7 and all washes are pooled, dialyzed against H₂O, lyophilized, and the total amount of unbound protein is determined by the methods of Bradford.[12]

Step 2: Immunoaffinity Chromatography. MGSA can be purified from acetic acid extracts of Hs0294-conditioned medium with an FB2AH7 immunoaffinity column.[16] An acetic acid extract from 1 liter of Hs0294 CM is dissolved in 30 ml of 0.1 M NaC₂H₃O₂, pH 7.2, and incubated for 48 hr with 10 ml of FB2AH7-bound Sepharose at 4° with gentle agitation. Unbound material is collected after filtration through a sintered

---

[16] G. Thomas and A. Richmond, manuscript submitted.

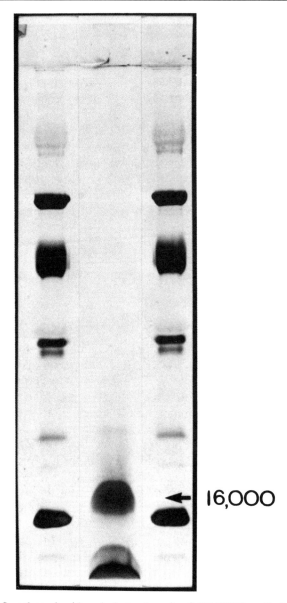

FIG. 4. SDS–polyacrylamide gel electrophoresis of RP-HPLC-purified MGSA under nonreducing conditions. Aliquots of lyophilized RP-HPLC-purified samples from MGSA/TGF pool II were tested for the presence of MGSA using the FB2AH7 ELISA. ELISA-positive fractions were then diluted in sample buffer and electrophoresed according to procedures of Laemmli,[14] in a 12% SDS–polyacrylamide gel. From left to right the lanes contain M_r standards, the ELISA-positive RP-HPLC fraction comprising the major MGSA peak, followed by M_r standards again. (Reprinted with permission of Richmond and Thomas.[9])

glass filter and the column is washed with 150 ml of 0.1 M NaC$_2$H$_3$O$_2$, pH 7.2. Bound protein is then eluted with 50 ml of 0.5 M NaC$_2$H$_3$O$_2$, pH 4.5, followed by a 150-ml wash with 0.1 M NaC$_2$H$_3$O$_2$, pH 7.2. Both the elution volume and the wash after elution are dialyzed exhaustively against 0.17 M acetic acid and then lyophilized. Approximately 1–3 μg of protein can be recovered from each liter of acetic acid-extracted CM which is applied to the column.

Step 3: Elution of Affinity Purified MGSA from Polyacrylamide Gels. MGSA purified by means of the Sepharose-bound monoclonal antibody FB2AH7 is electrophoresed on nonreducing polyacrylamide gels by a modification of the method of Laemmli.[14] After staining with silver, bands reproducibly appear at 16, 14, and <14 kDa, and occasionally bands are seen at 24–26 and 30 kDa. MGSA is subsequently eluted from the gels and assayed as described above in Method A, Step 4. When the gel proteins are assayed in this manner, the MGSA bioactivity can be eluted from the bands in a manner very similar to that described above for RP-HPLC-purified MGSA.[16] Gel slices containing no stained protein are included in each assay as a control.

Conclusions

MGSA is an autostimulatory monolayer mitogen for the Hs0294 human malignant melanoma cell line. This mitogenic activity resides in what is apparently a family of acid- and heat-stable polypeptides with molecular weights ranging from 9000 to 26,000. MGSA can be partially purified by gel exclusion chromatography followed by RP-HPLC. After the RP-HPLC step, the major peak of MGSA is separated from the TGFα.

An ELISA has been developed for MGSA using the monoclonal antibody FB2AH7. In this assay, FB2AH7 antigen-positive peaks are coincident with the major MGSA peaks. When these MGSA/FB2AH7 antigen-positive fractions are subjected to SDS–PAGE and the protein bands in the gel are removed, extracted, and eluted, the MGSA bioactivity is recovered from the 9- to 16-kDa and 24- to 26-kDa proteins.

An FB2AH7 immunoaffinity column can be used for rapid purification of MGSA. When the specifically bound eluant is analyzed by SDS–PAGE and proteins are eluted from the gel, the MGSA bioactivity can be recovered in a pattern very similar to that observed with the RP-HPLC-purified MGSA.

Acknowledgment

Supported by Veterans Administration Merit Award and National Cancer Institute Grant R23 Ca 34590.

[12] Solid-Phase Synthesis of Type-α Transforming Growth Factor

By James P. Tam

Type-α transforming growth factor (TGF-α) is a mitogenic peptide factor[1] produced extracellularly by tumor cells[2] and by virally[3] and chemically transformed cells in culture.[4] The primary structure of TGFα from a retrovirus-transformed Fisher rat embryo fibroblast (rTGFα, Fig. 1), deduced from Edman degradation and the predicted cDNA sequence,[5-7] contains 50 amino acid residues and differs from that of the predicted cDNA sequence of the human TGFα in 4 amino acid residues.[8] Both the rat and human sequences share striking structural similarities with human and mouse epidermal growth factors (EGF)[9,10] and are likely to have similar arrangements of disulfide linkages as EGFs. Like EGF, TGFα stimulates mitogenesis, binds, activates the EGF receptor tyrosine kinase, and induces a reversible change in the morphology of normal fibroblasts to the transformed phenotype.[2-7,11]

TGFα is obtained only with difficulty from the natural source and the synthesis of TGFα has been undertaken[12] with the objectives of providing sufficient material to enable more extensive studies on their mode of

[1] Abbreviations follow the tentative rules of the IUPAC–IUB Commission on Biochemical Nomenclature, published in *J. Biol. Chem.* **247**, 979 (1972). Others: Boc, *tert*-butyloxycarbonyl; BrZ, 2-bromobenzyloxycarbonyl; Bzl, benzyl; cHex, cyclohexyl; ClZ, 2-chlorobenzyloxycarbonyl; DCC, dicyclohexylcarbodiimide; DIEA, diisopropylethylamine; DMF, *N,N*-dimethylformamide; Dnp, dinitrophenyl; HOAc, acetic acid; HPLC, high-performance liquid chromatography; HOBt, 1-hydroxybenzotriazole; MBzl, 4-methylbenzyl; TFA, trifluoroacetic acid; TFMSA, trifluoromethanesulfonic acid.

[2] J. E. De Larco and G. J. Todaro, *Proc. Natl. Acad. Sci. U.S.A.* **75**, 4001 (1978).

[3] B. Ozanne, J. Fulton, and P. L. Kaplan, *J. Cell. Physiol.* **105**, 163 (1980).

[4] H. L. Moses and R. A. Robinson, *Fed. Proc., Fed. Am. Soc. Exp. Biol.* **41**, 3008 (1982).

[5] H. Marquardt, M. W. Hunkapiller, L. E. Hood, D. R. Twardzik, J. De Larco, and G. J. Todaro, *Science* **223**, 1079 (1983).

[6] J. Massagué, *J. Biol. Chem.* **258**, 13606 (1983).

[7] D. C. Lee, T. M. Rose, N. R. Webb, and G. J. Todaro, *Nature (London)* **313**, 489 (1985).

[8] R. Derynck, A. B. Roberts, M. E. Winkler, E. Y. Chen, and D. V. Goeddel, *Cell* **38**, 287 (1984).

[9] C. R. Savage, Jr., J. Inagami, and S. Cohen, *J. Biol. Chem.* **247**, 7612 (1972).

[10] H. Gregory, *Nature (London)* **257**, 325 (1975).

[11] A. B. Roberts, L. C. Lamb, D. L. Newton, M. B. Sporn, J. E. De Larco, and G. J. Todaro, *Proc. Natl. Acad. Sci. U.S.A.* **77**, 3494 (1980).

[12] J. P. Tam, H. Marquardt, D. R. Rosberger, and G. J. Todaro, *Nature (London)* **309**, 376 (1983).

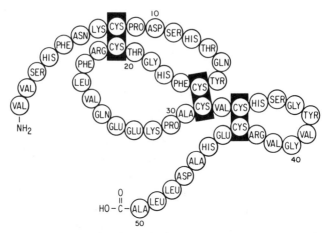

FIG. 1. Structure of rTGFα, disulfide pairing according to mEGF.[9]

action, and examining the role of their structures on their transforming activity. Since the theory and general procedures of solid-phase peptide synthesis[13] have been extensive and intensively reviewed in this series[14-16] and elsewhere,[17-19] this chapter will focus on methods necessary to achieve the manual synthesis of rTGFα. Several of the latest improvements of the solid-phase method, particularly the novel low–high HF to minimize strong acid-catalyzed side reactions,[20-22] are discussed in greater detail. However, readers are advised to consult previous articles and reviews for general understanding of the background mechanisms and operating procedures of the solid-phase method.

Synthetic Strategy

For the chemical synthesis of rTGFα, an improved version of the solid-phase method was used. Some of the salient features and rationale are outlined.

[13] R. B. Merrifield, *J. Am. Chem. Soc.* **85**, 2149 (1963).

[14] M. S. Doscher, this series, Vol. 47, p. 578.

[15] D. B. Glass, this series, Vol. 99, p. 119.

[16] J. M. Stewart, this series, Vol. 103, p. 3.

[17] G. Barany and R. B. Merrifield, *in* "The Peptides: Analysis, Synthesis, Biology" (E. Gross and J. Meienhofer, eds.), Vol. 2, p. 1. Academic Press, New York, 1980.

[18] J. Meienhofer, *in* "Hormonal Proteins and Peptides" (C. H. Li, ed.), Vol. 2, p. 45. Academic Press, New York, 1973.

[19] M. Bodanszky and J. Martinez, *Synthesis* 333 (1981).

[20] J. P. Tam, W. F. Heath, and R. B. Merrifield, *Tetrahedron Lett.* **23**, 2939 (1982).

[21] W. F. Health, J. P. Tam, and R. B. Merrifield, *J. Chem. Soc. Chem. Commun.* 896 (1982).

[22] J. P. Tam, W. F. Health, and R. B. Merrifield, *J. Am. Chem. Soc.* **105**, 6442 (1983).

1. Stable Peptide–Resin Linkage

The conventional linkage of the benzyl ester bond between the carboxyl-terminal amino acid and the solid support for solid-phase peptide synthesis is often derived from chloromethyl–resin and is not completely stable under the acidic conditions required for the removal of the Boc group. A new resin support, Boc-aminoacyl-4(oxymethyl)-phenylacetamidomethyl-resin (Boc-aminoacyl-OCH$_2$-Pam-resin), is 100-fold more stable than the conventional benzyl ester-resin linkage.[23,24] The increased stability of the Pam ester linkage has resulted in minimization of premature release of peptide chains from the resin, late initiation of peptides, and formation of deletion peptides lacking one or more residues at the carboxyl terminus. In short, the more acid-resistant Pam–resin support will reduce heterogeneity of peptide products, result in much higher yields, and is an absolute essential for the synthesis of a complex and long peptide such as TGFα.

2. Maximal and Stable Side Chain Protection

The synthetic strategy utilizes the acid-labile Boc group for temporary N^α-protection and more acid-stable groups for protection of the side chains. The complete protected rTGFα is shown in Fig. 2. The differential acid susceptibility between those two protection groups during repetitive deprotection of the Boc group should be sufficiently large to avoid the loss of side chain-protecting groups. Thus, side chain-protecting groups were selected to offer maximal stability and minimal loss during the synthesis. For example, N^ϵ-ClZ for Lys was used because it is much more stable than the benzyloxycarbonyl derivative to the acid deprotection steps and avoids side chain branching.[25] Other protecting groups were selected to minimize side reactions during the synthesis in addition to their acid stabilities. The β-cyclohexyl (cHex) ester of Asp was used because it minimizes aspartimide formation during the repetitive base treatments and because it is also more acid stable.[26] The N^{im}-dinitrophenyl for His was used because it offers stability and removes the possibility of racemization during the coupling step.[27] Finally, the BrZ for the phenolic protection of Tyr was

[23] A. R. Mitchell, B. W. Erickson, M. N. Ryabtsev, R. S. Hodges, and R. B. Merrifield, *J. Am. Chem. Soc.* **98**, 7357 (1976).
[24] A. R. Mitchell, S. B. H. Kent, M. Engelhard, and R. B. Merrifield, *J. Org. Chem.* **43**, 2845 (1978).
[25] B. W. Erickson and R. B. Merrifield, *J. Am. Chem. Soc.* **95**, 3750 (1973).
[26] J. P. Tam, T. W. Wong, M. W. Rieman, F. S. Tjoeng, and R. B. Merrifield, *Tetrahedron Lett.* 4033 (1979).
[27] S. Shaltiel and M. Fridkin, *Biochemistry* **9**, 5122 (1970).

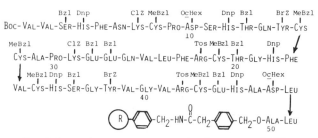

FIG. 2. Protecting group strategy and resin support of rTFα.

used because it greatly reduces the extent of alkylation side reaction during the final HF cleavage of all protecting groups.[28,29]

3. Quantitative Monitoring of the Coupling Step

To attain complete and efficient coupling reactions at each step and to avoid deletion and termination peptides, a double-coupling strategy using several coupling protocols was used. A typical coupling protocol was with the preformed symmetrical anhydride[30] in CH_2Cl_2 for the first coupling and in dimethylformamide for the second coupling. Coupling reactions of Gln and Asn was with preformed hydroxybenzotriazole ester[31] in DMF. However, coupling reactions to both Glu and Gln residues were carried out in preformed symmetrical anhydride in a DMF–NMP mixture to avoid pyroglutamyl formation and hence termination reaction.[32,33] Couping reactions to Gly and Arg were done in DCC in alone to avoid dipeptide and lactam formations.[19,34] The progress of the synthesis was monitored by three different methods. First, it was monitored with the new quantitative ninhydrin procedure.[35] The sensitivity of this test will allow the determination of the coupling efficiency to about 99.95%. Second, peptides, every 10 residues, while still attached to the solid support, were evaluated by solid-phase sequencing for correctness of sequence and for any indication of deletions by preview analysis.[36,37] Third, the peptide fragments were re-

[28] D. Yamashiro and C. H. Li, *J. Org. Chem.* **38**, 591 (1973).

[29] J. P. Tam, W. F. Health, and R. B. Merrifield, *Int. J. Pept. Protein Res.* **21**, 57 (1983).

[30] H. Hagenmaier and H. Frank, *Hoppe-Seyler's Z. Physiol. Chem.* **353**, 1973 (1972).

[31] W. Konig and R. Geiger, *Chem. Ber.* **103**, 788 (1970).

[32] R. D. DiMarchi, J. P. Tam, S. B. H. Kent, and R. B. Merrifield, *Int. J. Pept. Protein Res.* **19**, 88 (1982).

[33] R. D. DiMarchi, J. P. Tam, and R. B. Merrifield, *Int. J. Pept. Protein Res.* **19**, 270 (1982).

[34] R. B. Merrifield, A. R. Mitchell, and J. E. Clark, *J. Org. Chem.* **39**, 660 (1974).

[35] V. K. Sarin, S. B. H. Kent, J. P. Tam, and R. B. Merrifield, *Anal. Biochem.* **117**, 147 (1981).

[36] H. D. Niall, G. W. Tregear, and J. Jacobs, *in* "Chemistry and Biology of Peptides" (J. Meienhofer, ed.), pp. 695–699. Ann Arbor Press, Ann Arbor, MI, 1972.

[37] R. B. Merrifield, L. D. Vizioli, and H. G. Boman, *Biochemistry* **21**, 5020 (1982).

moved periodically from the resin support to be examined by HPLC for their integrity and purity. With these sensitive techniques, the progress and the level of homogeneity of the peptide products during the synthesis were known and decisions could be made to correct any serious mistakes.

4. A New Deprotection Strategy: Low–High HF Cleavage Method

The key to this synthesis is the new deprotection strategy at the final stage of the synthesis. The new deprotection method consists of a low and a high HF step.[20-22] The low HF step contains a low concentration of HF complexed with a weak base, dimethyl sulfide, to remove the benzyl protecting groups by the S_N2 mechanism as sulfonium salts to avoid serious alkylation side reactions. The high HF step is to remove the more acid-stable protecting groups such as Arg(Tos) and Cys(MBzl), which are not removable under the low HF step. The unusual latter property of the low-HF step also solves a major problem in the synthesis of complex peptides containing multiple cysteinyl residues which are protected as sulfides in the course of the synthesis. Since the protected cysteinyl sulfides are prone to oxidation to sulfoxides which are stable to the normal HF cleavage conditions, such a consequence leads to incomplete deprotection of the cysteinyl sulfides and, hence, to exceptionally low yield of the correct product. Furthermore, there is no satisfactory method to convert the sulfoxides at the fully protected peptide stage to the cysteinyl sulfide prior to the HF step. However, the low HF step is capable of concomitant reduction of sulfoxides to sulfides quantitatively during the cleavage reactions, thus allowing the smooth removal of the cysteinyl protecting groups during the subsequent high HF step.[38,39] Thus, the new low–high HF deprotection reduces many side reactions associated with the conventional HF cleavage reactions and provides a satisfactory solution to the synthesis of proteins containing multiple disulfide linkages.

Preparation of Boc-Ala-OCH$_2$-Pam-Resin

Preparation of the support is shown in Fig. 3. The synthetic route consists of a convergent preparation of the aminomethyl-resin[40] and the Boc-alanyl-4-(oxymethyl)phenylacetic acid.[41,42] Aminomethyl-resin is pre-

[38] S. Sakakibara, in "Chemistry and Biochemistry of Amino Acids, Peptides, and Proteins" (B. Weinstein, ed.), Vol. 1, p. 51. Dekker, New York, 1971.
[39] J. Lenard and A. B. Robinson, J. Am. Chem. Soc. 89, 181 (1967).
[40] A. R. Mitchell, S. B. H. Kent, B. W. Erickson, R. B. Merrifield, Tetrahedron Lett. 3795 (1976).
[41] J. P. Tam, S. B. H. Kent, T. W. Wong, and R. B. Merrifield, Synthesis 955 (1979).
[42] J. P. Tam, F. S. Tjoeng, and R. B. Merrifield, J. Am. Chem. Soc. 102, 6117 (1980).

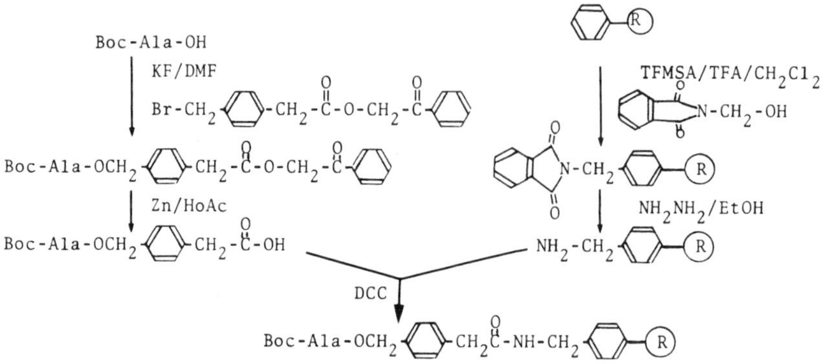

FIG. 3. Preparation of Boc-Ala-OCH₂-Pam-resin.

pared directly from polystyrene resin to allow precise control of the extent of substitution and is free from the undesirable side reactions. In the present synthesis of rTGFα, a substitution of 0.4 mmol/g is deemed suitable. Boc-alanyl-4-(oxymethyl)phenylacetic acid is prepared from a simple two-step synthesis and is coupled to aminomethyl-resins via DCC to give Boc-Ala-OCH₂-Pam-resin. The resulting Pam-resin is thus prepared from a chemically unambiguous route, without sacrificing HF cleavage yields and without significantly increasing the susceptibility to nucleophilic side reactions. The method described here is generally applicable to the preparation of other Boc-aminoacyl-OCH₂-Pam-resins.

Preparation of Boc-Ala-4-(oxymethyl)phenylacetic Acid

Reagents

1. Boc-Ala-OH (Protein Research Foundation, Japan, or her United States distributors)
2. Potassium fluoride (anhydrous, ultrapure, Alfa, Morton Thiokol, Inc.)
3. 4-Bromomethylphenylacetic acid phenacyl ester (RSV Corp., Ardsley, NY)
4. Zinc dust (7.1 μm, Alfa)
5. Dimethylformamide (Spectrophotometric grade, Aldrich Chemical Co.)
6. 85% acetic acid

Procedure

Boc-Ala-OH (3.78 g, 20 mmol) is dissolved in 120 ml of DMF in a 500-ml round-bottom flask equipped with magnetic stirrer. Grind KF

(3.48 g, 60 mmol) to a powder in a mortor under nitrogen or in glove bag, weigh, and add to Boc-Ala-OH. Stir vigorously and add bromomethylphenylacetic acid phenacyl ester (3.47 g, 10 mmol) and allow the reaction to go overnight. Monitor the reaction on thin-layer chromatography on silica gel GF 250-μm plates in petroleum ether:acetic acid (8:2, v/v). When all the starting material of bromomethylphenylacetic acid phenacyl ester is consumed, filter the suspension through a coarse-sintered glass funnel by sunction and wash 3× with 20 ml of DMF. Evaporate the combined filtrate in vacuum (use a mechanical pump). Redissolve the residue in 300 ml of ethyl acetate and extract with 5× with 200 ml each of pH 9.5 buffer (23 g K_2CO_3 and 26 g $NaHCO_3$ in 1 liter H_2O), then with 3× H_2O, 3× saturated NaCl solution, dry the ethyl acetate solution over $MgSO_4$ and examine by TLC for absence of Boc-Ala-OH. Evaporate ethyl acetate in a rotary evaporator and dry the resulting residue on vacuum pump overnight.

Dissolve the crude Boc-Ala-oxymethylphenylacetic acid phenacyl ester in 120 ml of 85% acetic acid in a 250-ml round-bottom flask with a magnetic stirrer at room temperature. Add zinc dust and monitor the reaction by TLC in chloroform:methanol (99:1, v/v) and in chloroform:acetic acid (95:5, v/v). Usually, the reaction will be completed in 4 hr. Filter off and wash zinc with 85% HOAc. Evaporate the combined filtrate to give a white solid. Remove the acetophenone (by-product of the reaction) by extraction from the solid with hexane or pentane (3×, 100 ml). Dissolve the solid in 0.1 N HCl (150 ml) and extract twice with an equal volume of ethyl acetate. Wash the combined ethyl acetate solution with 150 ml of 3× H_2O and saturated NaCl solution, dry with $MgSO_4$ and filter. Reduce the ethyl acetate to a small volume and add hexane to turbidity, cool to crystallize. Expect a yield of about 50%; the mp of Boc-Ala-oxymethylphenylacetic acid is 105–107°.

Preparation of Aminomethyl Resin

Reagents

1. Copoly (styrene-1% divinylbenzene) resin beads (Bio-Rad, 200–400 mesh)
2. N-(Hydroxymethyl)phthalimide (Aldrich Chemical Co.)
3. Trifluoromethanesulfonic acid (TFMSA, Aldrich Chemical Co.)
4. Anhydrous trifluoroacetic acid (TFA)
5. Hydrazine hydrate (Aldrich Chemical Co.)
6. Anhydrous CH_2Cl_2
7. Ethanol, 95%

Procedure

Weigh 10 g of washed and sized resin beads into a large synthesis reaction vessel (160 to 200 ml) and attach to a synthesis shaker in the hood. A 1-pint glass screw-capped bottle such as a well-washed and clean HCl reagent bottle can also be used as a substitute for the large reaction vessel. Mix 40 ml of trifluoroacetic acid and 40 ml of CH_2Cl_2 into a stoppered flask. Keep this and other solutions dry and in the hood. Quickly pipet with the aid of a rubber propipet 6.3 ml of trifluoromethanesulfonic acid into the 1:1 mixture of TFA–CH_2Cl_2. Put the used pipet in a graduated cylinder filled with H_2O and later rinse carefully. (Warning! TFA is a strong acid and TFMSA is an even stronger acid; handle with gloves and wear glasses.) Immediately pour the TFMSA solution in 10-ml portions into the resin suspension every 5 min. Shake and mix well at room temperature. Continue the reaction for 4 hr after the completion of the TFMSA additions. Filter by applying vacuum and wash with TFA–CH_2Cl_2(1:1) 3× 120 ml, CH_2Cl_2 5× 120 ml, CH_3OH 4× 120 ml, and discard all filtrates. Dry under vacuum. Estimate phthalimide content by infrared spectroscopy (IR). Make a pellet from 2 mg of ground resin + 200 mg KBr. Calculate the ratio pf peak heights at 1725/1601 cm^{-1} and divide by 5.90 to determine millimole per gram. Expect 0.45 mmol/g (90% yield) based on *N*-hydroxymethylphthalimide. Check by elemental nitrogen analysis. Place the dried resin into a 500-ml round-bottom flask fitted with a water-cooled condenser. Add 250 ml of ethanol (95%)–hydrazine hydrate (9:1) and reflux the mixture without stirring for 8 hr. (Don't use a magnetic stirring bar, the resin will be pulverized!) Filter the hot mixture by vacuum and wash with 3× 100 ml of hot ethanol (95%). Dry under vacuum and check the completeness of the removal of phthalimide by IR. The peak at 1725 cm^{-1} should completely disappear. Check also by quantitative ninhydrin test (see below).

Boc-Ala-4-(oxymethyl)-Pam-Resin

Aminomethyl resin (5 g, 2.0 mmol) is suspended in 100 ml of CH_2Cl_2 in a 150-ml reaction vessel on a solid-phase peptide synthesis shaker. Equilibrate the resin in this solvent for 2 hr so that the resin is fully solvated. Boc-Ala-OCH$_2$-phenylacetic acid (1.35 g, 4.0 mmol) is added to the resin. Shake 2 min. Add dicyclohexylcarbodiimide (0.82 g, 4.0 mmol) to the reaction vessel and shake for 2 hr. Filter, wash 100 ml each with 3× DMF, and 3× CH_2Cl_2. Determine the free amino groups by ninhydrin reaction and the substitution by hydrolyzing a portion of the resin sample (5 mg) in 12 *N* HCl–propionic acid (1:1, v/v) at 135° for 4.5 hr for amino acid analysis. If necessary (free amino groups higher than 5%), acetylate the residual free amino groups with 50 ml of a 1:1 mixture of premixed and cooled mixture of acetic anhydride and pyridine in 50 ml of DMF with shaking for 0.5 hr.

Stepwise Solid-Phase Peptide Synthesis

Reagents

1. CH_2Cl_2 (Eastman) is distilled from anhydrous Na_2CO_3. DMF (Spectrophotometric grade, Aldrich Chemicals) is stored over molecular sieve 4A for a week and filtered over fine-sintered glass filter before use
2. Deprotection reagents: 50% TFA (Halocarbon Products, NJ) in CH_2Cl_2 (1:1 mixture of TFA and CH_2Cl_2, v/v) containing 0.03% of ethanedithiol (30 μl in 1 liter of TFA-CH_2Cl_2 solution). Ethanedithiol is used to protect the cysteinyl moieties during the TFA treatments. Do not use indole as scavenger. Ethanedithiol should be added to the TFA solution prior to the deprotection reaction. TFA should also be distilled from concentrated sulfuric acid (100 ml of concentrated sulfuric acid/liter of TFA) to eliminate H_2O and other extraneous substances. It is important to have a TFA deprotection solution that is clear and well protected from moisture
3. Neutralization reagent: 5% DIEA (diisopropylethylamine) in CH_2Cl_2 (5 ml in 95 ml of CH_2Cl_2). DIEA is distilled twice from CaH_2 and then once through a fractionating column (Vigreaux)
4. Coupling reagents: DCC (dicyclohexylcarbodiimide) (Tridom-Fluka) should be readily soluble in CH_2Cl_2 and should give a clear solution. Prepare a 0.48 M (2.4 mmol/5 ml) solution in CH_2Cl_2 (494.4 mg/5 ml). Hydroxybenzotriazole (HOBt, Aldrich Chemicals) is recrystalized from 70% ethanol at least once before use. Prepare a 0.48 M solution in DMF (324 mg/5 ml). Boc-amino acids (Protein Research Foundation, Japan) are to be checked by TLC for purity before use. Prepare a 0.48 M solution in CH_2Cl_2. Several Boc-amino acids are not readily soluble in CH_2Cl_2. These are Boc-Arg(Tos), Boc-Gln, Boc-Asn, and Boc-Leu·H_2O. Dissolve these derivatives first in a minimal amount of DMF first and dilute to the right concentrations by CH_2Cl_2.
5. Direct DCC coupling: Place 5 ml of the 0.48 M solution in CH_2Cl_2 into the resin containing 10 ml of CH_2Cl_2 (including the hold-up volume of the resin). Shake and mix for 2 min. Add 5 ml of the 0.48 M solution of DCC solution in CH_2Cl_2 and mix for 1 hr. This

protocol is for a 2-g resin reaction and the total coupling volume should be maintained at about 20 ml

6. Preformed symmetrical anhydride coupling: Place 10 ml of the 0.48 M solution of Boc-amino acid in CH_2Cl_2 in a 50-ml round-bottom flask equipped with stirrer, and cooled in an ice bath. Add 5 ml of the 0.48 M solution of DCC solution and react for 15 min. Filter off dicyclohexylurea and wash with 10 ml of CH_2Cl_2. Add the combined filtrate to the drained resin. It is best to prepare the anhydride reaction during the TFA deprotection cycle. For the coupling in DMF, reduce CH_2Cl_2 of the filtrate in a stream of N_2 and add 10 ml of DMF

7. Preformed HOBt-active ester coupling: Place 5 ml of the 0.48 M solution of Boc-amino acid in CH_2Cl_2 in a 50-ml round-bottom flask equipped with stirrer and cooled in an ice bath. Add successively 5 ml each of the 0.48 M solution of DCC and the HOBt solution and react for 30 min. Filter off dicyclohexylurea and wash with 5 ml of CH_2Cl_2. Add the combined filtrate to the drain resin.

Procedure

4-(Boc-alanyloxymethyl)phenylacetamidomethyl-resin (2.0 g, 0.40 mmol/g), prepared as just described, was placed in a silanized reaction vessel and carried manually through 49 synthetic cycles in a mechanical shaker. All amino acids were protected with N^α-tert-butyloxycarbonyl (Boc). Side-chain protecting groups were as follows: Arg(Tos), Asp(OcHex), Cys(4-MeBzl), Glu(OBzl), His(Dnp), Lys(2-ClZ), Ser(Bzl), and Tyr(BrZ). Each synthetic cycle consisted of the following steps:

1. TFA-CH_2Cl_2 prewashes (3×, 2 min) and a 20-min deprotection with 50% trifluoroacetic acid/CH_2Cl_2
2. CH_2Cl_2 washes (5×, 1 min)
3. Neutralization with 5% diisopropylethylamine/CH_2Cl_2 (3×, 2 min)
4. Coupling with preformed symmetrical anhydrides (2.4 mmol, 3 Eq of the Boc-amino acid) for 1 hr in CH_2Cl_2
5. CH_2Cl_2 washes (3×, 1 min)
6. Neutralization with 5% DIEA (2×, 5 min)
7. CH_2Cl_2 washes (3×, 1 min)
8. DMF wash (1×, 2 min)
9. Recoupling in preformed symmetrical anhydride in dimethylformamide (DMF) for 1 hr
10. CH_2Cl_2 washes (3×, 1 min)

Coupling of Boc-Asn-OH and Boc-Gln-OH was mediated by the preformed hydroxybenzotriazole active ester in DMF. Boc-Gly-OH and Boc-

Arg(Tos)-OH were coupled with dicyclohexylcarbodiimide alone. All couplings were monitored by the quantitative ninhydrin test. A third coupling of symmetrical anhydride in N-methylpyrrolidinone at 50° for 2 hr was used when necessary to give >99.8% completion. [^{14}C]Boc-Leu and [^{3}H]Boc-Val were incorporated at positions 2 and 49 of the synthesis. Portions of protected peptide resin were removed after cycles 17, 28, and 37 for Edman degradation of preview analysis and HF deprotection to examine the progress and the results of the synthesis.

Monitoring Coupling Reactions: Quantitative Ninhydrin Test

The use of the Pam-resin also allows the quantitation of the completeness of the coupling reaction by removing a weighed amount of resin after each coupling. In essence, the quantitative test differs from the qualitative Kaiser test in only procedural manipulations. Both tests use the same reagents; however, the former produces specific and reliable determinations of the degree of coupling to replace qualitative estimates. In general, a background of 0.1 to 0.3% due to nonspecific reaction with the peptide resin is observed.

Reagents

1. Phenol (reagent grade)
2. KCN (reagent grade)
3. Pyridine (distilled from ninhydrin to remove primary amines)
4. Ninhydrin (Pierce Chemical)
5. Amberlite mixed-bed resin MB-3 (Bio-Rad)
6. Absolute ethanol

Prepare two standard reagents A and B. Both reagents can be kept for months when kept away from light.

Reagent A: Prepare separately solutions 1 and 2. Solution 1 contains 40 g of phenol with 10 ml of ethanol. Warm to dissolve. Stir with 4 g of Amberlite resin MB-3 for 45 min. Filter. Solution 2: Dissolve 65 mg of KCN in 100 ml of H_2O. Dilute 2 ml of the KCN solution to 100 ml with pyridine. Filter. Mix solutions 1 and 2 to give reagent A.

Reagent B: Dissolve 2.5 g of ninhydrin in 50 ml of ethanol. Stopper and store in the dark under nitrogen.

Procedure

Samples of peptide-resin after coupling should be washed twice either with 5% diisopropylethylamine in CH_2Cl_2 and three times with CH_2Cl_2 and dried under vacuum before the test. Weigh a 2- to 5-mg sample of dry resin into a 10×75 mm test tube. Be sure all the resin beads are deposited at the bottom of the test tube. A convenient way is to use the weighing paper as a funnel. Add 100 μl of reagent A (6 drops using a disposable glass pipet) and 25 μl of reagent B (2 drops). As a control, add only the reagents. Carefully and gently mix the reagents (never use a vortex, which will definitely splatter resin beads all over the test tube) and place both tubes, covered, in a heating block preadjusted to 100°. After 10 min place the tubes in ice-cold water. The characteristic intense blue color of a positive reaction will develop immediately. However, if the coupling step is quantitative, a brownish color will develop. Add 1 ml of 60% ethanol in water and mix thoroughly. Filter through a Pasteur pipet containing a tight plug of glass wool. Make the solution to 2.0 ml with 60% ethanol. Measure the absorbance of the sample filtrate against the reagent blank at 570 nm. The observed α-amino groups can be calculated by the following equation:

$$\mu mol/g = [(A_{570}V_{ml})/(\epsilon_{570}W_{mg})]10^6$$

where A is the observed absorbance at 570 nm, V is the dilution volume in milliliters, W is the weight of the resin, and ϵ is the extinction coefficient of the amino acid in the peptide resin being determined (see comment).

Comment on the Assay. For routine monitoring, an effective extinction coefficient ϵ of 1.5×10^4 M^{-1} cm^{-1} is used except for the following amino acid on the peptide-resin as the N^{α}-amino groups: those with Glu(OBzl) and Gln, use 1.2×10^4 M^{-1} cm^{-1}; with Gly, Asp(OBzl), Asp(OcHzl), and Asn, use 1.0×10^4 M^{-1} cm^{-1}; with β-amino acids such as β-alanine, use 0.5×10^4 M^{-1} cm^{-1}. Proline and secondary α-amino acids do not give the blue color reaction and should be used as background correction.

Low–High HF Cleavage to Remove all Protecting Groups

Reagents
1. HF (Matheson)
2. HF line (Protein Research Foundation, Japan)
3. *p*-Cresol (Gold-label, Aldrich Chemical Co.)
4. *p*-Thiocresol (Tridom-Fluka)
5. 2-Mercaptoethanol (Tridom-Fluka)
6. Dimethyl sulfide (Tridom-Fluka)
7. Thiophenol (Alrich Chemical Co.)
8. Dithiothreitol (Chemical Dynamic, NJ)
9. 8 M urea (ultrapure grade, Schwarz/Mann, Cambridge, MA) in 0.1 M Tris–HCl buffer (Schwarz/Mann), pH 8.0

Procedure

Protected rTGFα-resin (0.7 g) was first treated three times with 1 M thiophenol in DMF for 8 hr to remove the N^{im}-dinitrophenyl protecting group of His and then with 50% trifluoroacetic acid/CH_2Cl_2 (10 ml) for 5 min to remove the N^{α}-tert-butyloxycarbonyl group. The dried peptide-resin was treated with the low–high HF method of cleavage. For the low HF treatment, the peptide-resin was premixed with dimethyl sulfide, p-thiocresol, and p-cresol. Liquid HF at $-78°$ C was added to give a final volume of 10 ml (65:2.5:7.5:25, v/v). The mixture was equilibrated to 0° with stirring in an ice bath. After 2 hr, the HF and dimethyl sulfide were removed *in vacuo*. The high HF treatment was initiated by recharging the reaction vessel at $-78°$ with 14 ml of fresh liquid HF to give a total volume of 15 ml of HF–p-cresol–p-thiocresol. The reaction was carried out at 0° for 1 hr. After evaporation of HF at 0° and washing with cold ethermer-captoethanol (99:1, v/v, 30 ml) to remove p-thiocresol and p-cresol, the crude reaction mixture was extracted with 100 ml of 8 M urea–0.2 M dithiothreitol in 0.1 M Tris buffer, pH 8.0. The cleavage yield was 97% based on the back hydrolysis of the HF-treated resin by 12 N HCl–phenol–acetic acid (2:1:1, v/v, 2 ml) and by the ^{3}H-labeled peptide found in the 8 M urea extracts.

Comments on Low–High HF Cleavage. Several precautions and good laboratory practice will usually produce satisfactory results. These procedures are recommended:

1. Clean and disassemble the HF line periodically to prevent contaminations of previous cleavages. Very often, a heterogeneous and complex peptide mixture after HF cleavage can be explained by a dirty and contaminated HF line.

2. Check the HF line for leaks. One of the most troublesome problems is a leaky HF line, which can lead to introduction of water, CO_2, or acetone (from the dry ice–acetone bath) into the reaction vessel. Consequently, the HF reaction mixture will be diluted, with uncertain results. Furthermore, HF or DMS can be lost during the reaction, also leading to unsatisfactory results. To avoid this problem, it is best to check the HF line, with the reaction vessel in place prior to the actual experiment for leaks. This can be simply carried out by evacuating the HF line and closing the system to the outside. The vacuum should hold for at least 4 hr. If it fails to hold, replace O-rings or dismantle the HF line for thorough cleansing and replacement of worn-out parts.

3. Avoid incomplete mixing during the reaction. Incomplete mixing of the HF reagent and the peptide-resin or peptide can occur in several ways. The commonest one is due to the magnetic stirring bar being frozen by the

p-cresol or *p*-thiocresol mixture during the introduction of the reagents. This can be alleviated simply by adding the reagents in the following order: (1) peptide or peptidyl-resin, (2) *p*-cresol or *p*-thiocresol or both, in a melted form, placed carefully on top of the resin using a warm pipet, (3) after cooling the *p*-cresol mixture solidifies; then introduce the magnetic stirring bar and (4) DMS. Note that, in this order, the HF will first encounter the *p*-cresol mixture before the peptide-resin and will not produce extremely high acidity in the mixture.

Purification

The 8 M urea solution containing the crude mixture of rTGFα was sequentially dialyzed (Spectra Por 6, MW cutoff 1000) at 0° for 72 hr against 4 liters each of deaerated and N_2-purged 8 M urea-0.1 M dithiothreitol, 8 M urea, 4 M urea, and 2 M urea, all in 0.1 M Tris buffer, pH 8.0. No precipitate should occur during the dialysis, and 80% of the product (by ^3H counts) was recovered. Oxidation and regeneration of rTGAα was performed in 1 M urea, pH 8.0, Tris buffer (200 ml) in 1.5 mM oxidized and 0.75 mM reduced glutathione for 16 hr.[43] The clear solution was dialyzed against 8 liters each of 0.1 M Tris, pH 8.0, and 1 M HOAc. Lyophilization of the clear solution gave 125 mg of peptide. A portion of the lyophilized material (75 mg) was submitted to gel filtration on a 2.5 × 75 cm P-10 column eluted with 1 M HOAc (4-ml fractions). Detection by absorbance at 280 nm and radioactivity showed a broad peak of polymeric material eluting at the position of the lysozyme marker (Fig. 4A) and a more defined peak eluting at the position of the insulin marker (monomer). Fractions corresponding to the polymeric and monomeric material were pooled separately. The polymer-to-monomer ratio was 11 to 9 (by radioactivity). Lyophilization gave 41 mg of polymer and 31 mg of monomer. The polymeric material was again reduced, oxidized, and applied on a P-10 column to give an additional amount of monomer (24 mg). The combined crude monomer fraction was purified by ion-exchange chromatography on a CM–Sephadex column (2.5 × 15 cm) using a linear pH gradient of 5.6 to 8.0 in 0.1 M NH_4OAc and rTGFα eluted as a major peak (fractions 25–30)(Fig. 4B). Lyophilization gave 44 mg of protein, which accounted for 80% of the crude starting material. The peptide was further purified by reversed-phase C_{18} liquid chromatography (2.5 × 30 cm) eluted with 0.05% TFA–CH_3CN, with the major peak at 28% of CH_3CN being collected. Lyophilization gave 36 mg of highly pure rTGFα (Fig. 4C). The synthetic yield based on the starting loading of

[43] V. P. Saxena and D. B. Wetlaufer, *Biochemistry* **9**, 5015 (1971).

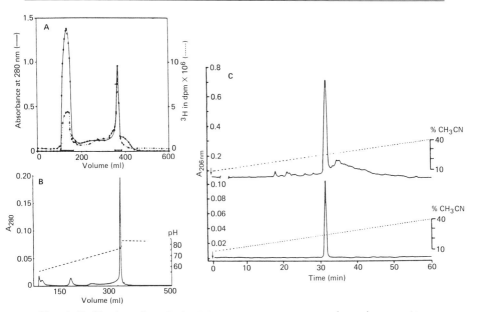

FIG. 4. Purification of synthetic rTGFα (A) P-10 gel permeation chromatography. (B) CM–Sephadex ion-exchange chromatography. (C) high-pressure liquid chromatography (HPLC) using C_{18} reversed-phase column. (D) HPLC of purified rTGFα.

alanine to the resin was 31% and a summary of the synthesis and work-up is given in Table I.

Characterization of Synthetic rTGFα

The purified synthetic rTGFα in sodium dodecyl sulfate–gel electrophoresis carried out in 15% polyacrylamide gels[44] and stained with Coomassie brilliant blue G-250 gave a single band with an apparent molecular weight of 6000. The purified material (50-μg sample) when examined in HPLC (Vydac column, C_{18}) with a 10 to 40% linear gradient of acetonitrile in 0.045% trifluoroacetic acid at a flow rate of 1 ml/min gave a single symmetrical peak (Fig. 4D) and coeluted with the native rTGF as a single symmetrical peak.[12] Amino acid analysis after 24 hr in 5.7 N HCl gave, respectively, in residues per molecules (theoretical values in paratheses): Asp, 3.02 (3); Thr, 1.85 (2); Ser, 2.76 (3); Glu, 5.00 (5); Pro, 1.86 (2); Gly, 3.00 (3); Ala, 2.79 (3); Val, 5.55 (6); Cys, 5.70 (6); Leu, 5.37 (6); Tyr, 2.83 (3); Phe, 2.97 (3); Lys, 2.02 (2); His, 4.72 (5); Arg, 1.91 (2).

[44] U. K. Laemmli, *Nature (London)* **227**, 680 (1970).

TABLE I
SYNTHESIS AND WORK-UP OF TGF

Step	Peptide–resin (mmol/g)	Yield per step (%)	Cumulative yield (%)
(1) Boc-Ala-resin	0.40	100	100
(2) Boc-Peptide resin	0.39	98	98
(3) HF cleavage	0.36	92	90
(4) Dialysis	0.29	80	72
(5) P-10 chromatography[a]	0.20	68	49
(6) CM-Sephadex	0.16	82	40
(7) C_{18} HPLC	0.12	76	31

[a] Polymeric materials recycled.

Biological Assays

When synthetic TGFα was compared for growth, transforming properties, and binding to the EGF-receptor with the putative native TGFα *in vitro*, both exhibited parallel and indistinguishable dose–response curves in these four different assays. Moreover, synthetic rTGFα also gave similar activities as mouse EGF in these assays. Dose–response curves for the mitogen assay of synthetic and natural rTGFα were performed on normal

TABLE II
COMPARISON OF SYNTHETIC TGF, NATURAL TGF, AND EGF BY DIFFERENT ASSAYS

Assay	Half-maximal activity [nM][a]		
	mEGF[b]	Natural TGF	Synthetic TGF
Stimulation DNA synthesis[c]	1.4	1.7	2.3
EGF-radioreceptor[d]	1.7	3.7	4.2
Phosphotyrosine kinase[e]	0.1	0.4	0.3
Soft agar growth[f]	0.5	0.5	0.3

[a] From the dose–response curves.
[b] m EGF from Collaborative Res., Inc.
[c] Performed on serum-deprived medium-conditioned NRK cells. [^{125}I]iododeoxyuridine incorporation was counted after 24 hr.
[d] 50% inhibition of [^{125}I]EGF (New England Nuclear to the EGF receptors of A431 carcinoma cells after 1 hr.
[e] EGF-receptor kinase from A431 cells and [Val^5]angiotensin is the peptide substrate.
[f] Performed in the presence of 10% serum and rTGFβ. Colonies were quantitated after 7 days.

rat kidney cells in Dulbecco's minimal essential medium with 10% heat-inactivated fetal bovine serum for 24 hr and maintained in medium containing 0.1% fetal bovine serum for 7 days before the assay as described.[2-4] Incorporation of [^{125}I]iododeoxyuridine was counted after a 24-hr exposure. Soft agar assay was performed on 49F fibroblast cells in the presence of 0.1 μg ml^{-1} of mouse TGFβ in phosphate-buffered saline as described[2-4] (also see Rizzino [32], this volume). Colonies were quantitated after 10 days by number and size using a Bausch and Lomb Analyzer. Dose–response curves of radioreceptor and stimulation of tyrosine-specific kinase assay were performed as described.[45] Inhibition of ^{125}I-labeled EGF binding to the EGF receptor of subconfluent monolayers of formalin-fixed A431 cells occurred after a 1-hr incubation at 22° with either synthetic rTGFα or natural mEGF. The kinase assay was performed using A431 membrane vesicles and [Val^5]angiotensin as substrate[46] (also see Pike [33], this volume). A summary of the specific activities of the half-maximal response is shown in Table II.

Acknowledgment

This investigation was supported by U.S. Public Health Service Grant CA36544, awarded by the National Cancer Institute, Department of Health and Human Services.

[45] H. Ushiro and S. Cohen, *J. Biol. Chem.* **255**, 8363 (1980).
[46] T. W. Wong and A. R. Goldberg, *J. Biol. Chem.* **258**, 1022 (1983).

[13] Identification of Receptor Proteins for Type-α Transforming Growth Factor

By JOAN MASSAGUÉ

Transforming growth factors (TGF) is a term originally used to designate hormonally active polypeptides that can induce the transformed phenotype when added to normal, nontransformed fibroblasts. TGFs were first discovered in culture fluids from retrovirally transformed cells[1] and later found in tumor-derived cells as well as in nonneoplastic sources. Full purification of the active components from culture fluids of retrovirally transformed rat fibroblast has shown that the transforming activity in these fluids is due to the simultaneous action of two otherwise unrelated factors

[1] J. E. De Larco and G. J. Todaro, *Proc. Natl. Acad. Sci. U.S.A.* **75**, 4001 (1978).

termed TGFα and TGFβ, respectively.[2,3] TGFα is a structural and functional analog of epidermal growth factor (EGF), the potent mitogen from submaxillary glands and possibly other endocrine sources. TGFβ is not related to EGF or any other known growth factor. It promotes anchorage-independent proliferation, the "transformed phenotype," in cultured rodent fibroblasts when they are simultaneously stimulated by mitogens including TGFα or EGF. TGFα and EGF are small polypeptides ($M_r \sim$ 6000) encoded by separate genes[4,5] but sharing about 40% of their primary amino acid structure.[3,6,7] Despite the extensive discrepancy in primary structure, TGFα and EGF bind with equivalent affinity to the same cellular receptor molecule in all the systems so far examined. The systems examined in detail include A431 human carcinoma cells which overexpress EGF receptors, NRK-49F rat fibroblasts which are sensitive to transformation by TGFs, as well as membranes isolated from these two cell lines and from human placenta[8,9] (J. Massagué, unpublished results). Chemically synthesized TGFα also has the same affinity for the receptor as EGF.[10] Based on these observations, it has been tentatively concluded that TGFα and EGF are two analogs equipotent *in vitro*. Indeed, TGFα has the same mitogenic potency as EGF, and EGF can substitute equipotently for TGFα in the induction of the transformed phenotype in the presence of homogenous TGFβ.[11]

This chapter describes methodology to characterize the interaction of TGFα with cellular and membrane receptors. Since TGFα and EGF interact with a common receptor type in all systems studies so far, the receptor assays included in this chapter are equally valid to characterize receptor interaction with EGF. Methodology to assay ^{125}I-labeled EGF binding to receptors in intact cells, isolated membranes, and solubilized receptor preparations has been recently reviewed in this series,[12] and is directly

[2] M. A. Anzano, A. B. Roberts, A. B. Smith, M. B. Sporn, and J. E. De Larco, *Proc. Natl. Acad. Sci. U.S.A.* **80**, 6264 (1983).

[3] J. Massagué, *J. Biol. Chem.* **258**, 13606 (1983).

[4] A. Gray, T. J. Dull, and A. Ullrich, *Nature (London)* **303**, 722 (1983).

[5] R. Derynck, A. B. Roberts, M. E. Winkler, E. Y. Chen, and D. V. Goeddel, *Cell* **38**, 287 (1984).

[6] H. Marquardt, M. W. Hunkapiller, L. E. Hood, D. R. Twardzik, J. E. De Larco, and G. J. Todaro, *Proc. Natl. Acad. Sci. U.S.A.* **80**, 4684 (1983).

[7] H. Marquardt, M. W. Hunkapiller, L. E. Hood, and G. J. Todaro, *Science* **223**, 1079 (1984).

[8] J. Massagué, *J. Biol. Chem.* **258**, 13614 (1983).

[9] J. Massagué, *J. Cell Biol.* **100**, 1508 (1985).

[10] J. P. Tam, H. Marquardt, D. F. Rosberger, T. W. Wong, and G. J. Todaro, *Nature (London)* **309**, 376 (1984).

[11] J. Massagué, *J. Biol. Chem.* **259**, 9756 (1984).

[12] G. Carpenter, this series, Vol. 109. p. 101.

applicable to study ^{125}I-labeled TGFα binding as well. Therefore, this chapter will be specifically concerned with methods to identify TGFα/EGF receptor proteins by affinity labeling.

Preparation of ^{125}I-Labeled TGFα

Homogeneous preparations of TGFα are obtained from Snyder–Theilen feline sarcoma virus-transformed rat embryo fibroblasts by the method of Marquardt and Todaro[13] with minor modifications as previously described (see [10], this volume). We label TGFα with ^{125}I using the gently oxidizing conditions of the solid-phase glucose oxidase: lactoperoxidase method (Enzymobeads, Bio-Rad). In this method, hydrogen peroxide slowly produced by the action of glucose oxidase is used to oxidize [125]I$^-$ to [125]I$_2$ which in turn reacts with tyrosine, and to a much lesser extent hystidine residues, in the protein. At the end of the reaction, ^{125}I-labeled TGFα is separated from unreacted ^{125}I by molecular filtration chromatography.

Materials

TGFα: 2.5 μg of lyophilized polypeptide, or 10 μl of 0.25 mg/ml TGFα, in 1 mM HCl

Reaction buffer: 0.3 M sodium phosphate, pH 7.4

Enzymobeads (Bio-Rad Laboratories): reconstituted in H_2O as per manufacturer's directions

Na[125]I (16 Ci of ^{125}I/mg of iodine, Amersham): 1 mCi/10 μl solution at neutral pH

D-Glucose: 20 mg/ml in sterile H_2O. Allow to racemize to β-D-glucose by incubation at room temperature overnight

Elution buffer: 0.1 M ammonium acetate containing 5 mg/ml of bovine serum albumin (BSA)

Sephadex G-50-80 (Pharmacia): a column of 40 × 0.8 cm equilibrated with elution buffer

Procedure

1. Mix the TGFα with 50 μl of reaction buffer in a 1.5-ml polypropylene microcentrifuge tube.
2. Add 25 μl of Enzymobeads and 1.5 mCi of Na[125]I.
3. Start the reaction by addition of 25 μl of D-glucose solution. Incubate 30 min at room temperature with occasional shaking to prevent settling of the beads.
4. Stop the reaction by addition of 0.2 ml of elution buffer followed by centrifugation at 12,000 g for 30 sec.

[13] H. Marquardt and G. J. Todaro, *J. Biol. Chem.* **257**, 5220 (1982).

5. Load supernatant on the column. Wash the Enzymobeads with a second 0.2-ml aliquot of elution buffer. After centrifugation, load the supernatant on the column.
6. Run the column by gravity at 0.5–0.6 ml/min. Collect 0.5-ml fractions.
7. Count the radioactivity in 2-μl aliquots from each fraction. Pool peak fractions of ^{125}I-labeled polypeptide. Measure the total volume of the pool and count radioactivity in one aliquot to determine the specific activity of the preparation.
8. Store the preparation in small aliquots at −20°.

Comments

[125]I-labeled TGFα elutes well after a minor peak of radiolabeled high-molecular-weight material which comes out with the void volume and probably represents radiolabeled BSA and/or aggregated TGFα. Sixty to 70% of the ^{125}I-labeled TGFα elutes in about 5.0 ml of peak fractions. The labeled material in this pool is 94–97% precipitable by 10% trichloroacetic acid. The specific activity is calculated assuming that the total area of the ^{125}I-labeled TGFα peak corresponds to 2.5 μg of TGFα. Preparations of ^{125}I-labeled TGFα obtained by this method have a specific activity of 150–220 Ci/g (0.6–0.8 mol of ^{125}I/mol of TGFα).[125]I-Labeled TGFα in this preparation copurifies with native TGFα when added to crude conditioned media extracts used as a source of TGFα.[8] ^{125}I-Labeled TGFα obtained by this method exhibits the same biological potency as native TGFα.[8]

Affinity Cross-Linking of TGFα/EGF Receptors

Photoreactive derivatives of ^{125}I-labeled EGF have been used to affinity label the EGF receptor from various sources.[14] In addition, EGF labeled with ^{125}I using chloramine-T as the oxidizing agent yields activated radiolabeled derivatives that become spontaneously attached to receptor upon photolysis of ligand–receptor complexes.[15] Although photoaffinity labeling has been a valuable approach for the identification of receptor proteins for polypeptide hormones, including EGF receptors, this technique has various drawbacks. First, the ligand must be chemically derivatized and subsequently repurified. Second, chemical derivatization may alter the receptor binding and other properties of the ligand. Third, exposure to intense ultraviolet light to photolyze ligand–receptor complexes may alter

[14] R. A. Hock, E. Nexo, and M. D. Hollenberg, *J. Biol. Chem.* **250,** 10737 (1980).
[15] P. G. Comens, R. L. Simmer, and J. B. Baker, *J. Biol. Chem.* **257,** 42 (1982).

the structural properties of the receptor system under study. An alternative methodology that obviates these problems is receptor affinity cross-linking, first introduced to label insulin receptors with ^{125}I-labeled insulin.[16,17] In this method, a bifunctional agent is added to preformed ^{125}I-labeled ligand–receptor complexes to achieve the covalent cross-linking between the components of these complexes. We have adapted this method to affinity label receptors for TGFα and EGF[8] using as the cross-linking agent disuccinimidyl suberate (DSS) which reacts divalently with free amino groups in proteins.[16]

Labeling of Receptors in Intact Cells

Materials

Binding buffer I: 128 mM NaCl, 5 mM KCl, 1.2 mM MgSO$_4$, 1.2 mM CaCl$_2$, 50 mM HEPES, pH 7.7, 2 mg/ml BSA

Binding buffer II: Dulbecco's modified Eagle's medium (Gibco), supplemented with 25 mM HEPES, pH 7.7, 2 mg/ml BSA

TGFα and ^{125}I-labeled TGFα: Appropriate dilutions prepared in binding buffer in polypropylene plastic test tubes

EGF and ^{125}I-labeled EGF: EGF is obtained as previously described[18] and appropriately diluted in binding buffer for the experiment. ^{125}I-Labeled EGF is labeled with ^{125}I using the solid-phase lactoperoxidase method exactly as described above for TGFα

Disuccinimidyl suberate (DSS, Pierce Chemical Co.): dissolved at 27 mM concentration in dimethyl sulfoxide immediately before use

Cell detachment buffer: 0.25 M sucrose, 10 mM Tris, 1 mM EDTA, pH 7.4, 0.3 mM phenylmethylsulfonyl fluoride (PMSF)

Cell solubilization buffer: 125 mM NaCl, 10 mM Tris, 1 mM EDTA, pH 7.0, 1% Triton X-100

Protease inhibition cocktail I: 1 mg/ml leupeptin, 1 mg/ml antipain, 5 mg/ml aprotinin, 10 mg/ml soybean trypsin inhibitor, 10 mg/ml benzamidine hydrochloride (all from Sigma) in H$_2$O

Protease inhibitor cocktail II: 1 mg/ml pepstatin, 30 mM PMSF (all from Sigma) in dimethyl sulfoxide

Electrophoresis sample buffer: 100 mM Tris, pH 6.8, 20% glycerol, 2% sodium dodecyl sulfate, 0.05% bromphenol blue, and 100 mM dithiothreitol

Cell monolayers: Near-confluent cell monolayers grown on 10-cm^2 tissue culture dishes are used. However, sparse cultures should be used in

[16] P. F. Pilch and M. P. Czech, *J. Biol. Chem.* **254**, 3375 (1979).
[17] J. Massagué and M. P. Czech, this series, Vol. 109, p. 179.
[18] C. R. Savage and S. Cohen, *J. Biol. Chem.* **247**, 7609 (1972).

studies involving cell lines with very high receptor numbers (for example, A431 human carcinoma cells) to prevent excessive depletion of the free ^{125}I-labeled ligand by binding to the cells. Studies involving cells that are weakly attached to the culture vessel should be performed with cell suspensions obtained by mechanical detachment (gentle scraping) or detachment by a 5- to 10-min incubation at 37° with 1 mM EDTA, 125 mM NaCl, 5 mM KCL, 5 mM glucose, 25 mM HEPES, pH 7.4, 2 mg/ml BSA

Procedure

1. Wash cell monolayers with binding buffer, and allow them to equilibrate with this buffer for 30 min at the temperature chosen for the experiment, normally 4°. Binding buffers I and II can be used interchangeably in experiments performed at 4°. However, only buffer II should be used in experiments at higher temperatures, for example 20 or 37°C (see comments below).

2. Aspirate buffer and add 1.5 ml ice-cold binding buffer to each well. Add ^{125}I-labeled TGFα (or ^{125}I-labeled EGF) in a small aliquot to achieve a final concentration in the 0.05–1 nM range. A 100-fold excess of native ligand is immediately added to duplicated wells to assess the specificity of the affinity-labeled cell components. Cultures are then incubated for 3.5–4 hr at 4°, on an oscillating platform set at about 120 cpm to achieve even distribution of the ligands throughout the experiment. Equilibrium binding is achieved with most cultured cell lines under these conditions.

3. Aspirate the medium and wash monolayers on ice three times with cold binding buffer.

4. Leave the dishes with 2.0 ml of binding buffer I lacking BSA. Add 10 μl of 27 mM DSS solution to each dish. To prevent precipitation of DSS in the ice-cold aqueous buffer, DDS is delivered with a micropipet directly inside the buffer solution, not on its surface, and the dish is swirled immediately after this addition. Incubate for 15 min at 4° with agitation.

5. Rinse the monolayers quickly with ice-cold detachment buffer. Then remove the cells by scraping them off with a Teflon scraper into 1.0 ml of this buffer. Rinse the dishes with 0.4 ml of the same buffer, centrifuge the cell suspensions for 2 min at 12,000 to obtain a cell pellet.

6. Remove the supernatant. Solubilize each cell pellet by resuspension with 60–100 μl of solubilization buffer supplemented with 10 μl/ml of protease inhibitor cocktails I and II. Incubate for 40 min at 4° with end-over-end mixing.

7. Remove insoluble material by centrifugation for 15 min at 12,000 g or equivalent. Mix the supernatant with 1 vol of electrophoresis sample buffer and heat at 100° for 2 min.

Comments

The protocol described above is designated to obtain affinity-labeled samples that will be subjected to SDS–PAGE analysis. Step 7 in this protocol may be modified to obtain samples of nondenatured, affinity-labeled receptor for other types of analyses. However, if rapid solubilization of the samples with sodium dodecyl sulfate is not performed, over-cross-linking of TGFα/EGF receptors can occur due to DSS that has partitioned into hydrophobic domains of cells and nonionic detergent micelles and has not been quenched by Tris buffer. To avoid over-cross-linking of the sample in these situations preliminary titration assays must be performed to determine the optimal concentration of DSS for each experiment. Alternatively, the hydrosoluble analog bis(sulfosuccinimidyl) suberate can be used in place of DSS.

If affinity labeling of TGFα/EGF receptors is performed with cell suspensions, polypropylene test tubes are recommended for all incubations. After detachment, cells are washed with binding buffer and adjusted to $2-10 \times 10^5$ cells/ml. Washes during the experiment are done by centrifugation at 2000 g for 5 min in the cold.

Using this TGFα/EGF receptor affinity-labeling methodology approximately 3–6% of the receptor-bound ligand becomes covalently cross-linked with the receptor. The criterium to assess the specificity of labeling of a cellular component using this protocol is the inhibition of the labeling by an excess of native polypeptide added during incubation of cells with ^{125}I-labeled TGFα or ^{125}I-labeled EGF.

The conditions described above can be manipulated to accommodate incubations with ^{125}I-labeled TGFα or ^{125}I-labeled EGF at 37° in experiments to characterize the internalization and intracellular degradation of these ligands and their receptors. However, the actual cross-linking reaction should be performed in the cold and in the presence of binding buffer I whenever possible to avoid cross-linking of the receptor with other cellular components.

Labeling of Receptors in Isolated Membrane Preparations

Materials

The following materials described for the affinity labeling of receptors in intact cells are used:

Binding buffer I
TGFα and ^{125}I-labeled TGFα and/or EGF and ^{125}I-labeled EGF
Disuccinimidyl suberate in dimethyl sulfoxide
Electrophoresis sample buffer

In addition:

Tris/EDTA buffer: 10 mM Tris, pH 7.0, 1 mM EDTA
Solubilization buffer: 1% Triton X-100, 10 mM Tris, pH 7.0, 1 mM
EDTA, 0.3 mM PMSF
Membranes: membrane preparations from primary tissues or culture
cells are suspended in binding buffer at 0.2–1.0 mg of membrane
protein/ml

Procedure

1. Add ^{125}I-labeled TGFα or ^{125}I-labeled EGF at 0.3–1 nM final concentration to 0.3 ml of membrane suspension in 1.5-ml microcentrifuge tubes. Add a 100-fold excess of native TGFα (or EGF) to assess the specificity of affinity-labeled receptor components in duplicate tubes.

2. Incubate for 40 min at 22° or 3.5 hr at 4° to reach equilibrim binding. Incubate with end-over-end mixing or frequent agitation to prevent settling of the membranes.

3. Sediment the membranes by centrifugation at 12,000 g for 10 min in the cold. Remove the supernatant. Wash the membrane pellet by resuspension in 1.4 ml of ice-cold binding buffer and centrifugate. Repeat washes twice.

4. Resuspend the membrane pellet in 0.3 ml of binding buffer lacking BSA. Add 3 μl of 27 nM DSS solution in dimethyl sulfoxide.

5. Incubate with agitation for 15 min at 4°. Quench unreacted DSS by addition of 1.0 ml of Tris/EDTA buffer. Sediment the membranes by centrifugation at 12,000 g for 10 min in the cold. Aspirate the supernatant and wash the membranes once more with Tris/EDTA buffer.

6. Solubilize the membrane pellet by incubation for 40 min at 4° with 60–100 μl of solubilization buffer. Sediment the unsoluble material by centrifugation at 12,000 g for 10 min.

7. Prepare the affinity-labeled, soluble extract for SDS–PAGE by heating at 100° for 2 min with 1 vol of electrophoresis sample buffer.

Comments

The step of membrane solubilization with Triton X-100 is included to optimize the visualization of specifically labeled receptor species, and to minimize the appearance of nonspecifically labeled material. If cells or membrane preparations rich in TGFα/EGF receptors (A431 cells, human placenta membranes, etc) are used this step can be omitted because the ratio of specifically labeled receptor components to nonspecifically labeled components is high.

Certain types of membrane preparations tend to aggregate during incubation in binding buffer, forming large clumps that may cause disturbances in the cross-linking reaction. Homogenization of the membrane preparation by repeated passage through a narrow-gauge syringe needle before the incubation with ^{125}I-labeled TGFβ and the addition of DSS is recommended in these cases.

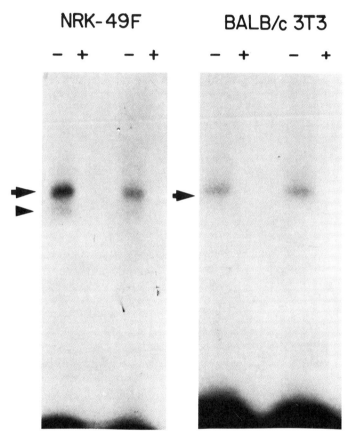

FIG. 1. Identification of TGFα/EGF receptor proteins in NRK-49F and BALB/c 3T3 cells by affinity labeling. Near-confluent cell monolayers in 10-cm² culture dishes were incubated for 3.5 hr at 4° in the presence of 1 nM ^{125}I-labeled TGFα or 1 nM ^{125}I-labeled EGF alone (−) or with 300 nM unlabeled EGF (+). Cells were cross-linked with cell-bound ligands by treatment with 0.15 mM disuccinimidyl suberate for 15 min at 4°. The affinity-labeled cells were scraped off the dishes and solubilized in the presence of Triton X-100. The Triton-soluble extracts from NRK-49F cells and BALB/c 3T3 cells labeled with ^{125}I-labeled TGFα (left lanes) or ^{125}I-labeled EGF (right lanes) were subjected to SDS–PAGE on 6% polyacrylamide gels. Shown are the autoradiograms (5 days) of the fixed, dried gels. *Arrow*: the 170-kDa labeled species; *arrowhead*: the 140-kDa labeled species.

Analysis of the Data

Figure 1 shows an autoradiogram from a sodium dodecyl sulfate – polyacrylamide electrophoresis slab gel on which samples of affinity-labeled NRK-49F rat fibroblasts and BALB/c 3T3 mouse fibroblasts were run. NRK-49F and BALB/c 3T3 cells contain relatively low numbers $(2 \times 10^4$ sites/cell) of TGFα/EGF receptors[9] and for this reason optimization of the labeling and receptor solubilization using the conditions of the

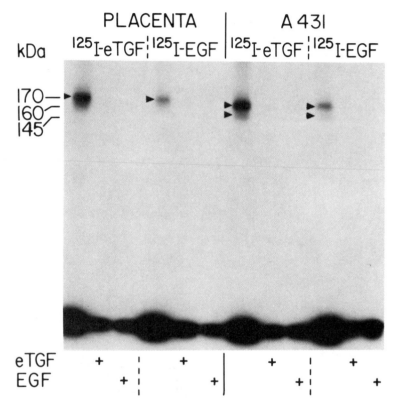

FIG. 2. Identification of TGFα/EGF receptor proteins in isolated membrane preparations. Human placenta membranes (120 μg of membrane protein) and A431 human carcinoma membranes (80 μg of membrane protein) were incubated in 200 μl of binding buffer I containing 1 nM ^{125}I-labeled TGFα (^{125}I-labeled eTGF) or 1 nM ^{125}I-labeled EGF with or without an excess (0.5 μM) of TGFα (eTGF) or EGF as indicated. After washing out the free ligands at 4°, membranes were resuspended in 300 μl of ice-cold binding medium and cross-linked to membrane-bound ligands by the addition of 0.27 mM disuccinimidyl suberate at 4°. Incubations were arrested 15 min later and affinity-labeled membranes were subjected to SDS – PAGE on an 8% polyacrylamide gel. A 3-day autoradiogram from this gel is shown. (From Ref. 8.)

above-described protocol were needed to eliminate the background of nonspecifically labeled bands. Cells affinity labeled by cross-linking with either ^{125}I-labeled TGFα or ^{125}I-labeled EGF exhibited one major labeled species of 170 kDA. The labeling of this species was somewhat more intense with ^{125}I-labeled TGFα than with ^{125}I-labeled EGF, even though the same number of counts per minute were loaded on both lanes. This difference in labeling is attributable to the fact that TGFα has three primary amino groups (amino terminus plus two lysine residues) available for reaction with DSS, whereas EGF, which lacks lysine, can react with DSS via the amino terminus only. The labeling of the 170-kDa receptor species is specific as it could be inhibited by an excess of native ligand added during incubation of cells with radioligand (Fig. 1). In addition to the 170-kDa labeled species, minor amounts of specifically labeled material of about 140 kDa were also detected in the autoradiograms. This labeled species most likely corresponds to TGFα/EGF receptors that have undergone limited proteolysis during scraping and solubilization of the cells.[19]

Figure 2 illustrates the affinity labeling of TGFα/EGF receptors in membrane preparations isolated from human placenta and A431 human carcinoma cells. Like in the case of intact NRK-49F and BALB/c 3T3 cells, the components labeled by ^{125}I-labeled TGFα and ^{125}I-labeled EGF exhibited the same apparent molecular mass, 170,000 in placenta membranes and 160,000 in A431 membranes. Furthermore, the labeling of these species could be completely inhibited by an excess of either TGFα or EGF, regardless of the radioligand or the membrane type used (Fig. 2). These results support the conclusion that TGFα and EGF interact with the same binding site on the same receptor molecule.

[19] P. S. Linsley and C. F. Fox, *J. Supramol. Struct.* **14**, 461 (1980).

[14] Purification of Type-β Transforming Growth Factor from Human Platelets

By RICHARD K. ASSOIAN

Type-β transforming growth factor (TGFβ)[1] is a bifunctional regulator of cell growth[2,3] that can be isolated, in low amounts (μg TGFβ/kg), from

[1] A. B. Roberts, C. A. Frolik, M. A. Anzano, and M. B. Sporn, *Fed. Proc., Fed. Am. Soc. Exp. Biol.* **42**, 2621 (1983).

many tissues, cells, and cell lines.[4-6] Human platelets contain relatively high amounts of TGFβ (mg TGFβ/kg),[6] making it the preferred source of the growth factor to date. Structural studies with TGFβ purified from several sources have yielded identical results and shown that the native growth factor (M_r 25,000) is composed of two apparently identical subunits (M_r 12,500) held together by disulfide bonds.[6-9] This chapter describes in detail our current procedure for purification of TGFβ from human platelets. The methodology yields the pure growth factor in $1\frac{1}{2}$ weeks with few days requiring more than 1–2 hr of labor.

Reagents and Solutions

Human Platelets. The amount and freshness of human platelets are the two major determinants of yield. Freshly drawn platelets yield about 2–5 times more TGFβ than do clinically outdated platelets (1–7 days clinically outdated), but even outdated platelets, when processed in batches of 25 units or more, will provide enough pure TGFβ for many experiments in cell culture. (Most experiments in cell culture require only 0.1–2 ng/ml TGFβ.) Less starting material results in proportionately greater losses of growth factor. In our hands, the following yields are typical: 25 units of outdated platelets yields 10 μg TGFβ; 100 units of outdated platelets yields 100 μg TGFβ; 100 units of fresh platelets yields 200 μg TGFβ. Under all circumstances, platelets should be washed and frozen upon receipt and stored at $-70°$ until quantities sufficient for extraction have been accumulated.

Platelet Washing Solution.[10] A solution contains 9 vol of Tris buffer (17 mM Tris base, 0.15 M NaCl, 0.1% glucose brought to pH 7.5 with

[2] R. F. Tucker, G. D. Shipley, H. L. Moses, and R. W. Holley, *Science* **226**, 705 (1984).

[3] A. B. Roberts, M. A. Anzano, L. M. Wakefield, N. S. Roche, D. F. Stern, and M. B. Sporn, *Proc. Natl. Acad. Sci. U.S.A.* **82**, 119 (1985).

[4] A. B. Roberts, M. A. Anzano, C. A. Frolik, and M. B. Sporn, *Cold Spring Harbor Conf. Cell Proliferation* **9**, 319 (1982).

[5] B. J. Childs, J. A. Proper, R. F. Tucker, and H. L. Moses, *Proc. Natl. Acad. Sci. U.S.A.* **79**, 5312 (1982).

[6] R. K. Assoian, A. Komoriya, C. A. Meyers, D. M. Miller, and M. B. Sporn, *J. Biol. Chem.* **258**, 7155 (1983).

[7] C. A. Frolik, L. L. Dart, C. A. Meyers, D. M. Smith, and M. B. Sporn, *Proc. Natl. Acad. Sci. U.S.A.* **80**, 3676 (1983).

[8] A. B. Roberts, M. A. Anzano, C. A. Meyers, J. Wideman, R. Blacher, Y. C. E. Pan, S. Stein, S. R. Lehrman, J. M. Smith, L. C. Lamb, and M. B. Sporn, *Biochemistry* **22**, 5692 (1983).

[9] J. Massagué, *J. Biol. Chem.* **259**, 9756 (1984).

[10] H. N. Antoniades, C. D. Scher, and C. D. Stiles, *Proc. Natl. Acad. Sci. U.S.A.* **76**, 1809 (1979).

HCl) and 1 vol of citrate buffer (38 mM citric acid, 88 mM sodium citrate, 2.2% glucose).

Extraction Solution (Acid/Ethanol). 375 ml ethanol (95%), 7.5 ml concentrated HCl, 33 mg phenylmethylsulfonyl fluoride.

Eluant for Gel Filtration Column 2 (1 M Acetic Acid Containing 8 M Urea). To preclude carbamylation of TGFβ during gel filtration in this solution, the urea should be of ultrapure grade and always maintained at acid pH. We use the following protocol: a solution of 8 M urea is prepared in 1 M acetic acid. The acetic acid concentration is then readjusted to 1 M by the addition of glacial acetic acid. The final solution is degassed under vacuum.

BioGel P-60 Equilibrated in 1 M Acetic Acid, 8 M Urea. Unusual care must be taken to prepare a column that will yield good flow rates for many runs. We swell two grades of BioGel P-60 (50–100 and 100–200 mesh) separately overnight in water, degas them under vacuum, and then wash each gel several times with a 10-fold excess (v/v) of column solution. About 50% of the original slurry volumes should be discarded during defining. The first tenth of the column is filled and packed with the 50–100 mesh gel. The remaining column volume is filled with 100–200 mesh gel by carefully layering it on top of the packed 50–100 mesh BioGel.

Assay Methods

Originally we monitored purification of TGFβ using growth of NRK fibroblasts (clone 49F) in soft agar as the assay. The procedure, described in Refs. 6 and 11 and elsewhere in this volume (Rizzino [32]), is extremely sensitive, detecting <0.05 ng/ml TGFβ. However, it is a laborious assay and requires a 7-day incubation period. Currently, we are monitoring purification by SDS–polyacrylamide gel electrophoresis and using the soft agar assay solely to confirm the biological activity of the final preparation.

SDS gels (prepared as described by Laemmli[12]) contain 12.5% acrylamide and are poured into slabs 0.75 mm thick. Samples of TGFβ (usually aliquots from column fractions; 1 M in acetic acid) are brought to dryness in a Speed Vac (Savant) or vacuum desiccator. The residue is redissolved in SDS sample buffer and electrophoresed at 120 V. (Our sample buffer for the SDS gels is made dense with 8 M urea rather than sucrose or glycerol. Use of urea improves recovery of proteins dried onto the walls of tubes.) The gels are fixed overnight at room temperature in a solution containing

[11] A. B. Roberts, L. C. Lamb, D. L. Newton, and M. B. Sporn, *Proc. Natl. Acad. Sci. U.S.A.* **77**, 3494 (1980).

[12] U. K. Laemmli, *Nature (London)* **227**, 680 (1970).

formaldehyde: 125 ml water, 55 ml 95% ethanol, and 30 ml reagent grade (35% solution) formaldehyde. The fixed gel is stained with silver using the Bio-Rad kit and methodology. Detection of TGFβ by this method is somewhat less sensitive (requiring 25–50 ng TGFβ for easy detection of bands) than the bioassay, but it reduces purification time by 2–3 weeks and permits more accurate pooling by allowing detection and analysis of contaminants as well as TGFβ.

Purification

Step 1: Washing Platelets. Human platelets (supplied as a suspension in plasma) are collected by centrifugation (3200 g, 30 min at 23°) and washed twice by gently resuspending the pellet into a small volume of washing buffer, diluting the suspension with washing buffer (500 ml/25 units platelets), and centrifuging the cells as described above. After the final wash, the supernatant should be clear and colorless. Washed platelets (about 1 g/unit) are either immediately extracted (see below) or frozen. We freeze platelets by suspending them in a small volume of washing buffer and adding the suspension dropwise into liquid nitrogen. The nitrogen is decanted and the platelet "balls" are transferred to a disposable beaker for storage at −70°. Freezing the platelets in this manner permits instantaneous homogenization of frozen samples.

Step 2: Extraction of Platelets with Acid/Ethanol. Freshly washed or frozen platelets are added to a solution of acid/ethanol (4 ml/g platelets) and extracted in a blender (1–2 min) until homogeneous. The volume of the extract is adjusted to 6 ml/g platelets by addition of water. After an overnight incubation at 4°, centrifugation is used to remove precipitated proteins. The pellet is discarded, and the supernatant is brought to pH 3 (as determined on the pH meter) by the careful addition of concentrated NH$_4$OH. Crude TGFβ is prepared by adding 2 vol of ethanol (95%) and 4 vol of freshly opened ethyl ether to the pH 3 solution; a white, flocculant precipitate should appear quickly. Precipitation of TGFβ activity is complete after overnight incubation of the mixture at 4°. Centrifugation can be used to collect the precipitate (about 4 mg/g washed platelets), but if large amounts of platelets have been extracted ($>$ 100 units), the volume at this stage can be difficult to handle by centrifugation. Vacuum filtration through two layers of Whatman #1 paper is an alternative to centrifugation for collecting large amounts of crude TGFβ. Regardless of the method of collection, crude TGFβ is suspended in 1 M acetic acid (0.5 ml/g platelets) and incubated at 4° overnight with stirring. This volume of acetic acid does not completely solubilize the precipitate, but it does dissolve about 90% of TGFβ activity. Centrifugation is used to clarify the solution, and the pellet is discarded. The supernatant can be dialyzed against 1 M acetic acid using a low-molecular-weight cutoff membrane or directly applied to the first gel filtration column (see below).

Step 3: Gel Filtrations. The following procedure describes gel filtration conditions suitable for the total purification of TGFβ from 25–50 units of platelets. These same conditions can be used for extracts from 50–100 units of platelets, but gel filtration alone will no longer yield homogeneous TGFβ, and further purification of the peptide by HPLC will be necessary. We routinely use HPLC to desalt the final preparation of TGFβ; the additional purification can be performed while desalting.

Column 1: Bio-Gel P-60 in 1 M acetic acid. The solubilized platelet extract (10 ml/25 units platelets) is gel filtered at 20 ml/hr on a column (4.4 × 115 cm) of BioGel P-60 equilibrated in 1 M acetic acid. Fractions containing 5 ml are collected. A typical elution profile for absorbance at 280 nm is shown in Fig. 1A. (There will also be a very large peak of absorbance at the internal column volume if the extract has not been dialyzed.) When aliquots (usually 25–50 μl) of selected column fractions are analyzed by SDS gel electrophoresis and silver staining, a profile similar to the inset in Fig. 1A is obtained. For the column run shown in Fig. 1, the minor band (M_r 25,000) in fractions 195–205 corresponds to TGFβ. Under the described column conditions, TGFβ consistently elutes with an aberrantly low molecular weight, as determined by calibration of the column, and usually coelutes with the third or fourth peak of absorbance at 280 nm (refer to Fig. 1). Thus, an overall absorbance profile at 280 nm can be used to identify, to a first approximation, the column fractions likely to contain TGFβ. Detailed analysis of those fractions by SDS gels will permit an accurate determination of TGFβ elution position. Fractions containing the peak of TGFβ should be pooled and freeze dried. We freeze dry the pool in portions so that a single test tube (15 × 100, polypropylene) will ultimately contain all of the TGFβ. Yield of total protein at this stage of purification is about 5 mg/25 units of platelets. The TGFβ is about 1–10% pure, and all of the contaminants are of smaller molecular weight (refer to Fig. 1, inset, fractions 105–205). The residue is carefully dissolved in 0.5 ml of 1 M acetic acid containing 8 M urea (see "Reagents and Solutions" for details of preparation) for further purification.

Column 2: Bio-Gel P-60 in 1 M acetic acid, 8 M urea. To remove the low-molecular-weight contaminants from the preparation of TGFβ, the partially purified growth factor is gel filtered (at 3 ml/hr) on a column (1.6 × 85 cm) of BioGel P-60 equilibrated in 1 M acetic acid, 8 M urea (see "Solutions and Reagents" for details of preparation). Fractions containing 0.5 ml are collected. Although there is too little TGFβ to detect by absorbance at 280 nm, analysis of selected column fractions (10-μl aliquots) by

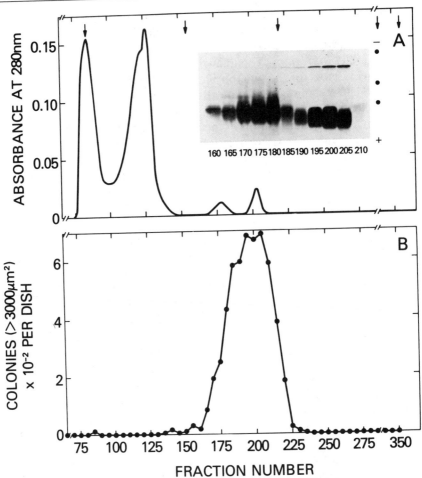

FIG. 1. Gel filtration of the platelet extract. An extract of platelets, prepared from 25 units, was purified on BioGel P-60 which had been equilibrated in 1 M acetic acid. Panel A shows the absorbance profile at 280 nm (solid line), and the elution positions of molecular weight standards [the vertical arrows indicate, from left to right, the positions taken by BSA (V_o), chymotrypsinogen A (M_r 27,000), lysozyme (M_r 14,000), insulin (M_r 6000), and [^{3}H]glycine (V_i), respectively]. The inset to Panel A shows a silver-stained SDS gel obtained after electrophoresis of aliquots (50 μl) from column fractions which were coincident with the region of TGFβ bioactivity. The closed circles at the far right of the inset indicate, from top to bottom, the positions taken by carbonate dehydratase (M_r 30,000), soybean trypsin inhibitor (M_r 21,000), and cytochrome c (M_r 13,500), respectively. The bioassay used for detection of TGFβ was colony formation of NRK cells grown in medium containing agar, EGF, and 10% calf serum[6,11] (see Panel B).

SDS gel electrophoresis and silver staining shows that gel filtration in the presence of urea prevents retarded elution of TGFβ and results in separation of the growth factor from the remaining contaminants (Fig. 2). The fractions containing the peak of TGFβ are pooled and desalted (see below). At this stage of purification, TGFβ should be 75–95% pure, and free of other known growth factors in platelets.

Note that the urea in these samples traps acetic acid while they are being brought to dryness in the Speed Vac. The trapped acetic acid interferes with accurate SDS gel electrophoresis and should be removed. We routinely dissolve the dried samples in 0.2 ml of water and redry them to remove the final traces of acetic acid. This cycle is repeated three times or until the pH of the redissolved samples is 5–6 (determined with pH paper). The dried sample is prepared for electrophoresis by dissolving it in 25 μl of SDS sample buffer lacking urea.

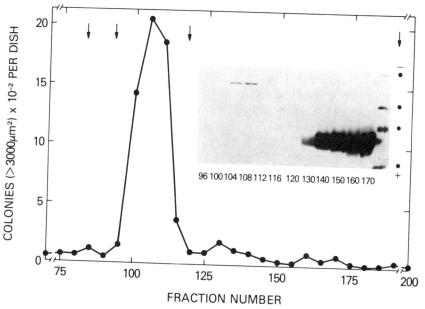

FIG. 2. Gel filtration of partially purified TGFβ. TGFβ (2–5 mg protein, pooled from the BioGel P-60 column) was gel filtered on BioGel P-60 equilibrated in 1 *M* acetic cid, 8 *M* urea. The solid lines show the elution of TGFβ as determined by bioassay (see the legend to Fig. 1 for details), and the vertical arrows show, from left to right, the elution positions of BSA, chymotrypsinogen A, lysozyme, and insulin, respectively. The inset to the figure shows a silver-stained SDS gel obtained after electrophoresis of aliquots (10 μl) from the indicated column fractions. The lane at the far right of the gel shows from top to bottom, the positions taken by carbonate dehydratase, soybean trypsin inhibitor, cytochrome *c*, and insulin, respectively.

Step 4: Desalting of Purified TGFβ. Urea can be removed from the solution of TGFβ either by dialysis[6] or HPLC.[13] Dialysis against 4 mM HCl can be performed in a microdialysis apparatus with a low-molecular-weight cut-off membrane (Bethesda Research Laboratories). The HCl solution is intrinsically sterile, so the dialyzed sample can be used in cell culture without filtration. Addition of albumin (0.1–1 mg) to the TGFβ prior to dialysis will minimize absorption of the growth factor to the membrane. We also minimize loss of TGFβ by dialyzing the sample in portions using a single well of the microdialysis apparatus.

Although dialysis is a simple way to remove urea from TGFβ, we find that recovery of the growth factor is better when HPLC is used to desalt, particularly when it is not desirable to add excess carrier to the TGFβ solution. An HPLC step also permits concomitant removal of any remaining protein contaminants; the purity of TGFβ after HPLC is routinely ≥95%.

For HPLC desalting, we use a procedure similar to that described by Tucker *et al.*[13] Peak fractions of TGFβ from the P-60/urea column are diluted with 1 vol of water (HPLC grade), and loaded (at 0.8 ml/min) onto an analytical C$_{18}$ column (Synchropak) equilibrated in 0.1% trifluoroacetic acid (Pierce, Sequanal Grade). Under these conditions, TGFβ binds to the support whereas urea and acetic acid do not (Fig. 3). The column is washed with equilibration solvent until the effluent no longer absorbs light at 210 nm. A gradient of acetonitrile (HPLC grade) in 0.1% trifluoroacetic acid (20–50% over 2 hr. 0.8 ml/min) is used to elute TGFβ and the remaining contaminants. Fractions containing 0.8 ml are collected. The major peak of absorbance at either 210 or 280 nm should elute at 35–40% acetonitrile and corresponds to TGFβ.

Large-Scale Purification of TGFβ from Human Platelets

The scale of the purification scheme described above can be readily increased to yield amounts of TGFβ sufficient for structural studies. A procedure for purification of TGFβ from 250 units of human platelets follows. Platelets are extracted with 1 liter of acid–ethanol and proportionately increased volumes of other solvents. The final ethanol–ether precipitate is collected by vacuum filtration and dissolved in 50 ml of 1 M acetic acid. Centrifugation is used to clarify the solution, and the resulting supernatant is gel filtered (125 ml/hr) on a column (10 × 100 cm) of BioGel

[13] R. F. Tucker, E. L. Branum, G. D. Shipley, R. J. Ryan, and H. L. Moses, *Proc. Natl. Acad. Sci. U.S.A.* **81**, 6757 (1984).

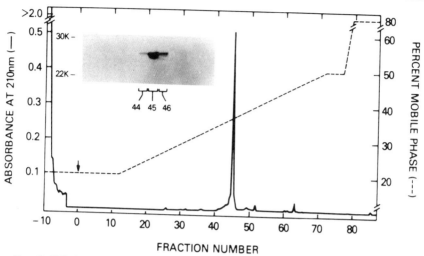

FIG. 3. HPLC desalting of purified TGFβ. Peak fractions of TGFβ from the P-60/urea column were pooled and desalted by reversed-phase HPLC as described in the text. The large amount of absorbance at 210 nm prior to fraction number zero represents urea and acetic acid, which are not retained by the C_{18} column. The vertical arrow at fraction number zero shows where the programmed elution of TGFβ began. The acetonitrile gradient used to elute TGFβ is represented by the dashed line. The inset to the figure shows the pattern obtained after silver staining of an SDS–polyacrylamide gel loaded with aliquots from fractions 44, 45, and 46 (the peak of absorbance at 210 nm). Molecular weight standards for the SDS–polyacrylamide gel are shown left of the inset.

P-60 equilibrated in 1 M acetic acid. Fractions containing 40 ml are collected. The absorbance profile for the column resembles that described above; comparison of the profiles should identify the region containing TGFβ. Aliquots (0.15 ml) of column fractions in the TGFβ region are used for SDS–polyacrylamide gel electrophoresis and detailed analysis of TGFβ elution position. Peak fractions, as determined by the gel, are pooled and freeze dried. The residue is dissolved in 1 M acetic acid, 8 M urea (20 ml) and applied to a column (5 × 90 cm) of BioGel P-60 equilibrated in the sample solvent. Gel filtration proceeds at 25 ml/hr, and 5-ml fractions are collected. TGFβ is localized by analyzing aliquots (25 µl) of selected column fractions on SDS gels; fractions containing TGFβ (now about 50% pure) are pooled, diluted with 1 vol of water (HPLC grade), and applied (0.8 ml/min) to an HPLC column (semipreparative C_{18}, Synchropak). Other HPLC conditions are as described above. Yield of pure TGFβ is 0.5–0.75 mg.

Other Procedures for Purification of TGFβ from Platelets

Massagué[14] has described a procedure for the purification of TGFβ from 40–50 units of platelets using acid/ethanol extraction, gel filtration on BioGel P-60 in 1 M acetic acid, and molecular exclusion HPLC of the P-60 pool. The extraction and gel filtration were performed similarly to the procedure described above. For molecular filtration HPLC, pooled TGFβ from the BioGel column was freeze dried, redissolved in a small volume of 25% acetonitrile, 1 M acetic acid in water, and chromatographed (at 0.5 ml/min) over two I-125 columns (Waters Associates) which had been set up in series and equilibrated with 25% acetonitrile, 1 M acetic acid. The same solvent system was used as eluant. The procedure yielded homogeneous TGFβ as determined by SDS gel elctrophoresis and N-terminal sequence analysis.

Quantitation of Purified TGFβ

TGFβ can be quantitated by amino acid analysis after hydrolysis of a sample in constant boiling HCl (Pierce) under standard conditions. Modern analyzers are sufficiently sensitive to permit chemical quantitation of TGFβ without sacrificing a large amount of material. Less than 0.5 μg of TGFβ is usually sufficient to yield accurate quantitation with these analyzers, but care in handling and preparation of glassware and solvents is essential to obtain accurate values for the amino acids (particularly glycine and histidine). The amino acid composition of human TGFβ as determined by sequence analysis of the purified protein and cDNA clones[15] is as follows (numbers in parentheses indicate residues of amino acid per mole of the native protein): Asp (20), Thr (6), Ser (16), Glx (18), Pro (18), Gly (10), Ala (14), half-Cys (18), Val (14), Met (2), Ile (10), Leu (20), Tyr (14), Phe (6), His (6), Lys (16), Arg (8), and Trp (6).

If an amino acid analysis cannot be used as the method of quantitation, purified TGFβ can be exhaustively dried and weighed,[13] or quantitated in a bioassay by comparison to dose–response curves from well-characterized systems. We routinely use growth of NRK fibroblasts (clone 49F) in soft agar as a bioassay for TGFβ. When assayed in the presence of EGF (0.5–1 nM) and as using TGFβ quantitated by amino acid analysis, we find that 10–20 pM TGFβ fully stimulates colony formation (colonies >60 μm in diameter), 4 pM TGFβ half-maximally stimulates colony formation, and <1 pM TGFβ has very little colony-stimulating activity.

[14] J. Massagué and B. Like, J. Biol. Chem. **260**, 2636 (1985).
[15] R. Derynck, J. A. Jarrett, E. Y. Chen, D. H. Eaton, J. R. Bell, R. K. Assoian, A. B. Roberts, M. B. Sporn, and D. V. Goeddel, Nature (London) **316**, 701 (1985).

Storage of Purified TGFβ

Purified TGFβ can be stored frozen in either 4 mM HCl (preferably with albumin carrier) or the solution from which it elutes during HPLC. When stored in the HPLC solvent at $-70°$ and at a concentration ≥ 0.2 mg/ml, purified TGFβ, even in the absence of carrier, is stable and can be quantitatively recovered for at least 6 months. We routinely store our concentrated stocks of TGFβ in the HPLC solvent and prepare working solutions for cell culture by dilution of the stock into 4 mM HCl containing 1 mg/ml BSA (0.5 μg/ml TGFβ final concentration). Working solutions can be stored at $-20°$ for at least 4 weeks with little loss of activity.

Acknowledgment

I thank Michael Sporn for support throughout the development of these procedures and Brigitte Wittmann-Liebold for helpful discussions concerning gel filtration in the presence of urea. These procedures were developed at the Laboratory of Chemoprevention, National Cancer Institute, National Institutes of Health.

[15] Isolation of the BSC-1 Monkey Kidney Cell Growth Inhibitor

By Robert W. Holley, Julia H. Baldwin, and Sybil Greenfield

The growth inhibitor is isolated from medium that has been conditioned by exposure to high-density cultures of BSC-1 (African Green monkey kidney) cells.[1-3] The BSC-1 cells are grown either in 15-cm tissue culture dishes or in plastic roller bottles, depending on the volume of conditioned medium that is to be processed. Although the yield of inhibitor is lower per unit volume of conditioned medium, the use of roller bottles is advantageous if volumes larger than a few liters are to be processed. The growth inhibitor is isolated from the conditioned medium by ultrafiltration, and is purified by chromatography on BioGel P-60 in 1 M acetic acid, followed by high-performance liquid chromatography.

[1] R. W. Holley, P. Bohlen, R. Fava, J. H. Baldwin, G. Kleeman, and R. Armour, *Proc. Natl. Acad. Sci. U.S.A.* **77**, 5989 (1980).

[2] R. W. Holley, R. Armour, J. H. Baldwin, and S. Greenfield, *Cell Biol. Int. Rep.* **7**, 525 (1983).

[3] R. W. Holley, J. H. Baldwin, S. Greenfield, and R. Armour, *Ciba Found. Symp.* **116**, 241 (1985).

Cell Cultures. Stock cultures are initiated from frozen BSC-1 cells at approximately 2-month intervals. The cells are grown in Dulbecco–Vogt's modified Eagle's medium (DMEM) with 0.45% glucose[4] and 10% calf serum at 37° in a CO_2 incubator. Cells are detached from the culture dishes with 0.025% trypsin in calcium- and magnesium-free Tris/saline buffer[5] (NaCl, 0.8%; KCl, 0.038%; Na_2HPO_4, 0.01%; glucose, 0.1%; Sigma 7–9, 0.3%; adjusted to pH 7.4 with 1 N HCl) with 0.5 mM Na_2EDTA.

High-density BSC-1 cell cultures, used for preparation of conditioned medium, are grown in either 15-cm Falcon tissue culture plates or in 490-cm² Corning plastic roller bottles in DMEM with 1% calf serum. The medium (25 ml for the tissue culture plates or 125 ml for the roller bottles) is changed at least once a week until the cells reach a high density, approximately 1.5×10^5 cells/cm². The cultures are maintained in DMEM with 0.2% calf serum for cells in 15-cm tissue culture plates, or in 1% serum for cells in roller bottles, with a medium change once a week.

Preparation of Conditioned Medium. If 15-cm tissue culture plates are used, the medium with 0.2% serum is aspirated off the plates and is discarded, and the plates are filled with 100 ml/plate of DMEM without serum. The tissue culture plates are returned to the CO_2 incubator for 24 hr.

If roller bottles are used, the medium with 1% serum is poured from the bottles and is discarded, and the bottles are filled with 1.33 liters per bottle of DMEM without serum. It is important for maintenance of the cultures in the roller bottles to have the pH of the medium between 7.2 and 7.4. The roller bottles are left standing upright, without rolling, with caps loose, in a CO_2 incubator at 37° for 24 hr. The conditioned medium is then removed for isolation of the inhibitor.

From tissue culture plates, the medium is removed by aspiration into a collecting flask. From roller bottles, the medium is simply poured from the bottles into the collecting flask. As soon as the conditioned medium is collected, it is acidified slightly by the addition of 1% (v/v) of 1 M acetic acid. The BSC-1 cell cultures are returned to DMEM, with either 0.2% serum (25 ml/15-cm tissue culture plate) or 1% serum (125 ml/roller bottle), for at least 1 day before the cultures are used again to prepare conditioned medium. Conditioned medium is prepared repeatedly from the same cultures for approximately 2 months.

Concentration of the Growth Inhibitor by Ultrafiltration. Serum-free medium, conditioned by exposure to cells for 1 day, as described above, is concentrated 100- to 200-fold by ultrafiltration at 4°. For 4–12 liters of

---

[4] H. J. Morton, *In Vitro* **6**, 89 (1970).
[5] J. D. Smith, G. Freeman, M. Vogt, and R. Dulbecco, *Virology* **12**, 185 (1960).

conditioned medium it is convenient to use a large (2.5 liter) Amicon unit with a 15-cm YM10 ultrafiltration membrane. After concentration, without stirring, to approximately 50 ml, the concentrate is diluted with 200 ml of 0.01 M acetic acid, and the solution is reconcentrated to 50 ml. The final 50 ml of 100- to 200-fold concentrate is removed and set aside; it contains most of the protein present in the conditioned medium but relatively little of the inhibitor. Most of the growth inhibitor is adsorbed to the YM10 membrane and is recovered by washing the membrane with three 10-ml portions of 1 M acetic acid, with agitation to remove attached protein; the first two washes are left on the membrane approximately 3 min and the last 30 min.

The inhibitor obtained from 20–24 liters of conditioned medium is combined (total of 60 ml of 1 M acetic acid solution; washes from two 15-cm YM10 concentrations) and the solution is concentrated in a 65-ml Amicon ultrafiltration unit (43-mm YM10 membrane) to approximately 5 ml. The final concentrate is removed from the ultrafiltration unit and the YM10 membrane is washed twice (15 min each) with 2 ml of 1 M acetic acid. The two washes and the concentrate are combined. This inhibitor solution is stable for several years at −20°.

P-60 Gel Chromatography. The inhibitor solution is chromatographed at 4° on a 2 × 30 cm column of BioGel P-60 (100–200 mesh; Bio-Rad Laboratories) previously equilibrated with 1 M acetic acid. Fractions of 4 ml are collected in polypropylene tubes, to which have been added 0.10 ml of a 0.1 mg/ml solution of bovine serum albumin in 1 M acetic acid. The absorption of the fractions at 280 nm is measured and the growth-inhibitory activity of the fractions is determined by bioassay. The growth-inhibitory activity elutes from the column after most of the absorption at 280 nm. Approximately a 40-fold increase in specific activity of the growth inhibitor is obtained with 60–80% recovery of activity. The yield of inhibitor is 500 to 1000 units per liter of conditioned medium.

Bioassay of the Kidney Epithelial Cell Growth Inhibitor.[6] BSC-1 cells are plated at approximately 2 × 10⁵ cells/5-cm plate and are grown 2 or 3 days in DMEM supplemented with 0.4 μg/ml of biotin and 0.2% calf serum. Fresh medium is placed on the cells before the assay additions are made. Aliquots (5 μl) of the P-60 column fractions are assayed (with the addition of 5 μl of 1 M sodium carbonate to neutralize the 1 M acetic acid). Incorporation of [*methyl*-³H]thymidine is measured from 20 to 25 hr after the assay additions. The cells are then washed with Tris/saline buffer[5] (see above, with CaCl₂, 0.01%; MgCl₂, 0.01%) and the cells are removed from the plates with 2 ml of 0.025% trypsin in calcium- and

[6] R. W. Holley and J. A. Kiernan, *Proc. Natl. Acad. Sci. U.S.A.* **71**, 2908 (1974).

magnesium-free Tris/saline buffer[5] (see above) with 0.5 mM Na$_2$EDTA. The cell suspension is combined with an equal volume of 10% trichloroacetic acid, and the acid-insoluble material is collected by filtration on glass fiber filter papers. The filter papers are rinsed with 95% ethanol, placed in scintillation vials, dried at 90°, and the radioactivity is determined by scintillation counting. The results are calculated as percentage inhibition of the thymidine incorporation observed in control cells. A unit of inhibitory activity is defined as the amount that inhibits thymidine incorporation 50%.

High-Performance Liquid Chromatography. The most active fractions of the peak of growth inhibitory activity from the P-60 column (typically, fractions 17–22) are combined. The combined fractions are lyophilized and the residue is redissolved in approximately 0.5 ml of 1 M acetic acid. This solution is injected on a Waters μBondapak CN column (3.9 mm × 30 cm) or on a Du Pont Zorbax PEP RP/1 column (6.2 mm × 8 cm) or an LKB Ultropac TSK ODS-120T 5μm column (4.6 × 250 mm), and chromatography is with Altex model 322 equipment.

When the Waters μBondapak CN column is used, buffer A is a pH 4 pyridine–formic acid buffer (29 ml pyridine, 19 ml 88% formic acid made up to 1 liter with water). Buffer B is a pH 4 pyridine–formic acid buffer with *n*-propanol (29 ml pyridine, 19 ml 88% formic acid, 600 ml *n*-propanol made up to 1 liter with water). Both buffers are filtered and degassed. The isocratic elution sequence is 0% buffer B for 4 min, 16.6% buffer B for 4 min, 41.7% buffer B for 20 min, 50% buffer B for 20 min, 66.6% buffer B for 20 min, and 66.6–100% buffer B for 1 min.[1] The column eluate is monitored with a fluorescamine stream sampling detector.[7] Three-minute fractions are collected, with a flow rate of 0.7 ml/min. The inhibitor elutes at approximately fraction 14.

With the Du Pont Zorbax PEP RP/1 column or the LKB Ultropac TSK ODS-120T column, the sample in 0.5 M acetic acid is diluted with an equal volume of 0.1% trifluoroacetic acid before injection. The elution program is 6 min with 0.1% trifluoroacetic acid followed by 90 min of a linear gradient from 0.1% trifluoroacetic acid to 99.9% acetonitrile–0.1% trifluoroacetic acid. The inhibitor elutes at approximately 50% acetonitrile in 0.1% trifluoroacetic acid. The column eluate is monitored at 280 nm with a Kratos model SF769 variable wavelength spectrophotometer. The flow rate is 0.7 ml/min and 3-min fractions are collected.

Before bioassay, the fractions are lyophilized with 10 μg BSA/fraction and residues are redissolved in 0.20 ml of 0.5 M acetic acid per fraction. A typical assay uses 0.2 μl of a fraction. Recovery of growth inhibitory activ-

---

[7] P. Bohlen, S. Stein, J. Stone, and S. Udenfriend, *Anal. Biochem.* **67,** 438 (1975).

ity from the column is approximately 50%, with a 10- to 20-fold increase in specific activity. Its specific activity is now 200–300 units/μg. The growth inhibitor has an apparent molecular weight of 24,000 on a nonreducing SDS–polyacrylamide gel, and, based on silver staining of the gel, gives a single major band, with traces of lower molecular weight material.

The purified growth inhibitor is extremely active in inhibiting the growth of BSC-1 cells and inhibits thymidine incorporation approximately 50% at 1 ng/ml. Maximum inhibition (80–90%) takes place at 10 ng/ml. The most sensitive cell that has been encountered is the CCL64 mink lung epithelial cell line. With CCL64 cells, thymidine incorporation is inhibited 50% by approximately 0.05 ng/ml and 95–99% by 1 ng/ml.[8] The inhibitor arrests the growth of BSC-1 and CCL64 cells in the G_1 (or G_0) phase of the cell cycle.[1,3]

The BSC-1 monkey kidney cell growth inhibitor is closely related to platelet type-β transforming growth factor (TGFβ).[9] For another discussion of the isolation of TGFβ from platelets see Assoian, this volume [14]. Assay of TGFβ is also performed using soft agar methods described by Rizzino, this volume [32].

Acknowledgments

This work was supported in part by Grant 59 from the American Cancer Society, and by Public Health Service Grant CA11176 awarded by the National Cancer Institute. R.W.H. is an American Cancer Society Professor of Molecular Biology.

[8] R. W. Holley, R. Armour, J. H. Baldwin, and S. Greenfield, Cell Biol. Int. Rep. 7, 141 (1983).
[9] R. F. Tucker, G. D. Shipley, H. L. Moses, and R. W. Holley, Science 226, 705 (1984).

[16] An Assay for Type-β Transforming Growth Factor Receptor

By LALAGE M. WAKEFIELD

Type-β transforming growth factor (TGFβ) is a bifunctional regulator of cell growth[1,2] that has been found in both normal and neoplastic tissues.[3–5] Specific high-affinity receptors for TGFβ have recently been

[1] A. B. Roberts, M. A. Anzano, L. M. Wakefield, N. S. Roche, D. F. Stern, and M. B. Sporn, Proc. Natl. Acad. Sci. U.S.A. 82, 119 (1985).
[2] R. F. Tucker, G. D. Shipley, H. L. Moses, and R. W. Holley, Science 226, 705 (1984).

METHODS IN ENZYMOLOGY, VOL. 146

described on fibroblastic and epithelial cells of human and rodent origin,[6-8] and on both normal and transformed cells.[7] The number of TGFβ receptors is reduced in normal cells infected by acute transforming retroviruses,[9] suggesting a possible down regulation of the receptor by endogenously produced TGFβ. In devising an assay for the cellular TGFβ receptor, two specific problems must be overcome. Since the platelet contains TGFβ at relatively high concentrations,[5] the whole blood-derived serum used in cell culture contains appreciable amounts of the growth factor, which may decrease the apparent number of assayable receptors. Also, TGFβ is a relatively hydrophobic molecule that adheres strongly to many surfaces, thereby increasing the background of nonspecific binding in any assay system. The assays for the TGFβ receptor described in this chapter were developed with the aim of minimizing both these effects.

Reagents and Solutions

TGFβ. TGFβ is isolated and purified to homogeneity from freshly drawn human platelets,[5] as described in detail in this volume (Assoian [14]). Since recoveries of labeled TGFβ from the iodination procedure are highest when relatively large quantities of TGFβ are used, a series of radioreceptor assays is best done with 5–10 μg TGFβ, using 2–5 μg for the iodination and 5–8 μg to determine nonspecific binding.

^{125}I-Labeled TGFβ. Since most cells have a very high-affinity receptor for TGFβ (1–50 pM), at relatively low copy number per cell (1000–50,000), it is important to use label of high specific activity in order to detect the receptor. TGFβ can be iodinated to high specific activity (2–4 μCi/pmol; 80–160 μCi/ug) with retention of 50–100% of the biological activity using a modified chloramine-T method,[6] described in detail elsewhere in this volume (Frolik and De Larco [9]). We typically find that >99% of the label is precipitable by trichloroacetic acid immediately after

---

[3] A. B. Roberts, M. A. Anzano, C. A. Frolik, and M. B. Sporn, *Cold Spring Harbor Conf. Cell Proliferation* **9**, 319 (1982).

[4] C. B. Childs, J. A. Proper, R. F. Tucker, and H. L. Moses, *Proc. Natl. Acad. Sci. U.S.A.* **79**, 5312 (1982).

[5] R. K. Assoian, A. Komoriya, C. A. Meyers, D. M. Miller, and M. B. Sporn, *J. Biol. Chem.* **258**, 7155 (1983).

[6] C. A. Frolik, L. M. Wakefield, D. M. Smith, and M. B. Sporn, *J. Biol. Chem.* **259**, 10995 (1984).

[7] R. F. Tucker, E. L. Branum, G. D. Shipley, R. J. Ryan, and H. L. Moses, *Proc. Natl. Acad. Sci. U.S.A.* **81**, 6757 (1984).

[8] J. Massagué and B. Like, *J. Biol. Chem.* **260**, 2636 (1985).

[9] M. A. Anzano, A. B. Roberts, J. E. De Larco, L. M. Wakefield, R. K. Assoian, N. S. Roche, J. M. Smith, J. E. Lazarus, and M. B. Sporn, *Mol. Cell. Biol.* **5**, 242 (1985).

iodination, of which 95% is capable of binding specifically to the receptor.[6] The specific bindability of the label decreases linearly with time by ~0.35% per day thereafter, and we routinely correct the measured free TGFβ concentration to take this into account. After ~30 days, although the label still shows appreciable specific binding to the cell surface, the results obtained on Scatchard analysis of the data are never very reproducible. It cannot be overemphasized that TGFβ adheres strongly to most surfaces and precautions must be taken to minimize losses of growth factor during storage and handling. Of the various plastics available, TGFβ sticks least to polypropylene, and if glass is used, it must be siliconized. Losses are also reduced in the presence of an excess of a carrier protein and in acid solution. We store our ^{125}I-labeled TGFβ as 10- to 20-μCi aliquots from a 100 μCi/ml stock solution in 4 mM HCl, 1% bovine serum albumin (BSA) at −20°, and only do the serial dilutions for the binding assay immediately before use. A full binding analysis will typically require 4–10 μCi label.

Binding Buffer. Dulbecco's modified Eagle's medium (DMEM) containing 0.1% BSA and 25 mM HEPES, pH 7.4, is used. Alternatively, bicarbonate-free Minimum Essential Medium (Eagle) containing 0.1% BSA and 25 mM HEPES pH 7.4 can be used, which obviates the need to perform the binding assay in a CO_2 atmosphere. Binding of TGFβ to its receptor seems relatively insensitive to changes in pH over the range pH 6–8.

Wash Buffer. Hanks' buffered salt solution, containing 0.1% BSA, is used. This should be kept at 4° until use. Other buffered saline solutions may be substituted, but note that the presence of Ca^{2+} and Mg^{2+} helps prevent cell detachment during washes.

Solubilization Buffer. Combine 1% Triton X-100, 10% glycerol, 20 mM HEPES, pH 7.4.

Radioreceptor Assay

Two forms of the radioreceptor assay have been developed, one suitable for cells that grow in monolayer culture and the other for cells in suspension culture. We find that if cells that normally grow in monolayer culture are treated with Versene and assayed in suspension, there is an apparent decrease in the affinity of the receptor, suggesting that cell shape may have an important effect on receptor parameters. It is therefore preferable to assay cells in as near as possible their normal growth state.

Monolayer Assay

The monolayer assay is a modification of that described by Frolik et al.[6] Cells are seeded in their normal growth medium at ~10^5 cells/well in 24-well cluster plates and allowed to grow for 24 hr. Ideally, the cells should cover 80–90% of the well surface at the time of assay. The monolayers are then washed once with 1 ml/well binding buffer and incubated in this medium for 2 hr at 37° to allow dissociation of bound endogenous TGFβ, either secreted by the cells or bound from the serum. A single 2-hr wash at 37° will remove 85–95% of prebound TGFβ depending on the concentration of label added (200 or 10 pM), while two 1-hr washes with a medium change in between will remove 90–100%.

After removal of bound endogenous ligand, cells are washed once more with 1 ml/well of binding buffer and 200 μl of ^{125}I-labeled TGFβ in binding buffer is added to each well, over a concentration range of 1–500 pM (1200–600,000 cpm/well). Nonspecific binding is determined at four to six different labeled ligand concentrations in the presence of 10 nM unlabeled TGFβ. Binding buffer alone is added to 4 wells which will be used for cell counts.

Cells are then incubated for 2 hr at 25°, or for 3 hr at 4°, in a 5–10% CO_2 atmosphere for binding to occur. We find there is little internalization or degradation of ligand at 25° for most cell types. However, if it is necessary to do the assay at 4°, the plates should be gently agitated to overcome problems of slow ligand diffusion, particularly at low ligand concentrations. A satisfactory arrangement that we employ is to place the plates in an air tight plastic box that has been briefly flushed with CO_2 and to place this on a Clinical Rotator (Arthur H. Thomas Co., Philadelphia, PA) in a 4° cold room. The CO_2 atmosphere is unnecessary if the bicarbonate-free MEM-based binding buffer is used. It should be noted that many cells do not fare well in serum-free conditions at 4° for 3 hr, and will detach from the wells. For comparison of receptor properties between a large number of cell types, it is therefore preferable to work at 25°.

To quantitate free ligand concentrations, the media are removed from the wells at the end of the incubation and counted in a gamma counter. This is a critical step since the free ligand concentration is generally only about half of the added concentration due to high levels of ligand binding to the plastic of the wells. If any cells have detached from the plate during the course of incubation, these should be removed by centrifugation before the media samples are counted. The TCA precipitability of the labeled ligand (final TCA concentration of 12%) is determined for two or more of the media samples, and compared with that for the label before incubation with the cells to check for ligand degradation.

To quantitate bound ligand, the cell monolayers are then washed four times with ice-cold wash buffer, and the cell-associated radioactivity is solubilized by incubation with 0.75 ml of the Triton-based solubilization buffer for 20–30 min at room temperature. After trituration of the solubilized sample, an aliquot is counted on a gamma counter. This solubiliza-

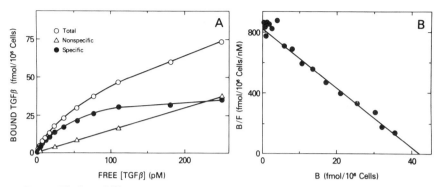

FIG. 1. Binding of ^{125}I-labeled TGFβ to normal rat kidney fibroblasts. (A) Total, nonspecific, and specific binding curves. (B) Scatchard analysis of the binding data in (A). See text for details.

tion protocol is used rather than NaOH- or formic acid-based extraction procedures because it reduces the recovery of label nonspecifically bound to the plastic (~30% of added counts). Collagenase may be used instead of Triton,[7] but not all cell lines are completely removed from the plate by this method and many have to be scraped as well as collagenase treated. The cells in the four wells that contained no label are trypsinized after washing with the ice-cold wash buffer and the number of cells/well is determined by counting on a Coulter counter or hemocytometer.

The amount of TGFβ specifically bound is determined by subtracting the nonspecific from total TGFβ bound, and the receptor affinity and number of receptors/cell can be determined from a plot of B/F (specifically bound ligand/free ligand concentration) vs B according to the method of Scatchard[10] (see, for example, Fig. 1). Where binding data yield a curvilinear Scatchard plot, the contributions of the different receptor sites can be computer resolved using the "Ligand" program of Munson and Rodbard.[11] Nonspecific binding typically ranges from ~1% of the total binding at the lowest added ligand concentrations to up to 50% at high ligand concentrations when the receptor is saturated. The counts specifically bound usually range from 100 to 20,000 cpm/well, and the limit of detection of the assay is ~1000 receptors/cell. The sensitivity of the assay can be increased if necessary by using more cells, in larger 12- or 6-well cluster plates, and scaling up reagent volumes accordingly.

[10] G. Scatchard, *Ann. N.Y. Acad. Sci.* **51**, 660 (1949).
[11] P. J. Munson and D. Rodbard, *Anal. Biochem.* **107**, 220 (1980).

Suspension Assay

This assay is used for cells that will not grow in monolayer, and is based on a method developed by Cantrell and Smith[12] for binding of interleukin-2 to lymphocytes.

Cells (ideally $10-50 \times 10^6$) are incubated in two changes of serum-free growth medium (50 ml/10^7 cells) at 37° for 2 hr to remove bound endogenous TGFβ. After washing, cells ($0.5-2.0 \times 10^6$/0.2 ml) are placed in siliconized tubes with serial dilutions of ^{125}I-labeled TGFβ, from 1 to 500 pM, in binding buffer. Nonspecific binding is determined in the presence of 20 nM unlabeled TGFβ. The cells are then incubated at 25° for 2 hr in a $5-10\%$ CO_2 atmosphere, with constant agitation (using a Clinical rotator as for the monolayer experiments) to keep the cells in suspension. Again, the need for a CO_2 atmosphere can be circumvented by the use of the bicarbonate-free binding assay, in which case better mixing is obtained if the tubes are rocked. At the end of 2 hr, the reaction is stopped by addition of 1 ml ice-cold binding buffer and the cells are spun down (9000 g, 15 sec). The supernatant is removed and counted in a gamma counter to determine free ligand concentration. The pellet is resuspended in 100 μl of ice-cold binding buffer and spun through a 200-μl layer of a mixture of 84% silicone oil (Dexter-Hysol 550 fluid; Dexter Corp., Olean, NY) and 16% Paraffin oil (No. 0-119; Fisher Scientific Co., Fair Lawn, NJ) in 400-μl polyethylene microfuge tubes (9000 g, 90 sec). The tip of the tube containing the cell pellet is cut off using a small guillotine, and counted in a gamma counter to determine bound radioactivity. The binding data are then analyzed as for the monolayer assay, and the limit of detection of this assay is ~100 receptors/cell.

Results

A typical set of curves for TGFβ binding to normal rat kidney fibroblasts are shown in Fig. 1A, and the corresponding Scatchard analysis of the data is shown in Fig. 1B. This cell line has a single class of binding sites, with a dissociation constant (K_d) of 40 pM and ~25,000 sites/cell. In our laboratory we have assayed over 50 different cell types to date, and all express similar high-affinity receptors for TGFβ. The receptor parameters of representative cell types are shown in Table I. In general, the receptor has a K_d of $1-50$ pM and there are $1000-50,000$ receptors/cell. Epithelial cells tend to have higher affinity receptors than fibroblastic cells but at lower copy number, and the human tumor cell lines generally have receptor properties similar to those of normal cells. Peripheral blood lympho-

[12] D. A. Cantrell and K. A. Smith, *J. Exp. Med.* **158**, 1895 (1983).

TABLE I
TGFβ RECEPTOR PARAMETERS ON VARIOUS CELL TYPES

Cell type	Receptor parameters	
	K_d (pM)	Number/cell
Human embryonic lung fibroblast (WI-38)	27	17,000
Human adult lung fibroblast (Explant)	19	19,000
Rat kidney fibroblast (NRK 49F)	40	25,000
Normal human bronchial epithelial	10	10,000
Normal rat kidney epithelial (NRK 52E)	5	9,000
Bovine pulmonary artery endothelial	26	9,000
Normal human mesothelial	10	20,000
Human tonsillar T lymphocytes (PHA stimulated)	5	400
Human rhabdomyosarcoma (A673)	10	3,000
Human lung adenocarcinoma (A549)	13	10,000
Human fibrosarcoma (HT1080)	6	7,000

cytes, however, have exceptionally low numbers of TGFβ receptors (100–600/cell, depending on culture conditions) in the unstimulated cell, which may be related to the small size of these cells. We have observed for both lymphocytes and monocytes that the number of detectable TGFβ receptors appears to be 5- to 10-fold higher if the cells are assayed immediately after isolation rather than first being placed in culture overnight. Another important observation relating to the design of these binding assays is the consistent 2- to 5-fold decrease in affinity of the TGFβ receptor in quiescent compared with actively growing cell cultures, seen in the three cell lines assayed for this effect to date (NRK, A549, and Calu I). Thus for the monolayer assay, although it is desirable to have a large proportion of the well covered by cells to minimize the binding of ^{125}I-labeled TGFβ to the exposed plastic, the cells should not be allowed to become confluent and quiescent unless this is also done for all the other cells to be compared in the same set.

Acknowledgments

I thank Dr. Michael Sporn for support during the development of these procedures, much of which was done in collaboration with Dr. Charles Frolik, and I thank Diane Smith for excellent technical assistance. NRK 49F and NRK 52E cells were generously provided by Dr. Joseph De Larco. I acknowledge the receipt of a NATO/SERC Overseas Postdoctoral Fellowship.

[17] Identification of Receptors for Type-β Transforming Growth Factor

By JOAN MASSAGUÉ

Type-β transforming growth factor (TGFβ) is a 23- to 25-kDa polypeptide that consists of two apparently identical subunits of 11–12 kDa linked by disulfide bonds.[1-3] TGFβ is relatively abundant in blood platelets and is also found in a variety of normal tissues and transformed cell lines. It was originally discovered together with TGFα as one of the active components from retrovirally transformed cells that induce anchorage-independent growth in certain types of normal rodent fibroblasts. Despite sharing the same nomenclature, TGFβ and TGFα are now recognized as two widely different molecules that belong to distinct polypeptide hormone families. Unlike TGFα, which is structurally related to epidermal growth factor (EGF), the portion of TGFβ sequenced so far does not share primary amino acid structure with any other known protein.[2,4] TGFβ binds to specific cellular receptors. We have radiolabeled homogeneous preparations of TGFβ from human platelets under conditions in which the biological properties of the native polypeptide are retained. These preparations are suitable to characterize the binding properties of cellular receptors for TGFβ, and to probe the structure of these receptors by ligand-directed affinity labeling. The methods employed in these studies are the subject of this chapter. Another chapter in this volume (see [13]) deals with the identification of receptor proteins for TGFα.

Source of TGFβ

TGFβ has been isolated from a variety of sources, but human blood platelets are by far the richest known source of this growth factor. We use TGFβ purified to homogeneity from acid–ethanol extracts of recently outdated platelet concentrates prepared as previously described.[1] After lyophilization, the extract from 50 units of platelets are reconstituted in 1 M acetic acid and chromatographed over a column (5 × 90 cm) of Bio-Gel P-60 equilibrated and eluted at 30 ml/hr with 1 M acetic acid at room

---

[1] R. K. Assoian, A. Komoriya, C. A. Meyers, D. M. Miller, and M. B. Sporn, J. Biol. Chem. **258**, 7155 (1983).

[2] A. B. Roberts, M. A. Anzano, C. A. Meyers, J. Wideman, R. Blacker, Y.-C. E. Pan, S. Stein, S. R. Lehrman, J. M. Smith, L. C. Lamb, and M. B. Sporn, Biochemistry **22**, 5692 (1983).

[3] J. Massagué, J. Biol. Chem. **259**, 9756 (1984).

temperature. Fractions (25 ml) are collected. TGFβ elutes from this column in fractions 48–55, as determined by bioassay (growth of NRK-49F cells in DME containing 10% calf serum, 0.35% agar, and 0.3 nM EGF), radioreceptor assay, and silver staining of nonreducing sodium dodecyl sulfate–polyacrylamide electrophoresis (SDS–PAGE) gels to which 100 μl of each chromatographic fraction is applied after lyophilization.[4]

TGFβ elutes from this column in a characteristically retarded position that corresponds to 7- to 10-kDa polypeptides.[1,3] This property has been used advantageously to further purify TGFβ.[1,3,4] We use one single step that yields homogeneous TGFβ in a salt-free form. The pooled, lyophilized material (4–6 mg protein) from the BioGel P-60 step is reconstituted in 0.2 ml of 25% acetonitrile, 0.5 M acetic acid, and applied to two I-125 molecular filtration high-pressure liquid chromatography columns (Waters Associates) connected in series, and eluted at 1 ml/min with 25% acetonitrile, 0.5 M acetic acid. Fractions (0.5 ml) are collected in siliconized glass test tubes. Fractions containing TGFβ are identified by electrophoresis and silver staining of 5-μl aliquots, and by bioassay. TGFβ elutes immediately after the void volume, well separated from lower molecular weight impurities. TGFβ obtained after this step is homogeneous as determined electrophoretically and by N-terminal amino acid sequence analysis.[4] It can be lyophilized in siliconized glass tubes and stored at −20° for months without detectable loss of biological activity.

The use of siliconized glassware in the purification and storage of TGFβ is mandatory, as the purified peptide binds irreversibly to untreated glass or plastic surfaces. Surfasil (Pierce Chemical Co.) is satisfactory as a siliconizing agent for this purpose.

[125]I-Labeling of TGFβ

The solid-phase glucose oxidase:lactoperoxidase method is used for iodination of TGFβ under gently oxidizing conditions. In this method the hydrogen peroxide needed as a substrate in the lactoperoxidase-catalyzed reaction to oxidize $^{125}I^-$ to $^{125}I_2$ is slowly generated by a coupled glucose oxidase reaction that uses β-D-glucose as the substrate. The $^{125}I_2$ generated reacts with tyrosine residues in TGFβ At the end of the reaction the labeled TGFβ is separated from the enzymes, which are covalently linked to polyacrylamide beads, and from the unreacted iodine by molecular filtration chromatography. The described chromatography conditions have been designed to minimize losses of ^{125}I-labeled TGFβ during the process.

---

[4] J. Massagué and B. Like, J. Biol. Chem. 260, 2636 (1985).

Materials

TGFβ: 10 μg lyophilized in the bottom of a siliconized glass test tube

Reaction buffer: 0.3 M sodium phosphate, pH 7.4

Enzymobeads (Bio-Rad Laboratories): reconstituted in H_2O as per manufacturer's directions

Na[125]I: (16 Ci of ^{125}I/mg of I, Amersham): dissolved at neutral pH, at 1 mCi/10 μl

D-Glucose: dissolved at 20 mg/ml in sterile H_2O and allowed to racemize to β-D-glucose by incubation at room temperature overnight

Acetic acid: 2.0 M solution

BioGel P-10 (100–200 mesh, Bio-Rad Laboratories): a column of 20 × 0.8 cm equilibrated and eluted with 2 mg/ml bovine serum albumin (BSA, Sigma) in 0.5 M acetic acid at room temperature

Glassware: siliconized glass test tubes for fraction collection

Procedure

1. Add 50 μl of reaction buffer to the tube containing TGFβ. Agitate to dissolve TGFβ.
2. Add 25 μl of Enzymobeads and 1.5 mCi of Na[125]I.
3. Start the reaction by addition of 25 μl of D-glucose solution. Incubate for 30 min at room temperature with frequent shaking to prevent sedimentation of the beads.
4. Stop the reaction by addition of 0.2 ml of 2.0 M acetic acid. Sediment the beads by a brief centrifugation at 1000 g.
5. Load the supernatant on the BioGel P-10 column. Wash the beads with 0.2 ml of 2.0 M acetic. Remove the beads by centrifugation, and load the supernatant on the column.
6. Run the column by gravity at 0.5–0.6 ml/min. Collect 0.5-ml fractions in siliconized glass test tubes.
7. Count the radioactivity in 2 μl from each fraction. Pool the peak fractions of ^{125}I-labeled polypeptide. Count radioactivity in one aliquot to determine the specific activity of the preparation.
8. Distribute the preparation into small aliquots and store at 4° or −20°.

Comments

[125]I-Labeled TGFβ elutes close to the void column in BioGel P-10, in the presence of 0.5 M acetic acid, whereas free iodine and NaI elute at the salt volume. Approximately 60% of the ^{125}I-labeled TGFβ elutes in 2.0 ml of peak fractions. The radioactivity in this pool is over 95% precipitable by addition of 10% trichloroacetic acid in the cold followed by centrifugation

for 10 min at 2000 g. The calculated specific activity of the resulting ^{125}I-labeled TGFβ is usually between 50 and 100 Ci/g or 0.6–1.2 mol of I/mol of TGFβ. Like the native peptide, ^{125}I-labeled TGFβ prepared by this method migrates on SDS–PAGE gels as a 25-kDa species in the absence of reductant, and as an 11 to 12-kDa species in the presence of dithiothreitol.[4] The biological activity of ^{125}I-labeled TGFβ in mitogenic and anchorage-independent cell proliferation assays is 70–90% that of native TGFβ.[4] This limited decrease can be accounted for by peptide loss and, perhaps, minor structural damage during iodination. The radiolabeled preparations can be used in receptor assays for a period of 4–6 weeks after iodination.

Binding of ^{125}I-Labeled TGFβ to Intact Cells

Many types of cultured cells exhibit cell surface receptors that interact specifically with TGFβ. Like receptors for other polypeptide hormones, the TGFβ receptors are subject to complex dynamics in the cell. They internalize and direct ligand bound at the cell surface toward a lysosomal compartment in which degradation of TGFβ takes place. These properties of the receptor must be taken into consideration in the design of assays to determine the presence, number, and affinity of TGFβ receptors on cell surfaces. Binding assays to determine these parameters are performed in the cold, under conditions in which internalization of TGFβ–receptor complexes occurs to a very limited extent if at all. However, variations of the standard procedure described below involving incubation at 37° are useful in studying the internalization and fate of TGFβ and its receptor upon binding at the cell surface.

Materials

Binding buffer I: 128 mM NaCl, 5 mM KCl, 5 mM MgSO$_4$, 1.2 mM CaCl, 50 mM HEPES, pH 7.5, 2 mg/ml BSA

Binding buffer II: Dulbecco's modified Eagle's medium (Gibco) supplemented with 25 mM HEPES, pH 7.5, and 2 mg/ml of BSA

[125]I-Labeled TGFβ: An appropriate dilution is prepared in binding medium immediately before each experiment. This dilution should be made in a siliconized glass test tube

TGFβ: Diluted appropriately using the same buffer and conditions as for ^{125}I-labeled TGFβ

Cell solubilization buffer: 1% (v/v) Triton X-100, 10% (v/v) glycerol, 25 mM HEPES, pH 7.5, 1 mg/ml BSA

Cells: Depending on the number of TGFβ binding sites per cell, 2- or 10-cm^2 confluent monolayers preferably seeded in multiwell plates are used. Assays involving cell types that detach easily from the substratum should be performed with cell suspensions. Single cell suspensions can be obtained by mechanical detachment (by pipetting) or brief exposure at 37° to detachment buffer containing 1 mM EDTA, 128 mM NaCl, 5 mM glucose, 25 mM HEPES, pH 7.4, 2 mg/ml BSA. See below for comments on binding assays using cell suspensions

Procedure

1. Wash cell monolayers with binding buffer. Binding buffers I or II can be used indistinctly in assays performed at 4°, but binding buffer II should be used in assays involving incubation steps at 37°. After washing once, monolayers are allowed to equilibrate with binding buffer for 30 min at 4°.

2. Aspirate buffer and add ice-cold binding buffer to plates on ice. The volume of buffer is 0.67 ml/2-cm^2 well, or 2 ml/10-cm^2 well.

3. Add ^{125}I-labeled TGFβ diluted appropriately in a small volume of binding buffer. Concentrations of ^{125}I-labeled TGFβ in the 10–50 pM range are appropriate for most cell lines. If desired, the ^{125}I-labeled TGFβ may be incorporated in the buffer used in Step 2 before addition to the wells. Immediately before or after addition of ^{125}I-labeled TGFβ, add unlabeled TGFβ to those wells that require it. TGFβ may be used for two purposes. Increasing concentrations of TGFβ are used to supplement the tracer concentration of ^{125}I-labeled TGFβ in experiments designed to establish the saturation curve of TGFβ binding to cells. Additionally, a 50-fold or larger excess of unlabeled TGFβ is used in all experiments to determine the amount of ^{125}I-labeled TGFβ which is nonspecifically bound. This determination is based on the assumption that ^{125}I-labeled TGFβ and TGFβ compete for binding to relevant high-affinity sites, but not to nonspecific sites on the surface of cells or dishes.

4. Incubate assays for 3.5–4.0 hr at 4° on a platform oscillating at 120 cpm.

5. Aspirate the medium. Rinse cultures five times with ice-cold binding buffer I.

6. Add solubilization buffer (0.5 or 1.5 ml for 2- or 10-cm^2 wells, respectively). Incubate 40 min at 4°.

7. Count radioactivity in the soluble extracts. To determine specific binding, subtract the counts per minute obtained in wells incubated with ^{125}I-labeled TGFβ in the presence of excess TGFβ from the counts per minute in wells incubated with ^{125}I-labeled TGFβ alone. Count an aliquot of each ^{125}I-labeled TGFβ dilution used in the experiment to determine specific radioactivity values. Count the number of cells per well to determine the amount of TGFβ bound per cell.

Comments

TGFβ and ^{125}I-labeled TGFβ bind to nonsiliconized glass and various types of plastic surfaces. Significant nonspecific adsorption of ^{125}I-labeled TGFβ plastic culture dishes is observed in the absence of cells. However, adsorption to the dishes is markedly diminished by the presence of a confluent cell monolayer. Thus, the amount of radioactivity nonspecifically bound to culture dishes incubated with 170 pM ^{125}I-labeled TGFβ was decreased from about 4% of the input radioactivity in 10-cm^2 dishes without cells to 0.5% in dishes containing a confluent monolayer of BALB/c 3T3 cells. Obviously, the relative contribution of nonspecific ^{125}I-labeled TGFβ binding to the total ^{125}I-labeled TGFβ binding is an inverse function of the binding capacity of each cell type. For example, 35, 21, and 8% of the total ^{125}I-labeled TGFβ binding is nonspecific in cell types that contain 0.8×10^4 sites/cell (WI-38 fibroblasts), 1.9×10^4 sites/cell (NRK fibroblasts), or 16×10^4 sites/cell (3T3-L1 fibroblasts), respectively.

If the assay is to be performed with cell suspensions, siliconized glass test tubes are recommended for all incubations. After detachment from culture vessels, cells are washed twice by centrifugation with binding buffer, and then adjusted to about $2-10 \times 10^5$ cells/ml. Frequent shaking during incubation with ^{125}I-labeled TGFβ is needed to prevent cell sedimentation. Washes of the cells with binding buffer before and adter incubation with ^{125}I-labeled TGFβ can be done by centrifugation at 2000 g for 5 min at 4°.

Figure 1 shows a typical saturation curve of TGFβ binding to NRK-49F rat fibroblasts at 4°. The kinetics of ^{125}I-labeled TGFβ binding at 4° are relatively slow; equilibrium binding is achieved usually after 3–4 hr of incubation at this temperature.[4] Binding at low temperature probably involves cell surface receptors only, as internalization and exocytosis processes are greatly inhibited at 4°. Meaningful values for binding affinity constants and cell surface receptor numbers can therefore be derived from binding data obtained at 4° (Fig. 1). Binding at 37° is more complex, and it is not an equilibrium process because bound ^{125}I-labeled TGFβ undergoes rapid internalization, intracellular degradation, and release[5] (J. Massagué, unpublished results) while the cointernalized receptor may cycle intact back to the cell surface (J. Massagué, unpublished results). Binding of ^{125}I-labeled TGFβ to mouse fibroblasts at 37° reaches a steady state after 45–60 min of incubations, and remains constant thereafter for at least 3 hr unless extensive degradation of ^{125}I-labeled TGFβ occurs in these cultures.

[5] C. A. Frolik, L. M. Wakefield, D. M. Smith, and M. B. Sporn, *J. Biol. Chem.* **259**, 10995 (1984).

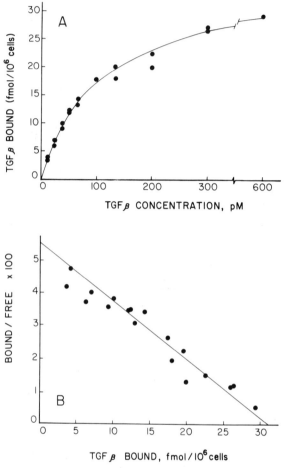

FIG. 1. Saturation curves of ^{125}I-labeled TGFβ binding to NRK-49F cells. (A) Near-confluent (1.7 × 10⁵ cells/well) monolayers of NRK-49F cells in 10-cm² culture wells were incubated for 4 hr at 4° in the presence of ^{125}I-labeled TGFβ alone (12–60 pM) or 60 pM ^{125}I-labeled TGFβ plus various amounts of unlabeled TGFβ to achieve the indicated final concentrations. The amount of TGFβ specifically bound in each condition was then determined. (B) Scatchard plot corresponding to the data in (A). (From Ref. 4.)

[125]I-Labeled TGFβ binding to monolayers of various cell types at 4° is optimal over a relatively broad pH range, 6.0–8.5 (J. Massagué, unpublished results).

The solubilization buffer recommended above minimizes the release of ^{125}I-labeled TGFβ nonspecifically bound to cells and culture vessels, while it completely solubilizes ligand bound specifically at 4° (J. Massagué, unpublished results). However, the ^{125}I-labeled TGFβ specifically associated with the cells after incubation at 37° may not be completely extractable by solubilization buffer at 4°. Complete extraction of ligand bound at 37° can be achieved with solubilization buffer at room temperature, or with 1% sodium dodecyl sulfate solution at 4°. While the nature of the Triton X-100-resistant cell compartment at 4° is not known, this finding may have operational interest in the study of internalization and intracellular sorting of ^{125}I-labeled TGFβ–receptor complexes.

Radioreceptor Assay for TGFβ

Cell monolayers which bind ^{125}I-labeled TGFβ can be used as a radioreceptor assay component to investigate the presence of TGFβ in biological samples.[6] We have used confluent monolayers of BALB/c 3T3 cells for this purpose, although other cell types may also be used specially if they display high cell surface binding capacity for tracer ^{125}I-labeled TGFβ. The radioreceptor assay is carried out with 2-cm^2 cell monolayers using the conditions described above for the ^{125}I-labeled TGFβ binding assay. A standard competition curve is constructed with 0.5–200 ng/ml of purified TGFβ, which the range of sensitivity of the radioreceptor assay. Samples to be analyzed can be added directly to the assays, immediately before addition of the tracer (25 pM) ^{125}I-labeled TGFβ. Crude samples that contain large amounts of impurities that may interfere with the assays should be subjected to a preliminary treatment to decrease the amount of impurities. Dialysis against 0.1 M acetic acid will precipitate some proteins but not TGFβ. After ultracentrifugation, TGFβ can be recovered from the supernatant by lyophilization, or by precipitation with ether:ethanol (2:1) in the cold. The dried sample is then reconstituted with 0.1 M acetic acid, neutralized, and subjected to radioreceptor assay. The results of the assay may be expressed in TGFβ nanogram equivalents, one nanogram equivalent being the amount of radioreceptor active material that competes with ^{125}I-labeled TGFβ with the same potency as 1 ng of TGFβ purified from human platelets.

To eliminate "false positives" in the radioreceptor assay, control determinations should be carried out in which cell monolayers are preincubated with the sample under study for 3.5 hr at 4°, under standard binding conditions. After five washes with ice-cold medium, the pretreated monolayers are exposed to ^{125}I-labeled TGFβ to measure the remaining binding capacity. The purpose of this control determination is to show that any decrease in ^{125}I-labeled TGFβ binding occurring in the assays is due to true

[6] J. Massagué, B. Kelly, and C. Mottola, *J. Biol. Chem.* **260**, 4551 (1985).

competition by TGFβ in the sample, and not due to the action of contaminating proteases that degrade ^{125}I-labeled TGFβ during the assay, or the potential presence of ^{125}I-labeled TGFβ binding proteins.

Binding of ^{125}I-Labeled TGFβ to Solubilized Receptors

Efforts in our laboratory to develop an assay for solubilized TGFβ receptors using various modifications of the polyethylene glycol precipitation method have failed due to substantial coprecipitation of free ^{125}I-labeled TGFβ which yields unacceptably high nonspecific background radioactivity values. However, we have taken advantage of the fact that TGFβ receptors from all sources examined bind specifically to the lectin, wheat germ agglutinin (WGA) (see below), to devise an alternative method for the assay of solubilized TGFβ receptors. In this assay, TGFβ receptors solubilized from whole cells or membrane preparations are incubated with ^{125}I-labeled TGFβ and then separated from free ^{125}I-labeled TGFβ by binding to Sepharose-coupled WGA. This assay is based on the rationale previously described for the soluble receptor assay for glucagon receptors.[7]

Materials

Source of receptor: The receptor is solubilized from its sources (intact cells, isolated membranes) by incubation for 40 min at 4° in the presence of 1% n-octylglucoside (Sigma), 125 mM NaCl, 1 mM EDTA, 25 mM HEPES, pH 7.4, and protease inhibitors [0.3 mM phenylmethylsulfonyl fluoride (PMSF), 50 μg/ml aprotinin, 10 μg/ml antipain, 10 μg/ml leupeptin, 10 μg/ml pepstatin, 100 μg/ml soybean trypsin inhibitor, 0.5 mM benzamidine hydrochloride]. The solubilized extract is clarified by centrifugation at 100,000 g for 60 min at 4°

Binding buffer: A solution appropriately formulated to yield the following concentrations after mixing with solubilized receptor samples: 128 mM NaCl, 5 mM KCl, 5 mM MgSO₄, 1.2 mM CaCl₂, 25 mM HEPES, pH 7.4, 0.8% n-octylglucoside, 2 mg/ml BSA

Washing buffer: 125 mM NaCl, 5 mM KCl, 5 mM MgSO₄, 25 mM HEPES, pH 7.4, 0.8% n-octylglucoside, 2 mg/ml BSA

Elution buffer: 125 mM NaCl, 25 mM HEPES, pH 7.4, 0.4 M N-acetylglucosamine, 0.8% n-octylglucoside, 2 mg/ml BSA

WGA-Sepharose: 1–2 mg WGA/ml of gel (P-L Biochemicals)

[125]I-labeled TGFβ

TGFβ

[7] R. Iyengar and J. T. Herberg, this series, Vol. 109, p. 13.

Procedure

1. Aliquots of the solubilized receptor extract are mixed with binding buffer, plus 10–100 pM ^{125}I-labeled TGFβ alone or with increasing concentrations of TGFβ in siliconized glass test tubes. Final volume is 0.3–0.5 ml.
2. After incubation for 4 hr at 4°, samples are transferred to 1.5-ml microcentrifuge plastic tubes that contain a small predetermined volume of packed, prewashed WGA-Sepharose beads. Incubations continue at 4° with end-over-end mixing for 2 hr or overnight.
3. Beads are sedimented by a brief (30 sec) centrifugation at 12,000 g, the supernatant is aspirated, and the pellet of beads is washed four times with washing buffer at 4°.
4. The washed beads are incubated for 1 hr at 4° with 0.5 ml of elution buffer.
5. After sedimentation of the beads, the radioactivity present in the supernatant is determined with a gamma counter.

Comments

Of various detergents tested, including n-octylglucoside, Triton X-100, Lubrol PX, sodium deoxycholate, and 3-[(3-cholamidopropyl) dimethyl-ammonio]-1-propane sulfate (CHAPS), we find the former to be the most effective for TGFβ receptor solubilization.

The volume of WGA-Sepharose beads in the assays depends on two parameters, the binding capacity of the particular batch of WGA-Sepharose used, and the amount of WGA-binding glycoprotein present in the solubilized receptor sample. The optimal volume of beads needed in a particular experiment must be determined in advance on the basis of these two variables. The optimal volume of WGA-Sepharose is that which provides TGFβ receptor binding capacity in small excess over that required to retrieve a maximal amount of receptor-bound ^{125}I-labeled TGFβ in the sample under study. This volume of WGA-Sepharose beads can be determined in parallel experiments using a fixed volume of beads and variable amounts of solubilized extract, and vice versa. Using WGA-Sepharose from the source listed above, we find that 10μl of packed beads is an optimal volume for assays of receptors solubilized from 50 μg of BALB/c 3T3 membrane protein. Sepharose beads sediment rapidly but the pellets can be easily disturbed and resuspended while the supernatant is removed. Therefore, care must be taken during the washes to avoid losing beads by aspiration.

The last step, namely elution of material specifically bound to WGA

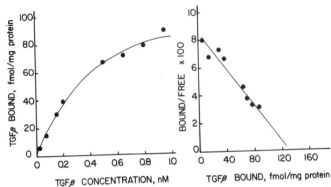

FIG. 2. Saturation curves of ^{125}I-labeled TGFβ binding to solubilized BALB/c 3T3 cell receptors. A membrane preparation from BALB/c 3T3 cells was solubilized in the presence of 1% *n*-octyl glucoside. Aliquots (50 μg of membrane protein) of the soluble extract were incubated at 4° in the presence of ^{125}I-labeled TGFβ alone (25–200 pM) or 200 pM TGFβ plus various amounts of unlabeled TGFβ to achieve the indicated final concentrations. After 4 hr, 10 μl of packed WGA-Sepharose beads was added to each sample and incubations continued at 4° overnight with mixing. The beads were then sedimented, washed, and eluted as described in the text. The amount of TGFβ specifically bound and eluted was then determined, and is plotted on the left panel. Right panel: Scatchard analysis of these data.

with the competing sugar, N-acetylglucosamine, is designed to eliminate radioactivity contributed by nonspecific adsorption of ^{125}I-labeled TGFβ to the Sepharose beads. ^{125}I-Labeled TGFβ adsorbed to the beads is not displaceable by N-acetylglucosamine. The background of nonspecifically adsorbed ^{125}I-labeled TGFβ is unacceptably high if the radioactivity bound to WGA-Sepharose beads during the assay is directly determined by counting the uneluted beads.

Using this binding assay, receptors solubilized from BALB/c 3T3 membranes exhibit a $K_d = 5 \times 10^{-10}$ M for ^{125}I-labeled TGFβ under conditions in which about 50% of the bound radioactivity at low concentrations of ^{125}I-labeled TGFβ is specific (Fig. 2). This method can be adapted to affinity label solubilized TGFβ receptors.[4]

This soluble radioreceptor assay is less sensitive and less accurate than the TGFβ binding assays performed with intact cell monolayers. Like in soluble receptor assays for other hormone receptors, there is no certainty that the TGFβ receptor is quantitatively retrieved by the WGA-Sepharose in every experiment.

Identification of TGFβ Receptor Proteins

Affinity cross-linking of ligand:receptor complexes using homobifunctional cross-linking agents was first developed for the identification of insulin receptor proteins,[8,9] and subsequently applied to other systems including the receptors for insulin-like growth factors I and II,[9] platelet-derived growth factor,[10] epidermal growth factor,[11,12] TGFα,[12] and others. The method is based on cross-linking receptors in intact cells, isolated membranes, or soluble extracts with receptor-bound ^{125}I-labeled ligand. The affinity-labeling reaction in this method does not require prederivatization of the ^{125}I-labeled ligand with a photoreactive group, as in standard photoaffinity-labeling methods. Rather, affinity labeling is achieved by addition of the cross-linking agent *after* the ^{125}I-labeled ligand:receptor complexes are formed. This treatment results in the formation of covalent complexes between the receptor protein and the labeled ligand. The structural properties of the labeled complexes can be analyzed by electrophoresis, chromatography, or other methods. The validity of this approach is illustrated by the fact that the receptor size and subunit structure proposed on the basis of receptor affinity cross-linking data for insulin, IGF-I, IGF-II, PDGF, EGF, and TGFα receptors has been eventually corroborated by purification or molecular cloning of the corresponding receptor proteins.[13-18]

Of several cross-linking agents currently available, disuccinimidyl suberate (DSS) has been the first and most widely used. It reacts with primary amino groups in proteins. This reagent has been used successfully in this laboratory to affinity label membrane components that have the properties expected from TGFβ receptors. The general protocol described below is designed for the affinity labeling of TGFβ receptors in cultured cell monolayers. The basic protocol can be modified to affinity labeling TGFβ receptors in cell suspensions, membrane preparations, and solubilized receptor samples.

[8] P. F. Pilch and M. P. Czech, *J. Biol. Chem.* **254,** 3375 (1979).
[9] J. Massagué and M. P. Czech, this series, Vol. 109, p. 179.
[10] D. F. Bowen-Pope and R. Ross, this series, Vol. 109, p. 69.
[11] J. Massagué, M. P. Czech, K. Iwata, J. E. De Larco, and G. J. Todaro, *Proc. Natl. Acad. Sci. U.S.A.* **79,** 6822 (1982).
[12] J. Massagué, *J. Biol. Chem.* **258,** 13606 (1983).
[13] T. Siegel, S. Ganguly, S. Jacobs, O. M. Rosen, and C. S. Rubin, *J. Biol. Chem.* **256,** 9266 (1981).
[14] A. Ullrich, J. R. Bell, E. Y. Chen, R. Herrera, L. M. Petruzzelli, T. J. Dull, A. Gray, L. Coussens, Y.-C. Liao, M. Tsubokawa, A. Mason, P. H. Seeburg, C. Grundfeld, O. M. Rosen, and J. Ramachandran, *Nature (London)* **313,** 756 (1985).
[15] C. L. Oppenheimer and M. P. Czech, *J. Biol. Chem.* **258,** 9033 (1983).
[16] S. Cohen, H. Ushiro, C. Stoscheck, and M. Chinkers, *J. Biol. Chem.* **257,** 1523 (1982).
[17] A. Ullrich, L. Coussens, J. S. Hayflick, T. J. Dull, A. Gray, A. W. Tam, J. Lee, Y. Yarden, T. A. Libermann, J. Schlessinger, J. Downward, E. L. V. Mayes, N. Whittle, M. D. Waterfield, and P. H. Seeburg, *Nature (London)* **309,** 418 (1984).
[18] A. R. Frackelton, Jr., P. M. Tremble, and L. T. Williams, *J. Biol. Chem.* **259,** 7909 (1984).

Materials

They include the following materials described above for binding assays for ^{125}I-labeled TGFβ in intact cells:

Binding buffer I, or binding buffer II
[125]I-labeled TGFβ
TGFβ
Cell monolayers

In addition:

Disuccinimidyl suberate (Pierce Chemical Co.): dissolved at 27 mM concentration in dimethyl sulfoxide immediately before use

Cell detachment buffer: 0.25 M sucrose, 10 mM Tris, 1 mM EDTA, pH 7.4, 0.3 mM PMSF

Cell solubilization buffer: 125 mM NaCl, 10 mM Tris, 1 mM EDTA, pH 7.0, 1% Triton X-100

Protease inhibitor cocktail I: 1 mg/ml leupeptin, 1 mg/ml antipain, 5 mg/ml aprotinin, 10 mg/ml soybean trypsin inhibitor, 10 mg/ml benzamidine hydrochloride (all from Sigma) in H_2O

Protease inhibitor cocktail II: 1 mg/ml pepstatin (Sigma), 1 mg/ml bestatin (Peninsula), 30 mM PMSF (Sigma) in dimethyl sulfoxide

Electrophoresis sample buffer: 100 mM Tris, pH 6.8, 20% glycerol, 2% sodium dodecyl sulfate, 0.05% bromphenol blue, and 100 mM dithiothreitol. Dithiothreitol must be omitted if the affinity-labeled receptors are to be studied in unreduced form

Procedure

1. Treat cell monolayers according to steps 1–4 described in the protocol for binding of ^{125}I-labeled TGFβ to intact cells. This treatment will generate a monolayer of cells with surface receptors occupied by ^{125}I-labeled TGFβ.

2. After the washes at 4°, add ice-cold binding buffer I lacking BSA, 0.67 ml/2-cm² well, or 2.0 ml/10-cm² well. Add 5 μl of 27 mM DSS solution/ml to each dish. Dispense the solution of DSS with a micropipet directly inside the buffer solution, not on its surface, and swirl dishes immediately after addition of DSS to minimize precipitation of DSS. Leave agitating at 4° for 15 min.

3. Rinse briefly with ice-cold detachment buffer. Then scrape cells off the dish with a disposable Teflon scraper in the presence of 0.5–1.0 ml of detachment buffer. Rinse the dish with a second aliquot of detachment buffer.

4. Sediment the cells by centrifugation at 12,000 g for 2 min, in 1.5-ml microcentrifuge tubes. Remove the supernatant.

5. Solubilize cell pellets in a minimal volume (60–100 μl) of solubilization buffer supplemented with 10 μl/ml of protease inhibitor cocktails I and II. Incubate 40 min at 4° with end-over-end mixing.
6. Remove insoluble cell debris by centrifugation for 15 min at 12,000 g. Mix the supernatant with 1 vol of electrophoresis sample buffer and heat at 100° for 2 min.

Comments

This protocol is designed for the analysis of affinity-labeled samples by SDS–PAGE. If other analytical techniques should be applied, the last step can be omitted or modified accordingly.

The cross-linking reaction is in part arrested by addition of Tris-containing buffer. The primary amino groups in Tris added in excess quench effectively unreacted DSS that may remain free in solution. However, it is important to note that Tris alone will not quench effectively the unreacted DSS that has partitioned into hydrophobic compartments in the cells. Effective quenching of residual DSS is achieved when samples are solubilized in the presence of sodium dodecyl sulfate. It is therefore important to proceed through step 6 as rapidly as possible. In protocols in which this is not possible, over-cross-linking of the sample can be avoided by carrying out preliminary tests to tritrate the minimal amount of DSS needed for efficient affinity labeling of the receptor, or by using the hydrosoluble analog, bis(sulfosuccinimidyl) suberate.

Variations of this general method have been successfully used in our laboratory when specific experimental requirements could not accommodate the general protocol. For example, cell solubilization with Triton X-100 or sodium dodecyl sulfate can be performed directly on the dishes after the cross-linking step, without the cell detachment step. Alternatively, the cross-linking reaction may be performed using cell suspensions obtained either before or after incubation with ^{125}I-labeled TGFβ. Isolated membrane preparations and solubilized receptors can also be affinity labeled by modifying centrifugations to sediment membrane fractions, or using WGA-Sepharose to retrieve solubilized, cross-linked receptor complexes. However, the general protocol described above usually yields the highest efficiency (15–18%) and specificity of receptor cross-linking with bound ligand.

Of other cross-linking agents tested to affinity label the TGFβ receptor, N-hydroxysuccinimidyl azidobenzoate, a heterobifunctional, photoreactive agent useful in studies on glucagon[19] and nerve growth factor recep-

[19] G. L. Johnson, V. I. MacAndrew, and P. F. Pilch, *Proc. Natl. Acad. Sci. U.S.A.* **78,** 875 (1981).

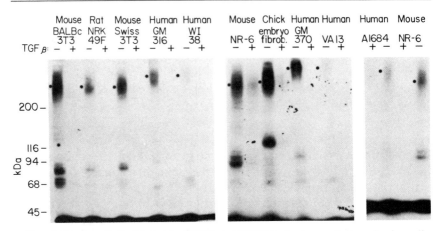

FIG. 3. Species affinity labeled by ^{125}I-labeled TGFβ in human, rodent, and avian cells. Near-confluent cell monolayers were detached by a brief exposure to buffer containing 1 mM EDTA, and were suspended in binding buffer. Aliquots (2×10^5 cells in 0.5 ml of buffer) were incubated in the presence of 150 pM ^{125}I-labeled TGFβ alone, or with 20 nM TGFβ as indicated, followed by incubation in the presence of 0.15 mM DSS. The affinity-labeled samples were solubilized with Triton X-100. The resulting extracts were subjected to SDS–PAGE in the presence of dithiothreitol. Shown are autoradiograms (4–6 days) from the resulting fixed, dried gels. Numbers at the left denote the positions corresponding to molecular weight markers. Black dots denote the major species labeled in human (330 kDa), rodent (280 kDa), and chick (290 kDa) cells, respectively. (From Ref. 22.)

tors[20] is very inefficient ($< 1\%$ cross-linking efficiency) to affinity label TGFβ receptors when added to preformed ^{125}I-labeled TGFβ: receptor complexes.[4] The sulfhydryl-reacting agent, m-maleimidobenzoyl-N-hydroxysuccinimide ester, which is effective in cross-linking ^{125}I-labeled cholecystokinin and ^{125}I-labeled insulin with their respective receptors,[21] is also ineffective with TGFβ receptors, even after reduction of the receptor to expose sulfhydryl groups (under conditions in which ^{125}I-labeled TGFβ binding to the receptor is not impaired) (J. Massagué, unpublished results). However, cleavable homologs of DSS such as ethylene glycol bis(succinimidyl succinate) and the hydrosoluble analog bis(sulfosuccinimidyl) suberate are comparable to DSS as cross-linking agents for TGFβ receptors (J. Massagué, unpublished results).

[20] J. Massagué, B. J. Guillette, M. P. Czech, C. J. Morgan, and R. A. Bradshaw, *J. Biol. Chem.* **256**, 9419 (1981).
[21] L. D. Madison, S. A. Rosenzweig, and J. D. Jamieson, *J. Biol. Chem.* **259**, 14818 (1984).

Analysis of the Data

Figure 3 shows an autoradiogram from an SDS–PAGE gel of various types of human, rodent, and chick cells affinity labeled with ^{125}I-labeled TGFβ. Cells were sequentially incubated with ^{125}I-labeled TGFβ (alone or with an excess of TGFβ) and DSS. The cross-linked samples were extracted with buffer containing Triton X-100 as described above and subjected to electrophoresis and autoradiography (with Kodak XAR film and Du Pont Lightning Plus screens). One major species of approximately 330 kDa in human cells, 280 kDa in rodent cells, and 290 kDa in chick cells can be

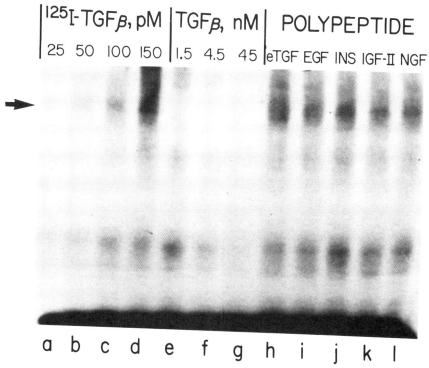

Fig. 4. Specificity of labeling of cellular components by ^{125}I-labeled TGFβ Aliquots (0.5 ml) of a suspension (4.1×10^5 cells/ml) of BALB/c 3T3 fibroblasts were incubated in the presence of increasing concentrations of ^{125}I-labeled TGFβ alone as indicated (lanes a–d), in the presence of 150 pM ^{125}I-labeled TGFβ plus various concentrations of native TGFβ as indicated (lanes e–g), or in the presence of 150 pM ^{125}I-labeled TGFβ plus TGFα (eTGF), EGF, insulin (INS), insulin-like growth factor-II (IGF-II), or nerve growth factor (NGF), all at 300 nM final concentration (lanes h–l). Cells were then treated with 0.25 mM disuccinimidyl suberate, extracted with 1% Triton X-100, and subjected to SDS–PAGE on 5–7% polyacrylamide gels. A 6-day autoradiogram from the fixed, dried gel is shown. The arrow points at the 280-kDa-labeled species. (From Ref. 4.)

observed. The labeling of this species is completely inhibited by a limited excess of TGFβ present during incubation of cells with ¹²⁵I-labeled TGFβ The intensity of labeling of this species in all cell types examined is proportional to the number of TGFβ receptors in each cell type, as determined by ¹²⁵I-labeled TGFβ binding assays. The electrophoretic migration of labeled TGFβ receptor species was not altered when the affinity-labeled samples were boiled in the presence of 5 M urea prior to electrophoresis. This observation suggests that the difference in apparent molecular mass

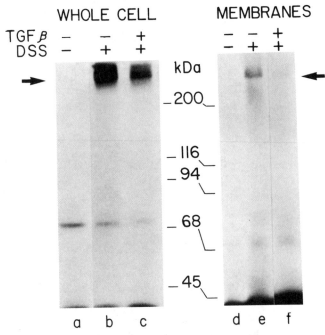

FIG. 5. SDS–PAGE and autoradiography of NRK-49F cells and membranes affinity labeled with ¹²⁵I-labeled TGFβ. Near-confluent monolayers of NRK-49F cells were detached from the culture vessels and either subjected directly to the affinity-labeling protocol or homogenized to obtain a crude membrane fraction. Suspensions of whole cells (2 × 10⁵ cells in 0.5 ml of buffer) (lanes a–c) or isolated membranes (150 μg of membrane protein in 0.3 ml of buffer) (lanes d–f) were affinity labeled by sequential incubations in the presence of 150 pM ¹²⁵I-labeled TGFβ and 0.25 mM DSS. Native TGFβ (1.5 nM) was added to some samples (lanes c and f) during incubation with ¹²⁵I-labeled TGFβ. Samples a and d were treated with dimethyl sulfoxide without the cross-linking agent, DSS. The affinity-labeled samples were extracted with a solution containing 1% Triton X-100 and subjected to electrophoresis on 7% (lanes a–c) or 6.5% (lanes d–f) polyacrylamide gels. Autoradiograms (3 days) from the resulting fixed, dried gels are shown. The positions of molecular size standards electrophoresed on parallel lanes are indicated. The arrows point at specifically labeled bands. (From Ref. 4.)

A TRITON SOLUBLE   B TRITON INSOLUBLE

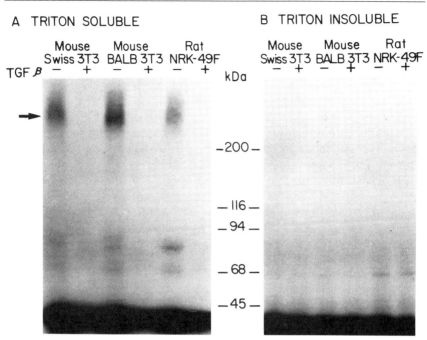

FIG. 6. Detergent solubility of affinity-labeled ^{125}I-labeled TGFβ receptors. Suspensions (1.8–2.2 × 10^5 cells in 0.5 ml of buffer) of mouse Swiss 3T3 and BALB/c 3T3 fibroblasts and rat NRK-49F fibroblasts were affinity labeled by incubation with 150 pM ^{125}I-labeled TGFβ alone (−) or in the presence of 40 nM TGFβ (+), followed by treatment with 0.25 mM DSS. The affinity-labeled cells were incubated in the presence 1% Triton X-100 for 40 min at 4°. The Triton-soluble extract was separated from insoluble material by centrifugation at 12,000 g for 15 min. After heating in the presence of 2% sodium dodecyl sulfate, the Triton-soluble (A) and Triton-insoluble (B) fractions were electrophoresed on 5–7% polyacrylamide gels in the presence of dodecyl sulfate. Four-day autoradiograms from the fixed, dried gels are shown. The positions corresponding to molecular size markers electrophoresed on parallel lanes are shown. The arrow points at 280-kDa-labeled species. (From Ref. 4.)

between TGFβ receptors from human and rodent sources is not due to incomplete unfolding of the molecule or noncovalent binding to other cellular components, but to some structural difference. However, comparative peptide maps of 330-kDa-labeled human receptors and 280-kDa-labeled rodent receptors have demonstrated a close structural relationship between receptors from these two sources.[22]

The specificity and high affinity of the labeled TGFβ receptor species is further illlustrated by the experiment in Fig. 4. BALB/c 3T3 cells were affinity labeled with increasing concentrations of ^{125}I-labeled TGFβ alone,

[22] J. Massagué, *J. Biol. Chem.* **260**, 7059 (1985).

or with 150 pM ^{125}I-labeled TGFβ in the presence of TGFβ or five other polypeptide growth factors. TGFα (EGF-like TGF, eTGF), EGF, insulin, IGF-II on nerve growth factor did not affect the labeling of the TGFβ receptor species, but 1.5 nM TGFβ completely inhibited it. Labeling of the 280-kDa species could be observed with as low as 25 pM ^{125}I-labeled TGFβ in the original autoradiograms. These observations indicate that the 280-kDa species corresponds to a TGFβ receptor with a $K_d \sim 10^{-10} M$.

In addition to the major labeled receptor species, minor labeled products in the 70- to 110-kDa range are frequently although not always observed in ^{125}I-labeled TGFβ-labeled samples. The concentration of TGFβ needed to completely inhibit the labeling of some of these minor species was 45 nM. This observation suggests that these labeled species correspond to binding sites with a $K_d \sim 10^{-8}$ M for TGFβ. However, high affinity TGFβ receptors of 65 and 85 kDa have been recently found (S. Cheifetz, B. Like, and J. Massagué, submitted for publication).

The 280-kDa TGFβ receptor component can also be labeled in isolated

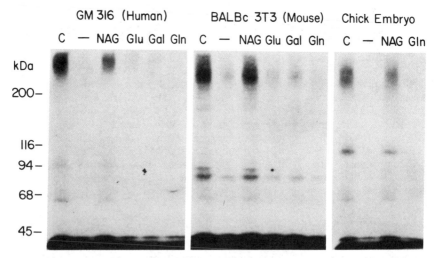

FIG. 7. Specific binding of affinity-labeled TGFβ receptors to WGA-Sepharose. Near-confluent monolayers of the indicated cell types were affinity labeled with ^{125}I-labeled TGFβ and solubilized in the presence of Triton X-100. Aliquots (90 μl) of the resulting extracts corresponding to 3×10^5 cells were incubated overnight at 4° in the presence of 15 μl of WGA-Sepharose beads alone ($-$), or with 0.1 M N-acetylglucosamine (NAG), 0.1 M glucose (Glu), 0.1 M galactose (Gal), or 0.1 M glucosamine (Gln). Control samples (C) were incubated in the presence of 15 μl of Sepharose CL-4B beads instead of WGA-Sepharose. After incubation, the beads were sedimented and 90-μl aliquots of the supernatant were subjected to SDS–PAGE and autoradiography. The resulting autoradiograms (6 days) are shown. (From Ref. 22.)

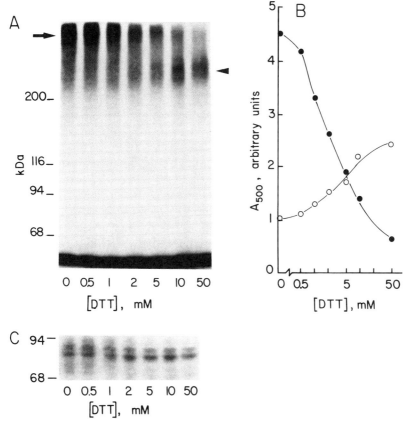

FIG. 8. Effects of dithiothreitol on the structure of affinity-labeled TGFβ receptors. Mono-layers of BALB/c 3T3 cells were affinity labeled with ^{125}I-labeled TGFβ and DSS, and then solubilized in the presence of 1% Triton X-100. Aliquots of the soluble extract corresponding to 3×10^5 affinity-labeled cells each were heated for 1 min at 100° in the presence of electrophoresis sample buffer containing the indicated concentrations of dithiothreitol (DTT). Samples were then electrophoresed on 5% polyacrylamide gels. (A) A 3-day autora-diogram from the fixed, dried, gels. The arrow shows the >500-kDa receptor complex; the arrowhead shows the reduced, 280-kDa component. (B) Densitometric quantitation of the >500-kDa species (●) and the 280-kDa species (○) on each lane in (A). (C) The 68- to 94-kDa region of a 9-day autoradiographic exposure of the same experiment. The experiment was repeated five times with similar results. (From Ref. 22.)

membrane preparations (Fig. 5). It is likely an intrinisc membrane protein because it cannot be extracted with 5 M urea (not shown) but is quantitatively solubilized by Triton X-100 (Fig. 6).

The ability of affinity-labeled TGFβ receptors to interact with plant lectins has been tested to examine whether the TGFβ receptor is glycosylated like other membrane glycoproteins that have extracellular domains. Figure 7 shows that detergent-soluble extracts containing affinity-labeled human, mouse, and chick TGFβ receptors can be depleted of this receptor type when they are incubated in the presence of WGA-Sepharose. The binding of TGFβ to WGA-Sepharose can be prevented by the presence of N-acetylglucosamine, the sugar specific for this lectin, but not by nonspecific sugars (Fig. 7). Affinity-labeled TGFβ receptors prebound to WGA-Sepharose can be eluted with N-acetylglucosamine. These properties of the TGFβ receptor are the basis of the binding assay for solubilized receptors as described above.

The 280- to 330-kDa TGFβ receptor species are observed in samples reduced by treatment with 50 mM dithiothreitol prior to electrophoresis. If reductant is eliminated from the electrophoresis buffer, the affinity-labeled material penetrates the gells as a complex of much higher (> 500 kDa) apparent molecular mass (Fig. 8), with only a residual amount of the label being present as diffusely migrating material in the 280-kDa region of the gels. As the concentration of reductant is increased, a progressive decrease in the amount of high-molecular-mass species and a concomitant increase in 280-kDa labeled species are observed. Increasing the concentration of dithiothreitol did not increase the intensity of labeling of minor labeled species (Fig. 8) indicating that they are not components of disulfide-linked receptor complexes. The overall decrease (by about one-half) in the labeling of receptor bands is attributed to reductant-mediated dissociation and loss of the ^{125}I-labeled TGFβ chain not cross-linked to the receptor. Similar observations have been made with human and aviant labeled TGFβ receptors electrophoresed with and without reductant (not shown). The molecular parameters of nondenatured affinity-labeled TGFβ receptor complexes have been determined by sucrose density gradient sedimentation and by molecular filtration chromatography, and are summarized in Table I.

TABLE I
PHYSICAL PARAMETERS OF MOUSE AND HUMAN FIBROBLAST TGFβ RECEPTORS[a]

Parameter	Mouse (BALB/c 3T3)	Human (GM316)
Stokes radius (nm)	8.2 ± 0.1 ($n = 3$)[b]	8.5 ± 0.2 ($n = 2$)
Sedimentation coefficient, $s_{20,\omega}$	12.6 ± 0.3 ($n = 3$)	13.0 ± 0.4 ($n = 2$)
Calculated M_r	565,000	615,000
Frictional ratio, f/f_0	1.4	1.4
M_r from SDS–PAGE	280,000	330,000

[a] Parameters are for ligand-occupied receptor complexes. (From Ref. 22.)
[b] Values are the average $\pm$ SD of the indicated number (n) of determinations.

The results indicate that the high-affinity TGFβ receptor from human, rodent, and avian cells consists of a disulfide-linked, glycosylated receptor complex of 565–615 kDa. This complex contains 280- to 330-kDa subunit that by its susceptibility to cross-linking with ^{125}I-labeled TGFβ is likely to contain the ligand binding site, and to be glycosylated. The nature of the other subunits in the receptor is unknown at present, but this receptor type could consist of two identical subunits of 280–330 kDa linked by disulfide bonds. Available data suggest that the oligomeric structure of the receptor is not induced by occupancy with the ligand.[22]

Section III

Somatomedin/Insulin-Like Growth Factors

[18] Solid-Phase Synthesis of Insulin-Like Growth Factor I

By DONALD YAMASHIRO and CHOH HAO LI

Divergent studies on somatomedin (Sm; formerly "sulfation factor"), insulin-like growth factor (IGF; formerly nonsuppressible insulin-like activity, NSILA), and multiplication-stimulating activity (MSA) have recently been unified as a consequence of the elucidation of the structures of these growth factors. Historical perspectives on this unification have recently been summarized.[1-3] The structural identity of somatomedin C and IGF-I has been established[4,5] and the entity has been designated as either Sm-C/IGF-I or IGF-I (Fig. 1). The structure of MSA from rat liver cell cultures is homologous to human IGF-II and differs in only five conservative amino acid substitutions.[6,7] Therefore, the name rat IGF-II has been proposed.[7] The nucleotide sequence of a human liver cDNA which encodes the complete amino acid sequence of IGF-I has been reported.[8] Chemical synthesis of human IGF-I by the solid-phase method[9] has shown that the biological activity associated with this factor is an intrinsic property of the peptide.[10]

Protected Peptide Resin of IGF-I

The linkage of the peptide chain to the insoluble polymer was through the so-called "Pam" attachment[11]: 4-(Boc-alanyloxymethyl)phenylaceta-

[1] R. E. Humbel, *in* "Hormonal Proteins and Peptides" (C. H. Li, ed.), Vol. 12, p. 57. Academic Press, New York, 1984.
[2] J. J. Van Wyk, *in* "Hormonal Proteins and Peptides" (C. H. Li, ed.), Vol. 12, p. 81. Academic Press, New York, 1984.
[3] S. P. Nissley and M. M. Rechler, *in* "Hormonal Proteins and Peptides" (C. H. Li, ed.), Vol. 12, p. 127. Academic Press, New York, 1984.
[4] E. Rinderknecht and R. E. Humbel, *J. Biol. Chem.* **253**, 2769 (1978).
[5] D. G. Klapper, M. E. Svoboda, and J. J. Van Wyk, *Endocrinology* **112**, 2215 (1983).
[6] E. Rinderknecht and R. E. Humbel, *FEBS Lett.* **89**, 283 (1978).
[7] H. Marquardt, G. J. Todaro, L. E. Henderson, and S. Oroszlan, *J. Biol. Chem.* **256**, 6859 (1981).
[8] M. Jansen, F. M. A. van Schaik, A. T. Ricker, B. Bullock, D. E. Woods, K. H. Sabbay, A. L. Nussbaum, J. S. Sussenbach, and J. L. Van den Brande, *Nature (London)* **306**, 609 (1983).
[9] R. B. Merrifield, *J. Am. Chem. Soc.* **85**, 2149 (1963).
[10] C. H. Li, D. Yamashiro, D. Gospodarowicz, S. L. Kaplan, and G. Van Vliet, *Proc. Natl. Acad. Sci. U.S.A.* **80**, 2216 (1983).

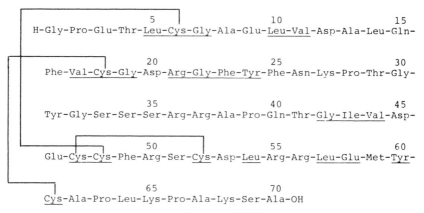

FIG. 1. Structure of IGF-I.

midomethyl-resin (see Tam [12], this volume, for use of Pam attachment polymer in the solid-phase synthesis of TGFα). 4-[N^α-Boc-alanyloxymethyl]phenylacetic acid (1.7 mmol) was coupled to aminomethyl-resin[12] (2.0 g, 0.84 mmol) in CH_2Cl_2 (20 ml) with dicyclohexylcarbodiimide (1.28 mmol) in 75 min at 24°. The starting material (1.60 g, 0.60 mmol) was placed in a Beckman model 990 peptide synthesizer and the amount of CH_2Cl_2 held by resin after drainage was determined (imbibed volume, about 6.4 ml). The synthetic protocol followed a two-stage coupling procedure[13] as follows: (1) 4 × CH_2Cl_2; (2) 1 × 55% trifluoroacetic acid in CH_2Cl_2; (3) 1 × 55% trifluoroacetic acid in CH_2Cl_2, 15 min; (4) 2 × CH_2Cl_2; (5) 3 × 25% dioxane in CH_2Cl_2; (6) 2 × CH_2Cl_2; (7) 1 × 5% diisopropylethylamine in CH_2Cl_2, 2 min; (8) 2 × CH_2Cl_2; (9) repeat step (7); (10) 5 × CH_2Cl_2; (11) 3 Eq of the symmetrical anhydride of Boc-amino acids[14] in CH_2Cl_2, 20 min; (12) 1 Eq of N-methylmorpholine and sufficient trifluoroethanol to give a concentration of 20%, 20 min; (13) 3 × CH_2Cl_2; (14) 3 × 33% ethanol in CH_2Cl_2. Unless otherwise noted volumes of solvents and reagents for each application were 2.5 × the imbibed volume and the mix time 1 min. The imbibed volume was periodically determined and the volumes for reagents increased accordingly, with step (2) requiring greater increases to achieve a stirrable slurry.

[11] A. R. Mitchell, S. B. H. Kent, M. Engelhard, and R. B. Merrifield, *J. Org. Chem.* **43**, 2845 (1978).
[12] J. T. Sparrow, *J. Org. Chem.* **41**, 1350 (1976).
[13] D. Yamashiro, J. Blake, and C. H. Li, *Tetrahedron Lett.* **18**, 1469 (1976).
[14] H. Hagenmaier and H. Frank, *Hoppe-Seyler's Z. Physiol. Chem.* **353**, 1973 (1972).

Symmetrical anhydrides were prepared as follows: 3.6 mmol of Boc-amino acid in 7 ml CH_2Cl_2 (12 ml for Boc-Leu-OH) was cooled to 0°, mixed with 3.0 ml of CH_2Cl_2 containing 1.8 mmol of dicyclohexylcarbodiimide (DCC), stirred at 0° for 15 min, filtered, and the precipitate rinsed with 2 ml CH_2Cl_2. An amino acid reservoir was charged with this filtrate and kept at 0°. Boc-Met (d-sulfoxide)-OH,[15] Boc-Arg(Tos)-OH, and Boc-Gln-OH were dissolved in 4.5, 3.4, and 5 ml dimethylformamide before dilution to 7 ml with CH_2Cl_2 and treatment by the symmetrical anhydride protocol. The coupling time for these derivatives in step (11) was 40 min. An exception to the symmetrical anhydride protocol was the coupling of Boc-Asn-OH for which the DCC-1-hydroxybenzotriazole procedure was used.[16] Thus, Boc-Asn-OH (3.0 mmol) and 1-hydroxybenzotriazole (3.3 mmol) in 6 ml dimethylformamide were cooled to −10° and 2.7 mmol of DCC in 4.5 ml CH_2Cl_2 was added and stirred for 45 min at −10 to 0°. This mixture was filtered, the precipitate rinsed with 1 ml dimethylformamide, and the filtrate added directly to the reaction vessel. For the second coupling stage, step (12), 1 Eq of diisopropylethylamine in 3 ml CH_2Cl_2 was delivered for residues 69 through 59 and all subsequent couplings in which dimethylformamide was used. For all other couplings the protocol was followed. The following side chain blocking groups were used: Asp, cyclopentyl[17]; Thr, Ser, and Glu, benzyl; Cys, 3,4-dimethylbenzyl[18]; Met, d-sulfoxide; Tyr and Lys, 2-bromobenzyloxycarbonyl; Arg, tosyl. The yield of protected peptidyl resin was 7.92 g (weight gain was 87% of theory).

Cleavage and Deblocking of Peptide and Oxidation

Protected IGF-I resin (1.336 g, or 0.101 mmol) was treated with 50% trifluoroacetic acid/CH_2Cl_2 (15 ml) for 15 min and then filtered and washed with CH_2Cl_2 and absolute ethanol. One-half of the dried peptide resin was treated with 10 ml of liquid HF[19] in the presence of 2-mercaptopyridine (203 mg)[20] and anisole (1.4 ml) for 75 min at 0°. The HF was removed with a rapid stream of nitrogen at 0°. The residue was mixed with cold ethyl acetate (10 ml) and stirred at 24° until a finely divided powder was obtained, which was filtered off. The other half of the peptide resin was treated in the same way. The two batches were combined and mixed with

[15] K. Hofmann, W. Haas, M.-J. Smithers, and G. Zanetti, *J. Am. Chem. Soc.* **87**, 631 (1965).
[16] W. Konig and R. Geiger, *Chem Ber.* **103**, 788 (1970).
[17] J. Blake, *Int. J. Pept. Protein Res.* **13**, 418 (1979).
[18] D. Yamashiro, R. L. Noble, and C. H. Li, *J. Org. Chem.* **38**, 3561 (1973).
[19] S. Sakakibara, Y. Shimonishi, Y. Kishida, M. Okada, and H. Sugihara, *Bull. Chem. Soc. Jpn.* **40**, 2164 (1967).
[20] D. Yamashiro, *Int. J. Pept. Protein Res.* **20**, 63 (1982).

20 ml of 6 M guanidine–HCl. The pH was adjusted to approximately 7 with 1 M NaOH, and the mixture was triturated for 1 hr. Dithiothreitol (68 mg)[21] was then added and the pH was maintained at 7.0 to 7.5 over 30 min by addition of 1 M NaOH. The resin was filtered off and washed with 6 M guanidine–HCl. The filtrate was diluted with 6 M guanidine HCl to a volume of 100 ml. The solution was dialyzed (SpectraPor 3, 11.5-mm diameter, molecular weight cut-off 3500, Spectrum Medical Industries, Los Angeles, CA) against 1 M acetic and then 0.1 M acetic acid. A large precipitate formed during the dialysis; lyophilization of the mixture gave 684 mg.

The crude reduced form of IGF-I was partially purified in a 1.9 × 26 cm column of CM-cellulose. Starting buffer was 0.01 M NH$_4$OAc containing 10 mM dithiothreitol and 8 M urea, pH 4.6; the eluting buffer was 0.2 M NH$_4$OAc containing 10 mM dithiothreitol and 8 M urea pH 5.0. Both were degassed prior to use. A gradient was produced by use of a 250-ml constant volume mixing chamber. Collection of 8-ml fractions (detection at 280 nm) showed a small peak (fractions 6–14) followed by a broad large peak (fractions 15–40). The latter was thoroughly dialyzed against 0.1 M acetic acid and lyophilized to give 487 mg.

Oxidation was performed in 2 M guanidine–HCl in 0.05 M NH$_4$HCO$_3$, pH 8.4, at 24° for 6 hr at a peptide concentration of 1 mg/ml. Thiol content was followed by the method of Ellman.[22] The solution was dialyzed against 0.1 M acetic acid and water and lyophilized to yield 446 mg.

Purification and Characterization of Synthetic IGF-I

The synthetic IGF-I was purified by gel filtration, partition chromatography,[23] and chromatofocusing. Gel filtration performed in a 2.5 × 43 cm column of Sephadex G-50 in 1 M acetic acid in 90-mg batches (Fig. 2) gave a total of 188 mg of monomeric material. Partition chromatography was then performed in a 1.9 × 38 cm column of Sephadex G-50 (20- to 44-μm dry particle diameter). This column, initially packed in water, was equilibrated with the lower phase of the two-phase solvent system 1-butanol:pyridine:0.1% aqueous acetic acid (5:3:10, v/v) followed by equilibration with the upper phase. The sample was applied in the upper phase, which was used to develop the chromatogram (Fig. 2). Partitioning of the material in

[21] W. W. Cleland, *Biochemistry* 3, 480 (1964).
[22] G. L. Ellman, *Arch. Biochem. Biophys.* 82, 70 (1959).
[23] D. Yamashiro, *in* "Hormonal Proteins and Peptides" (C. H. Li, ed.), Vol. 9, p. 25. Academic Press, New York, 1980.

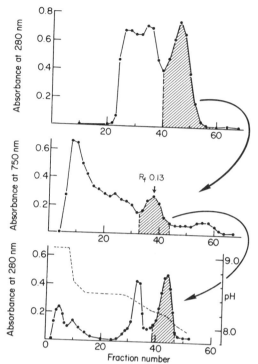

Fig. 2. Purification of synthetic IGF-I: Top: Gel filtration on Sephadex G-50. Middle: Partition chromatography on Sephadex G-50. Bottom: Chromatofocusing; broken line is the pH gradient.

two batches followed by lyophilization in the presence of added water gave 29 mg of product. This material was submitted to chromatofocusing (Pharmacia) in 1.0×26 cm column of Polybuffer exchanger PBE-94. Starting buffer was $0.025\ M$ ethanolamine–HCl, pH 9.4, and eluting buffer was Polybuffer 96–HCl, ph 7.0 (Fig. 2). For separation of the polybuffer the pooled fractions were mixed with 1-butanol (9 ml) and pyridine (5.4 ml) and subjected to partition chromatography in a 5×18 cm column of Sephadex G-50 in the same solvent system described above. The desired product emerged as a very broad peak with the polybuffer remaining on the column in the stationary aqueous phase. The fractions were mixed with water, evaporated to a low volume, and lyophilized. Final gel filtration in a 2.2×35 cm column of Sephadex G-50 in 1 M acetic acid gave one peak with a K_d of 0.32, and isolation by lyophylization gave 6.8 mg of highly purified IGF-I (approximately a 1% yield based on the initial loading of alanine on the solid support).

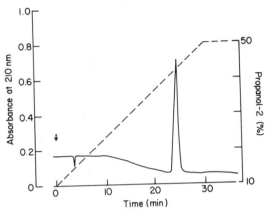

FIG. 3. HPLC of synthetic IGF-I

Amino acid analysis after 24-hr hydrolysis in constant boiling HCl gave, in residues per molecule (theoretical values in parentheses): Asp, 4.82 (5); Thr, 2.89 (3); Ser, 4.21 (5); Glu, 6.08 (6); Pro, 5.20 (5); ½-Cys, 5.28 (6); Gly, 6.73 (7); Ala, 6.19 (6); Val, 2.72 (3); Met, 0.95 (1); Ile, 0.69 (1); Leu, 5.90 (6); Tyr, 2.73 (3); Phe, 4.12 (4); Lys, 3.04 (3); Arg, 5.97 (6). Low values for Val and Ile can be accounted for by the acid-resistant Ile–Val sequence.

Analytical isoelectric focusing performed in a polyacrylamide gel (0.6 × 10 cm) in 8 M urea and 2% Ampholine, pI 9–5 (LKB), at 100 V for 16 hr showed a single band at pI 8.5 (staining was with Coomassie brilliant blue G-250 in perchloric acid).

Paper electrophoresis (50-μg sample) on Whatman 3 MM paper at pH 3.7 (pyridine acetate buffer) and pH 6.7 (collidine acetate buffer) for 4 hr at 400 V each gave a single spot (ninhydrin and Cl_2-toluidine detection) with R_f values relative to lysine of 0.51 and 0.25, respectively.

HPLC of a 40-μg sample was performed in a Vydac 201TP104 column with a 10 to 50% linear gradient of 2-propanol in 0.1% trifluoroacetic acid at a flow rate of 0.5 ml/min (detection at 210 nm) (Fig. 3).

High-performance partition chromatography[24] was performed in a 1.05 × 19 cm column of Superose 6B (Pharmacia #17-0489-01) under nitrogen pressures of up to 5 psi. The column, packed in water, was equilibrated with the lower phase of the two-phase solvent system 1-butanol:pyridine:1.5 M NaCl (1:2:4, v/v) and then with the upper phase. Samples were applied in the upper phase and the chromatogram developed with the upper phase. Detection was by the picrylsulfonic acid method (Fig. 4).[25]

[24] D. Yamashiro, Int. J. Pept. Protein Res. 22, 381 (1983).
[25] B. V. Plapp, S. Moore, and W. H. Stein, J. Biol. Chem. 246, 939 (1971).

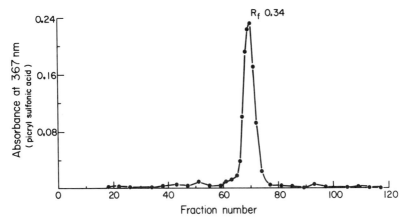

FIG. 4. High-performance partition chromatography of IGF-I on Superose 6B.

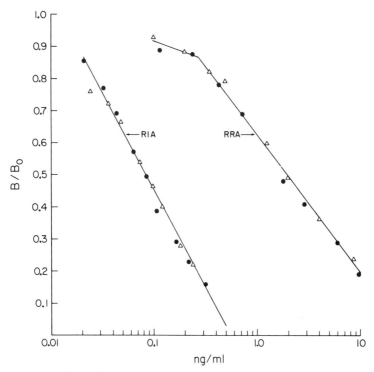

FIG. 5. Comparison of the displacement curves of synthetic and natural IGF-I in RIA (left) and RRA (right). B, Labeled IGF-I bound; B_0 labeled IGF-I bound in the absence of competitor. (●), Natural IGF-I; (△), synthetic IGF-I.

Biological Activities of Synthetic IGF-I[26]

Radioimmunoassay (RIA) was performed according to the method of Furlanetto *et al.*[27] and the radioreceptor assay (RRA) according to the method of D'Ercole *et al.*[28] The RRA and RIA methods of measurement of IGF-I have been described elsewhere in this volume. Figure 5 compares natural and synthetic IGF-I in RIA and RRA. The displacement curves produced by the two preparations are indistinguishable.

Mitogenic activities of natural and synthetic IGF-I were compared by assessing their ability to induce DNA synthesis in density-arrested BALB/c 3T3 cells. After exposure to platelet-derived growth factor (PDGF), these cells require only epidermal growth factor (EGF) and IGF-I to traverse G_1 and enter into the S phase of the cell cycle. In this system, EGF and IGF-I act synergistically to stimulate DNA synthesis. As shown in Table I, natural and synthetic IGF-I at concentrations of 16 ng/ml were indistinguishable in their stimulatory effect on DNA synthesis in the presence of EGF (10 ng/ml).

Comments

The linear sequence proposed for IGF-I was assembled by the all stepwise solid-phase method.[9] Coupling of amino acid residues was performed by a two-stage protocol in which preformed symmetrical anhydrides are prepared and coupled followed by use of trifluoroethanol to enhance coupling efficiency.[13] Cyclopentyl protection for the side chain of aspartyl residues was used to reduce base-catalyzed succinimide formation.[17] Sulfoxide protection of the side chain of the methionyl residue was used to avoid S-alkylation during synthesis, and it was removed in the final deblocking step by use of 2-mercaptopyridine in liquid HF.[20]

One of the critical steps in the synthesis of IGF-I was the correct formation of the disulfide bridges. Although the reduced form could be prepared in soluble form, oxidation under conventional conditions led mainly to insoluble products. Even when oxidation was performed under extremely dilute conditions (0.01 mg/ml) only low amounts of soluble monomer could be obtained. Earlier studies had indicated that reduced human growth hormone could be oxidized to the native form in 6 *M*

[26] J. J. Van Wyk, W. E. Russell, and C. H. Li, *Proc. Natl. Acad. Sci. U.S.A.* **81,** 740 (1984).

[27] R. W. Furlanetto, L. E. Underwood, J. J. Van Wyk, and A. J. D'Ercole, *J. Clin. Invest.* **60,** 648 (1977).

[28] A. J. D'Ercole, L. E. Underwood, J. J. Van Wyk, C. J. Decedue, and D. B. Foushee, *in* "Growth Hormone and Related Peptides" (A. Pecile and E. E. Muller, eds.), p. 190. Excerpta Medica, Amsterdam, 1976.

TABLE I

EFFECTS OF SYNTHETIC AND NATURAL IGF-I ON DNA
SYNTHESIS IN BALB/c 3T3 CELLS[a]

Condition	[³H]Thymidine incorporation (cpm)
EGF (10 ng/ml)	$18,681 \pm 2543$
EGF + natural (16 ng/ml)	$28,484 \pm 5961$
EGF + synthetic (16 ng/ml)	$25,770 \pm 3199$

[a] Data are reported as the mean $\pm$ SD of triplicate wells.

guanidine hydrochloride at 4° (T. Bewley, unpublished observations). Oxidation of reduced IGF-I under these conditions can be effected without precipitation, but no IGF-I is obtained. Successful oxidation was finally achieved in 2 M guanidine hydrochloride at pH 8.4 and 24° and a peptide concentration of 1 mg/ml. Complete solution was maintained throughout and the half-time of oxidation was about 1 hr. Similar success has been reported for phospholipase A_2[29] and β-bungarotoxin[30] at lower concentrations of denaturant. It is not self-evident that successful oxidation that requires denaturants implies that reduced IGF-I can spontaneously oxidize to the native form *in vivo*.

Acknowledgments

This work was supported in part by National Institutes of Health grants (GM-02907, AM-18677) and the Hormone Research Foundation.

[29] G. M. van Scharrenburg, G. H. Haas, and A. J. Slotboom, *Hoppe-Seyler's Z. Physiol. Chem.* **361,** 571 (1980).

[30] C.-J. Chen, S.-H. Wu, C.-L. Ho, and K.-T. Wang, *Proc. Natl. Sci. Counc., Repub. China* **6,** 411 (1982).

[19] Derivation of Monoclonal Antibodies to Human Somatomedin C/Insulin-Like Growth Factor I

By G. YANCEY GILLESPIE, JUDSON J. VAN WYK, LOUIS E. UNDERWOOD, and MARJORIE E. SVOBODA

The *in vitro* immunization of antigenically naive lymphocytes has provided a new dimension to the technology for the production of monoclonal antibody-secreting cell lines.[1-4] This technique involves incubation of immunocompetent lymphocytes with small amounts of antigen in medium that has been supplemented with thymocyte-generated growth factors. Splenic B lymphocytes that have been induced to proliferate in response to epitopes on the antigen are then fused to myeloma cells. The advantages of this approach are that the immunization can be performed in only a few days and smaller amounts of antigen are required. However, when *in vitro* challenge is used as the sole exposure of naive lymphocytes to antigen, the resulting antibodies have usually been of the IgM isotype.[4]

Somatomedin C (Sm-C), also known as insulin-like growth factor I (IGF-I), is a M_r, 7649 peptide that was isolated from human plasma and is structurally homologous with human proinsulin.[5,6] Its concentrations in plasma are growth hormone dependent and it stimulates *in vitro* growth of cartilage and a variety of other cell types derived from extraskeletal tissues.[7] Recently, Schimpff *et al.*[8] have shown that Sm-C/IGF-I promotes DNA synthesis in lectin-stimulated human peripheral blood lymphocytes in serum-free culture.

After two unsuccessful attempts in our laboratories to recover mouse monoclones to human Sm-C/IGF-I, we were able to produce monoclonal antibodies when spleen cells, immunized *in vivo,* were boosted by incubation *in vitro* with purified Sm-C/IGF-I for 5 days. These antibodies were of the IgG isotype and their specificity and affinity for Sm-C/IGF-I are characterized in this chapter.

Materials and Methods

Preparation and Iodination of Sm-C/IGF-I and Its Conjugation to Mouse Albumin

Sm-C/IGF-I was isolated from Cohn fraction IV of human plasma by previously described methods.[9] The preparation used for immunization

---

[1] H. Hengartner, A. L. Luzzati, and M. Schreier, *Curr. Top. Microbiol. Immunol.* **81,** 92 (1978).

[2] R. A. Luben and M. A. Mohler, *Mol. Immunol.* **17,** 635 (1980).

[3] R. L. Pardue, R. C. Brady, J. R. Dedman, and C. L. Reading, *J. Cell. Biol.* **91,** 81a (1981).

[4] C. L. Reading, *J. Immunol. Methods* **53,** (1982).

[5] E. Rinderknecht and R. E. Humbel, *J. Biol. Chem.* **253,** 2769 (1978).

[6] D. G. Klapper, M. E. Svoboda, and J. J. Van Wyk, *Endocrinology* **112,** 2215 (1983).

[7] D. R. Clemmons and J. J. Van Wyk, *in* "Handbook of Experimental Pharmacology" (R. Baserga, ed.), Vol. 57, p. 161. Springer-Verlag, New York, 1981.

[8] R.-M. Schimpff, A.-M. Repellin, A. Salvatoni, G. Thieriot-Prevost, and P. Chatelain, *Acta Endocrinol.* **102,** 21 (1983).

[9] M. E. Svoboda and J. J. Van Wyk, this series, Vol. 109, p. 798.

was a pool of side fractions of the terminal HPLC purification steps on an alkylphenyl column and was estimated to be 80% pure based on specific activity and polyacrylamide gel electrophoresis. This preparation was coupled to mouse albumin in a final molar ratio of 6:1 by the stepwise addition of glutaraldehyde. Analysis by HPLC and RIA revealed that over 98% of the original Sm-C/IGF-I had been coverted to large-molecular-weight forms or insoluble complexes.

Iodination of a more highly purified Sm-C/IGF-I was carried out using a fractional chloramine-T method. Purification of ^{125}I-labeled Sm-C/IGF-I (specific activity, 330–370 μCi/ug) was performed by affinity chromatography on a column of polyclonal anti-Sm-C/IGF-I antibody bound to glass beads.

In Vivo Immunization Schedules

Immunization of BALB/c male mice was initiated by intramuscular (im; thigh of each hind leg) injection of 100 μg of conjugated Sm-C/IGF-I and 37 μg of free Sm-C/IGF-I emulsified in an equal volume of Freund's complete adjuvant. After 6 weeks, all mice were injected intramuscularly with 80 μg of conjugated and 30 μg of free Sm-C/IGF-I, emulsified with Freund's incomplete adjuvant. Subsequent challenges utilized im injections at 6- to 8-week intervals with decreasing amounts of conjugated and free Sm-C/IGF-I in incomplete Freund's adjuvant. Two weeks after the fourth injection, the mouse with the highest titer of serum antibodies was given 50 μg of conjugated Sm-C/IGF-I and 18 μg of free Sm-C/IGF-I intraperitoneally. Four days later, the mouse was decapitated and its spleen removed aseptically.

Preparation and in Vitro Immunization of Spleen Cells

The spleen was disaggregated to produce a single cell suspension and the erythrocytes were lysed in ice-cold 0.15 M NH$_4$Cl in 0.05 M potassium phosphate buffer, pH 7.8. The immune splenocytes were washed twice by centrifugation (300 g, 20°, 8 min) in 50 ml of Dulbecco's modified Eagle's medium (DMEM) and were partitioned as follows: The bulk of the cells (85 × 10^6) were fused directly with an equal number of P3X63Ag8.653 myeloma cells[10] as described below. The remaining viable splenocytes, approximately 20 × 10^6, were cultured in the presence of Sm-C/IGF-I before fusing the survivors with the myeloma cells.

Splenocytes were cultured (5 days, 37°, 10% CO$_2$, 95% humidity) at 4 × 10^6 viable cells/ml in DMEM with 10% fetal bovine serum (FBS) and

[10] J. F. Kearney, A. Radbruch, B. Liesgang, and K. Rajewsky, *J. Immunol.* **123,** 1548 (1979).

mixed with an equal volume of the same medium that had been conditioned for 48 hr by a mixed culture of BALB/c and C57BL/6 thymocytes.[4] To this was added 12 μg of a sterile preparation of human Sm-C/IGF-I. The cells were incubated in an upright Falcon T-25 flask that was swirled twice daily. At the end of 5 days in culture, 8.9×10^6 cells survived (44.5%). These were fused with 11.6×10^6 "653" myeloma cells using the method described below.

Hybridization of Splenocytes and Myeloma Cells

The method that we routinely employ is a modification of that described by Galfre et al.[11] using 50% PEG 1500. The fused cells were washed and gently dispersed in 10 ml of selection medium [DMEM supplemented to 20% with FBS, mixed 1:1 with REVC-conditioned medium and containing hypoxanthine–aminopterin–thymidine–glycine (HATG) formulated according to Littlefield[12]]. By including aminopterin at the outset, greater than 50% of the unfused P3X63Ag.8.653 myeloma cells die within the first 24 hr. After overnight incubation, the surviving fused cells were harvested, washed, adjusted to 5×10^5 viable cells/ml of selection medium, and 100 μl of cell suspension was dispensed into each well of Costar 3596 microculture trays. Within 2 days, an additional 100 μl of fresh selection medium was added to each well. About half of the medium in each well was replaced with fresh medium every 3–4 days.

Wells with anti-Sm-C/IGF-I activity were cloned by limiting dilution and the monoclones were amplified by intraperitoneal injection of 3×10^6 viable cells into pristane-primed BALB/c mice. Ascites fluids were collected and held frozen ($-20°$) until used.

Radioimmunoassay for Immunoglobulin-Secreting Colonies and Anti-Sm-C/IGF-I

At the time of near confluence of the colonial outgrowths in the original microwell plates, supernates were screened for evidence of mouse immunoglobulin (Ig) using a semiquantitative radioimmunoassay that employed F(ab')$_2$ goat anti-mouse Ig immobilized in poly (vinyl chloride) microtest wells, rabbit anti-mouse Ig, and radioiodinated *Staphylococcus aureus* protein A. Radioactivity bound to the PVC well bottoms was quantified in a gamma scintillation counter.

Ig-positive microcultures were screened for anti-Sm-C/IGF-I activity by

[11] G. Galfre, S. C. Howe, C. Milstein, G. W. Butcher, and J. C. Howard, *Nature (London)* **266,** 550 (1977).
[12] J. W. Littlefield, *Science* **145,** 709 (1964).

incubating supernates with ^{125}I-labeled Sm-C/IGF-I and precipitating the complex with either PEG-6000 (Sigma Chemical Co., St. Louis, MO) at a final concentration of 12.5% or an appropriate dilution of rabbit anti-mouse Ig antiserum.

Affinity and Specificity of Monoclonals

Affinity constants of monoclonals were determined by the method of Scatchard[13] using the PEG-based RIA. Specificity was assessed using insulin-like growth factor II (9-SE-IV, gift of R. Humbel, Zurich, Switzerland), multiplication-stimulating activity (MSA, #82-235, Collaborative Research), chymotryptic digests of Sm-C/IGF-I prepared in our laboratory, porcine insulin (#815-D63-10, gift of E. Lilly Co.), bovine proinsulin (#615-07J-91-1, E. Lilly Co.), bovine B-chain insulin (Sigma Chemical Co.), rat insulin (R-169, Novo Research Institute, Copenhagen, Denmark), human growth hormone (HS1395, National Hormone and Pituitary Program), mouse EGF (#867-10, Collaborative Research), human TSH (WHO, 68/38), and mouse albumin (#29C-0165, Sigma Chemical Co.).

Results

Value of the Initial Screening for Immunoglobulin Production

Of the 85×10^6 in vivo-immunized splenocytes that were fused directly with an equivalent number of myeloma cells, 864 wells were plated at 5×10^4 each. Of these, 201 wells (23%) eventually had 1 or more colonial outgrowths. To reduce the number that had to be expanded for testing for anti-Sm-C/IGF-I antibody, supernates from near-confluent cultures were first screened for the presence of mouse Ig. The percentage of Ig-positive supernates was greatest in the wells containing the more rapidly growing colonies: 75% of those reaching near confluence by day 17 secreted Ig, whereas only 3% of those requiring 26 days to reach the same degree of confluency secreted Ig (Table I). Those that were positive were expanded to 16-mm-diameter wells to generate sufficient supernatant fluid to test in both types of anti-Sm-C/IGF-I RIAs.

Recovery of Anti-sm-C Monoclones

Seventy-one of the Ig-secreting colonies from the direct fusion were tested for antibodies to Sm-C/IGF-I. Using the double-antibody method for detecting antibody binding to ^{125}I-labeled Sm-C/IGF-I, only one super-

---

[13] G. Scatchard, *Ann. N.Y. Acad. Sci.* **51**, 660 (1949).

TABLE I

FREQUENCY OF IMMUNOGLOBULIN-SECRETING OUTGROWTHS AS A
FUNCTION OF GROWTH RATE

Days from plating to screening	Number of colonies tested for immunoglobin	Ig-secreting outgrowths	
		Number	%
Direct fusion: 864 wells			
16	84	63	75
20	78	7	9
26	39	1	3
Total	201 (23%)	71	35
Fusion after *in vitro* boost: 88 wells			
12	81	56	69
17	4	1	25
Total	85 (97%)	57	67

nate bound radiolabeled antigen in significant amounts. Two additional wells were positive by the PEG method (Table II). In 2 previous direct hybridization attempts in which over 540 wells were screened for antibody activity to Sm-C/IGF-I, none was unequivocally positive for antibody activity, despite the presence of detectable antibody in the sera of the splenocyte donor animals.

In striking contrast to this were the results of the *in vitro* boosting following *in vivo* immunization. Of the 88 wells that were plated (5.2×10^4 well) from this fusion, colonial outgrowths occurred in 85 wells (97%). We tested 81 supernatants for mouse Ig and 57 (69%) were positive. Forty-two of these outgrowths were successfully expanded *in vitro* and 24 of these (57%) were found by both methods to secrete anti-Sm-C/IGF-I antibody.

TABLE II

YIELD OF COLONIAL OUTGROWTHS SECRETING ANTI-Sm-C/IGF-I MONOCLONAL
ANTIBODIES

Method of immunization	Number screened for antibody to Sm-C/IGF-I	Number positive by precipitation with	
		Rabbit anti-mouse Ig	PEG 6000
In vivo	71	1(1%)	3(4%)
In vivo with *in vitro* boost	42	24(57%)	24(57%)

Characterization of Anti-Sm-C/IGF-I Monoclones

Two monoclones (sm1.2 and sm5.1) were selected for further characterization and were amplified by production of ascites fluid. Both proved to be $IgG_{1,\kappa}$ immunoglobulins that exhibited high titers, binding 40–60% of trace quantities of ^{125}I-labeled Sm-C at dilutions of 10^{-5} in the PEG RIA. Affinity constants were determined to be 1.09 and 1.01 × 10^{10} liter/mol for sm1.2 and sm5.1, respectively. Both clones were quite specific for Sm-C/IGF-I. Insulin-like growth factor II exhibited 5% cross-reactivity and multiplication-stimulating activity was approximately 1% as active as Sm-C/IGF-I (Fig. 1). Neither of the chymotryptic fragments of Sm-C/IGF-I, the insulin preparations, hGH, hTSH, mEGF, or mouse albumin exerted activity at concentrations as high as 10^{-6} M.

Discussion

Using a combination of *in vivo* immunization followed by a brief *in vitro* incubation in the presence of small quantities of purified antigen, we have generated 24 hybrid cell lines secreting monoclonal antibodies to human Sm-C/IGF-I. Two of these, selected on the basis of highest reactivity to Sm-C/IGF-I, were determined to be $IgG_{1,\kappa}$ and exhibited highly

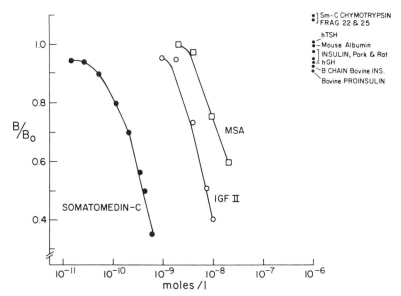

FIG. 1. Specificity and affinity of monoclone sm1.2 for human Sm-C/IGF-I were determined by competitive binding PEG-RIA as described in the text.

specific binding to Sm-C/IGF-I with affinities >10⁹ liter/mol. Including the parallel attempt reported here, we had virtually no success in producing monoclonal antibodies to Sm-C/IGF-I using direct fusions of myeloma cells with splenocytes from *in vivo* immunized animals, despite the presence of numerous immunoglobulin-producing colonial outgrowths. However, two other groups[14,15] have reported production of hybrids from *in vivo* immunizations. In the latter report the yield of Sm-C/IGF-I antibody-producing hybrids was 1.4%, compared with our 27.6% yield of specific antibody producers/well plated. The antibodies described in the two published reports appear to be similar in specificity and affinity to the monoclones that we have examined in detail.

Reports by several investigators suggest that primary *in vitro* immunization predisposes to the production of IgM antibodies. IgM is also the most prevalent isotype early after *in vivo* immunization. By using *in vitro* exposure to evoke an anamnestic response in cells previously sensitized by *in vivo* exposure to antigen, it is possible to combine the advantages of IgG antibodies with the efficiency of *in vitro* exposure. Although the data presented here represent our experience in only 1 comparison between the 2 techniques, our generation of 24 stable cell lines secreting specific antibody from 88 different wells compares favorably with yields reported by other investigators using the *in vitro* culture method without prior immunization.[4]

Hybridization of myeloma cells with immunized splenocytes to create specific antibody-secreting cell lines is a random process with a low order of success; fusion efficiencies of 10⁻⁵ to 10⁻⁸ are usually observed.[16-18] Techniques that significantly improve the rate of successfully generating antibody-producing hybrid cells may eventually find general utility. We have observed that an *in vitro* antigen challenge of *in vivo* immunized spleen cells markedly enhanced (1) fusion efficiency as defined by the ratio of immunoglobulin-secreting hybrid outgrowths/10⁶ splenocytes fused; (2) the percentage of Ig-secreting cells that produced specific antibody; and (3) the total number of specific antibody-secreting colonies. Moreover, due to the preferential survival of specific antibody producers, the total number of

[14] R. C. Baxter, S. Axiak, and R. L. Raison, *J. Clin. Endocrinol. Metab.* **54**, 474 (1982).
[15] U. K. Laubli, W. Baier, H. Binz, M. R. Celio, and R. E. Humbel, *FEBS Lett.* **149**, 109 (1984).
[16] F. Melchers, M. Potter, and N. Warner, *Curr. Top. Microbiol. Immunol.* **81**, (1978).
[17] C. Stahli, T. Staehelin, V. Miggiano, J. Schmidt, and P. Haring, *J. Immunol. Methods* **32**, 297 (1980).
[18] D. E. Yelton, D. H. Margulies, B. Diamond, and M. D. Scharff, *in* "Monoclonal Antibodies" (R. H. Kennett, T. J. McKearn, and K. B. Bechtol, eds.), p. 3. Academic Press, New York, 1980.

original microcultures that had to be handled was reduced considerably. It must be pointed out, however, that we have not established whether the 24 anti-Sm-C/IGF-I hybridomas are idiotypically distinct or recognize different epitopes. Indeed, the possibility exists that our hybridomas were derived from a few primary clones that were expanded during the 5-day *in vitro* incubation with the antigen.

Another criterion by which to evaluate the overall efficacy of the *in vitro* boosting technique is the utility of the antibodies generated. We have found monoclone sm1.2 to be highly useful as an affinity reagent for the purification of Sm-C/IGF-I.[19] When a 15% Na_2SO_4 precipitate of sm1.2 was coupled to cyanogen bromide-activated Sepharose 4B in ratios ranging between 1.7 and 6.8 mg of protein/ml of gel, we were able to extract up to 60 μg of Sm-C/IGF-I/ml of gel from crude extracts of human plasma containing several grams of protein. All of the Sm-C/IGF-I activity could be eluted in 0.2 M ammonium acetate, pH 3.2. Furthermore, our monoclonals have proven of great value in *in vitro* immunoneutralization studies directed at delineating the biological role of Sm-C/IGF-I in growth control.[20,21]

Summary

Somatomedin C, also called insulin-like growth factor I (Sm-C/IGF-I), is a highly conserved polypeptide required for the proliferation of many cell types. Since several attempts in our laboratory to recover monoclonal antibody-secreting hybrids to this peptide by the direct fusion of hyperimmunized splenocytes with myeloma cells had been unsuccessful, we modified our approach by coculturing hyperimmunized BALB/c splenocytes and a small amount of the antigen for 5 days prior to fusion with the P3X63Ag.8.653 myeloma cell line. Of 88 microcultures at risk, specific antibody was detected in 24. Two clones were expanded in ascites fluid and characterized as to isotype, affinity, and specificity. Both were $IgG_{1,\kappa}$ and bound human Sm-C/IGF-I with affinity constants of 1.09 and 1.01 × 10^{10} liter/mol, respectively. Both clones were quite specific for Sm-C/IGF-I with inconsequential binding to insulin-like growth factor II, multiplication-stimulating activity, any of the chymotryptic fragments of Sm-C/IGF-I, insulin preparations, hGH, hTSH, mEGF, or mouse albumin. *In vitro* boosting after primary *in vivo* immunization appears to provide

---

[19] S. D. Chernausek, P. G. Chatelain, M. E. Svoboda, L. E. Underwood, and J. J. Van Wyk, *Biochem. Biophys. Res. Commun.* **126**, 282 (1985).

[20] E. B. Leof, W. Wharton, J. J. Van Wyk, and W. J. Pledger, *Exp. Cell Res.* **141**, 107 (1982).

[21] W. E. Russell, J. J. Van Wyk, and W. J. Pledger, *Proc. Natl. Acad. Sci. U.S.A.* **81**, 2389 (1984).

monoclones of an IgG isotype in contrast to primary *in vitro* immunization, which reportedly favors an IgM isotype. The antibodies produced in this study have proved to be extraordinarily useful in defining the physiologic role of Sm-I/IGF-I with immunoneutralization techniques and in the purification of human Sm-C/IGF-I by affinity chromatography.

Acknowledgments

We thank Jane M. Bynum, Martha Pacilio, Mary Murphy, and Sylvia White for expert technical assistance. This work was supported by research Grants CA 30479, AM 01022, and HD 08299 from the NIH, by USPHS Training Grant AM 07129, and by the Neurosurgery Research Trust Fund. Judson Van Wyk is a recipient of a USPHS Research Career Development Award 5K06 AM 14115.

[20] Radioimmunoassay of Somatomedin C/Insulin-Like Growth Factor I

By RICHARD W. FURLANETTO and JEAN M. MARINO

Introduction

Of the assay methods which have been described for the measurement of somatomedin C/insulin-like growth factor I (Sm-C/IGF-I), radioimmunoassay methods are the most sensitive and specific. A number of immunoassays which are highly specific for Sm-C/IGF-I and which are capable of measuring picogram quantities of the hormone have been described[1-3] and Sm-C/IGF-I measurements are now an important diagnostic and investigative tool. Because Sm-C/IGF-I circulates in serum tightly bound to a specific carrier protein, most of these assays employ an extraction procedure when serum or plasma samples are being examined. We have found, however, that heparin promotes the dissociation of Sm-C/IGF-I from its binding protein and that the addition of heparin to the assay allows the quantitative measurement of the hormone in unextracted human serum.

This chapter describes the heparin assay procedure in detail and discusses its important features. The chapter also describes the methods used in our laboratory for the iodination of Sm-C/IGF-I and for production and

[1] R. W. Furlanetto, L. E. Underwood, J. J. Van Wyk, and A. J. D'Ercole, *J. Clin, Invest.* **60,** 648 (1977).

[2] R. M. Bala and B. Bhaumick, *J. Clin. Endocrinol. Metab.* **49,** 770 (1979).

[3] J. Zapf, H. Walter, and E. R. Froesch, *J. Clin. Invest.* **68,** 1321 (1981).

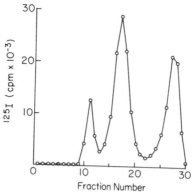

FIG. 1. Elution pattern of Sm-C/IGF-I iodination mixture on 1.5×50 cm column of Sephadex G-50 fine. Elution buffer was 0.03 M phosphate, 0.25% BSA, 0.02% sodium azide, pH 7.4. Fraction size was approximately 2 ml; 5-μl aliquots were counted. Peak 1 is iodinated albumin and damaged hormone; peak 2 is the radiolabeled Sm-C/IGF-I; peak 3 is unreacted sodium [^{125}I]iodide. Fractions 16–19 were combined, brought to 4% with BSA, aliquoted, and stored at $-20°$.

characterization of antisera to this hormone. Since the purification of Sm-C/IGF-I is the subject of two recent chapters in another volume in this series,[4,5] discussion of this topic has been omitted.

Preparation of Radiolabeled Sm-C/IGF-I

Sm-C/IGF-I is iodinated to a specific activity of 150–250 μCi/μg using limited quantities of chloramine-T as described by Freychet et al.[6] Hormone iodinated to higher specific activity is useable in the radioimmunoassay but is less stable. The iodinated hormone is separated from unreacted sodium iodide by chromatography on Sephadex G-50. Routinely 2.0 μg Sm-C/IGF-I is iodinated. The 2.0-μg aliquots are stored lyophilized in polypropylene microfuge tubes as $-70°$. We have stored Sm-C/IGF-I under these conditions for 2 years without significant loss of activity. All iodination procedures are performed at room temperature.

Prior to iodination the Sm-C/IGF-I is dissolved in 50 μl 0.5 M sodium phosphate buffer, pH 7.4. To this is added 1.0 mCi (10 μl) of carrier-free sodium [^{125}I]iodide, followed by 10 μl of a freshly prepared solution of 50 μg/ml chloramine-T. The solution is mixed gently and after 5 min an

[4] P. P. Zumstein and R. E. Humbel, this series, Vol. 109, p. 782.
[5] M. E. Svoboda and J. J. Van Wyk, this series, Vol. 109, p. 798.
[6] P. Freychet, J. Roth, and D. M. Neville, Jr., *Biochem. Biophys. Res. Commun.* **43**, 400 (1971).

aliquot is removed and the incorporation of iodide into protein is measured by precipitation with 10% trichloroacetic acid. Additional chloramine-T is added, if necessary, to give 30–50% incorporation. The iodination is terminated by the addition of 500 μl of 0.03 M sodium phosphate buffer containing 0.02% sodium azide, 0.25% BSA, pH 7.4. This solution is applied to a 1.5 × 50 cm column of Sephadex G-50 fine equilibrated in the same buffer and the column is developed at a flow rate of approximately 1 ml/min. Fractions of 2 ml are collected and aliquots counted (Fig. 1). The fractions containing the iodinated Sm-C/IGF-I (K_{av} 0.3–0.4) are combined and sufficient BSA is added to give a final concentration of 4%. Aliquots are stored at −20°. Iodinated Sm-C/IGF-I is remarkably stable when stored under these conditions and can undergo repeated freeze/thaw cycles without significant loss of immunoreactivity. We routinely use the iodinated hormone for 6 weeks without repurifying. If necessary the iodinated material can be repurified by chromatography over Sephadex G-50 as described above.

Preparation of Antiserum to Sm-C/IGF-I

Antisera to Sm-C/IGF-I have been prepared in a number of laboratories using a variety of immunization techniques.[1-3,7,8] One of us (R.W.F.) has used the method described below on two occasions to produce antisera suitable for use in the Sm-C/IGF-I radioimmunoassay.

Preparation of Sm-C/IGF-I Ovalbumin Complex

To enhance its immunogenicity the Sm-C/IGF-I is cross-linked to ovalbumin using glutaraldehyde as the coupling agent. For a typical conjugation 2.5 mg of partially purified hormone preparation (approximately 20% pure) and 1.0 mg ovalbumin (Sigma A5503) are dissolved in 700 μl 0.1 M sodium phosphate, pH 7.4. To this is added 300 μl of freshly prepared solution of 13 mM glutaraldehyde in phosphate buffer. The glutaraldehyde is added dropwise over 30 min. During the addition the solution turns a pale yellow color. The solution is allowed to stand at room temperature for 16 hr and is then dialyzed extensively against normal saline.

The purity of the material used for immunization is an important consideration in antiserum preparation. We have obtained highly specific antisera using only partially purified Sm-C/IGF-I preparations. Care must be taken, however, to use preparations which are relatively free of contami-

[7] R. L. Hintz, F. Liu, L. B. Marshall, and D. Chang, *J. Clin. Endocrinol. Metab.* **50,** 405 (1980).

[8] R. C. Baxter, A. S. Brown, and J. R. Turtle, *Clin. Chem.* **28,** 488 (1982).

nation with insulin-like growth factor II (IGF-II), since this peptide might cross-react in the immunoassay. Another important consideration in preparing the immunogen is the ratio of reagents used in the coupling procedure. A high concentration of glutaraldehyde results in the formation of an insoluble precipitate and is undesirable. Similarly, a low ratio of Sm-C/IGF-I to ovalbumin favors the formation of highly immunogenic ovalbumin complexes which might adversely affect production of antibodies to Sm-C/IGF-I. In our hands a 4:1 molar ratio of Sm-C/IGF-I to ovalbumin and a 10:1 molar ratio of glutaraldehyde to total protein has worked well.

Immunization of Rabbits

The Sm-C/IGF-I ovalbumin complex is diluted to a concentration of 1–2 mg protein/ml with normal saline and thoroughly emulsified with an equal volume of complete Freund's adjuvant. Mature New Zealand white rabbits of approximately 4 kg are immunized with this preparation using the multiple site immunization procedure described by Vaitukaitis *et al.*[9] except that simultaneous immunization with *Bordetella pertussis* vaccine is omitted. Each animal receives 500 to 1000 μg of the complex in a volume of 1 ml adjuvant distributed over approximately 30 sites in the axillary and femoral skin fold regions and along the back. Six weeks later the animals are bled and their sera tested for Sm-C/IGF-I binding activity. Animals with detectable Sm-C/IGF-I antibody are rechallenged with 100–250 μg of the complex emulsified in incomplete Freund's adjuvant and given intramuscularly in four to six sites. The animals are bled approximately 1 month later and at biweekly intervals thereafter. In responding animals the Sm-C/IGF-I antibody titer decreases slowly over the ensuing 4–6 months. Subsequent reimmunization with 50 μg of the complex dissolved in normal saline and given subcutaneously produces an anamnestic response, with binding activity increasing 5 to 10-fold within 10 days.

Characterization of Antiserum to Sm-C/IGF-I

Each antiserum must be analyzed to determine its useable titer, i.e., the dilution of antibody at which approximately 50% of a tracer amount of iodinated hormone is bound. This should be done using conditions similar to those to be employed in the radioimmunoassay. Antisera with acceptable titers are examined to determine the apparent affinity constant (K_a) for Sm-C/IGF-I. High-affinity antisera are desirable. The specificity of the antiserum is determined using human IGF-II, proinsulin, insulin, growth

[9] J. Vaitukaitis, J. B. Robbins, E. Nieschlag, and G. T. Ross, *J. Clin. Endocrinol. Metab.* **33**, 988 (1971).

hormone, and other available peptide hormones and growth factors. If the assay is to be used to measure Sm-C/IGF-I in nonhuman species, homologous hormones from those species should also be examined. Cross-reactivity with IGF-II is the most common problem. A good antiserum will have a useable titer of 1 to 5000 or greater, a K_a of 10^{10} or higher, and will show less than 5% cross-reactivity with IGF-II. A number of antisera fulfilling these criteria have been described,[1-3,8] including the antiserum available from the NIADDK.

Radioimmunoassay of Sm-C/IGF-I

Sm-C/IGF-I circulates in serum tightly bound to one or more large-molecular-weight carrier proteins.[1,10] This has a number of important effects on the immunoassay when it is used to quantitate Sm-C/IGF-I in blood samples.[11] In order to measure the total concentration of Sm-C/IGF-I in blood, steps must be taken to free the hormone from its carrier protein. One approach to this has been to physically separate the Sm-C/IGF-I from its binding protein by chromatographing the sample under acidic conditions[1] or by extracting the sample with acid–ethanol.[2,7,12] These procedures are tedious and introduce an additional source of error into the assay. We have found that the inclusion of heparin in the assay buffer promotes the dissociation of Sm-C/IGF-I from its binding protein and allows quantitative measurement of the hormone in unextracted human serum.[11] The following provides a detailed description of this modified assay as performed in our laboratory and compares its features to those of the more traditional assays involving prior separation of Sm-C/IGF-I from its binding protein.

Reagents and Supplies

We have found that the following reagents give satisfactory results in the Sm-C/IGF-I radioimmunoassay: bovine serum albumin (BSA; Sigma A4503), polyethylene glycol 8000 (PEG; Sigma P2139), human γ-globulin (hIgG; Sigma HGII) and sodium heparin (Organon Inc., West Orange, NJ). Contamination of BSA with Sm-C/IGF-I has not been a problem but we have found that certain batches of bovine γ-globulin contain relatively large quantities of Sm-C/IGF-I and are not suitable for use in the assay. We have not observed this with human γ-globulin. While it is important that

[10] R. W. Furlanetto, *J. Clin. Endocrinol. Metab.* **51**, 12 (1980).
[11] R. W. Furlanetto, *J. Clin. Endocrinol. Metab.* **54**, 1084 (1982).
[12] W. H. Daughaday, I. K. Mariz, and S. L. Blethen, *J. Clin. Endocrinol. Metab.* **51**, 781 (1980).

the radioimmunoassay be performed in polystyrene or polypropylene tubes, we have found no difference between polystyrene tubes obtained from a number of different suppliers.

Standards

There is no international standard for Sm-C/IGF-I, although a reference serum preparation is available from the NIADDK. Because of the scarcity of purified hormone, most laboratories have adopted a human serum or plasma sample as their working standard and express values relative to this. Unfortunately, the standards used by different laboratories have often varied widely in their Sm-C/IGF-I content, making interlaboratory comparisons difficult. We have adopted as our standard a commercially available pooled plasma sample, the Gilford QCS normal control serum, assayed (Gilford, Irvine, CA), which we arbitrarily define as containing 1 U/ml of Sm-C/IGF-I. This pool appears to be representative of a normal adult population, as 124 normal adults assayed in our laboratory had a mean serum Sm-C/IGF-I concentration of 0.98 ± 0.36 U/ml relative to this standard. One unit of Gilford corresponds to approximately 140 ng purified Sm-C/IGF-I. We include the Gilford standard, a purified hormone standard, and serum pools with low, normal, and high Sm-C/IGF-I values in all assays.

Radioimmunoassay Method

The Sm-C/IGF-I radioimmunoassay is a nonequilibrium assay using PEG precipitation[13] to separate antibody-bound and free hormone. It is performed in disposable polystyene or polypropylene culture tubes. The buffer used for the assay is 0.03 M sodium phosphate, 0.02% sodium azide, 0.25% BSA, pH 7.4, containing 1.0 U/ml sodium heparin. This buffer is used for all dilutions and additions except the IgG and PEG, which are dissolved in 0.05 M sodium phosphate, 0.15 M sodium chloride, pH 7.4.

The radioimmunoassay is performed as follows: to 300 μl buffer, containing the appropriate quantity of standard or unknown, is added 50 μl of a 1 to 1800 dilution of Sm-C/IGF-I antiserum. The tubes are incubated at 4° for 96 $\pm$ 5 hr after which approximately 15,000 cpm (10 fm) radiolabeled Sm-C/IGF-I, in a volume of 150 μl, is added. The incubation is continued for an additional 16 hr at 4°; 100 μl of 1% human IgG is then added, followed by 600 μl of 25% PEG. PEG is added at room temperature using a syringe-type pipet, such as a Cornwall multiple dispensing syringe. The tubes are vortexed thoroughly and centrifuged for 15 min at 1500 g at

---

[13] B. Desbuquois and G. D. Aurbach, *J. Clin. Endocrinol. Metab.* **33**, 732 (1971).

4°. The supernatant is removed by aspiration, the precipitate is washed once with 1 ml of 12.5% PEG, and the pellet is counted for 1 min in a gamma spectrophotometer.

Characteristics of the Sm-C/IGF-I Radioimmunoassay

The Sm-C/IGF-I RIA is quite sensitive. Fifty percent displacement of binding occurs at 1.0 mU/ml of hormone (140 pg/ml) and the working range of the assay is between 0.15 and 4.5 mU/ml. Because serum contains relatively large quantities of Sm-C/IGF-I (140 ng/ml), very small volumes are required: for example, for samples with Sm-C/IGF-I values in the normal range, 50% displacement occurs with 0.5 μl of serum per tube. A similar sensitivity has been reported for the antiserum distributed by the NIADDK (R. L. Hintz, personal communication, 1985). We routinely assay all samples in duplicate at two doses which differ by 3-fold or greater. For binding between 20 and 80%, the within assay coefficient of variation is approximately 6% and the between assay coefficient of variation is approximately 9%.

This heparin–polystyrene assay quantitatively measures Sm-C/IGF-I in unextracted human serum. This is shown in Fig. 2, where immunoreactive Sm-C/IGF-I concentrations in blood samples before and after acid gel filtration are compared. Chromatography under acid conditions is the standard method which has been used to separate Sm-C/IGF-I from its binding protein and to validate immunoassay results.[1] There is no signifi-

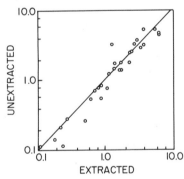

Fig. 2. Scattergram comparing immunoreactive Sm-C/IGF-I concentrations in serum samples before (unextracted) and after (extracted) separation of Sm-C/IGF-I from its binding protein by chromatography on Sephadex G-50 fine in 1 M acetic acid. Using this method the recovery of purified hormone averaged 86% ± 3% in three experiments. The extracted values have not been corrected for this. Sm-C/IGF-I concentrations are in units per mililiter. Note the logarithmic scales. The solid line is the least-squares fit to the data and has a slope of 1.05 and intercept of −0.02. $R = 0.96$.

cant difference in Sm-C/IGF-I concentrations between the two groups of samples ($p < 0.005$ using two-tailed t test). The recovery of Sm-C/IGF-I in the acid gel filtration procedure averaged 86%. The Sm-C/IGF-I immunoassay therefore measures greater than 85% of the hormone in unextracted serum samples.

Sample Type and Sample Handling Conditions

Table I compares immunoreactive Sm-C/IGF-I values in simultaneously drawn serum, heparinized plasma, and EDTA plasma samples. There is no significant difference between the serum and heparinized plasma samples but EDTA plasma consistently gives slightly lower results. For this reason serum or heparinized plasma samples are preferred. The data in Table I also indicate that immunoreactive Sm-C/IGF-I in serum is stable to incubation at room temperature for 72 hr and to repetitive freeze–thaw cycles.

Factors Affecting Assay Performance

A number of factors can affect the performance of the Sm-C/IGF-I radioimmunoassay. Of these, tube type and buffer composition deserve

TABLE I

EFFECT OF SAMPLE TYPE AND SAMPLE HANDLING CONDITIONS ON IMMUNOREACTIVE Sm-C/IGF-I

	Immunoreactive Sm-C/IGF-I (U/ml)[a]					
	Serum[b]				Plasma	
Sample	−20°[c]	4°[c]	22°[c]	Freeze–thaw[d]	Heparin[b]	EDTA[b]
1	0.31	0.36	0.34	0.34	0.29	0.29
2	0.50	0.46	0.50	0.47	0.46	0.40
3	0.68	0.74	0.77	0.70	0.64	0.60
4	1.02	1.11	1.16	1.05	0.96	0.90
5	2.12	2.01	2.17	2.03	2.20	1.91

[a] Assays were performed using the heparin–polystyrene assay system described in text. The values are expressed relative to the Gilford standard.

[b] Serum, heparinized plasma, and EDTA plasma samples were obtained from a single venipuncture using appropriate vacutainer tubes.

[c] Aliquots of serum samples were stored at −20, 4, and 22° for 72 hr.

[d] Samples were freeze-thawed nine times prior to assay.

special emphasis. We have found that assay reproducibility is much better when polystyrene or polypropylene tubes, rather than glass tubes, are used. The Sm-C/IGF-I binding protein adsorbs avidly to glass surfaces. This alters the nonspecific binding and adversely affects the dose–response behavior of the assay. This does not occur with plastic tubes. It is very important, however, that heparin be included in the radioimmunoassay buffer when serum or plasma samples are being assayed in plastic tubes.[11] Plastic stabilizes the Sm-C/IGF-I binding protein complex and, in the absence of heparin, leads to an underestimation of the total Sm-C/IGF-I content of the sample.[11] Protamine, by neutralizing heparin, has a similar effect on the assay but it is unlikely to be present in serum in sufficient quantities to be of significance.

Since heparin promotes the dissociation of Sm-C/IGF-I from its binding protein,[11,14] it should also be effective in other Sm-C/IGF-I radioimmunoassays. While we have not systematically tested other Sm-C/IGF-I antisera, we have obtained results similar to those reported here with another (high-affinity) antiserum available to us. It is possible, however, that with some antisera (particularly low-affinity antisera) an extraction procedure may still be required. Each laboratory must, therefore, determine the conditions necessary to quantitatively measure Sm-C/IGF-I in serum using the particular antiserum available. A number of extraction procedures have been described[2,7,8,12] and if necessary can be employed.

An important limitation of the heparin system relates to its use for measuring Sm-C/IGF-I in serum from other animal species. While the procedure works well with serum from certain species (i.e., chimpanzee, baboon, dog), rodent (rat, mouse, rabbit, and guinea pig) serum gives inconsistent results and an extraction procedure is necessary. It is therefore advisable to validate the procedure for each animal species whose serum is to be measured.

When assaying serum or plasma samples it is important that the total Sm-C/IGF-I be measured, even if this requires that an extraction procedure be used. Assays in which Sm-C/IGF-I is not quantitatively measured are subject to a number of variables which limit their usefulness and which can lead to erroneous results.[15] In nonquantitative systems variables which affect the binding of Sm-C/IGF-I to its binding protein (such as variations in sample type or sample handling conditions) alter the apparent immunoreactive Sm-C/IGF-I content of the sample.[11] These variables do not affect the quantitative assays since these assays do not depend on the integrity of

[14] D. R. Clemmons, L. E. Underwood, P. G. Chatelain, and J. J. Van Wyk, *J. Clin. Endocrinol. Metab.* **56**, 384 (1983).

[15] P. G. Chatelain, J. J. Van Wyk, K. C. Copeland, S. L. Blethen, and L. E. Underwood, *J. Clin. Endocrinol. Metab.* **56**, 376 (1983).

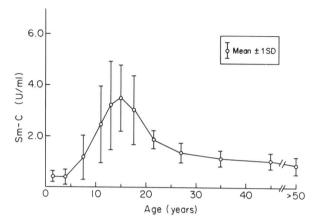

FIG. 3. Immunoreactive Sm-C/IGF-I concentrations as a function of age in humans. A total of 296 samples were analyzed using the heparin–polystyrene assay described in the text. The values are expressed relative to the Gilford standard.

the binding protein. We can find no justification, either practical or theoretical, for the use of nonquantitative methods.

Regulation of Sm-C/IGF-I Concentrations

Sm-C/IGF-I concentrations vary as a function of age,[1] sex,[16] and hormonal,[1-3] and nutritional[17,18] status. In humans the serum Sm-C/IGF-I concentration increases 10-fold between birth and adolescence and then decreases to more or less constant levels during adulthood (Fig. 3). While there is no significant difference in Sm-C/IGF-I levels between adult males and females, levels are slightly higher in young girls than in boys. This difference is apparent by 6 weeks of age and persists throughout childhood and adolescence. The pubertal peak in Sm-C/IGF-I values also occurs earlier in girls than in boys.[16]

Although a number of hormones are known to affect serum Sm-C/IGF-I concentrations, growth hormone is by far the most important hormone regulating Sm-C/IGF-I levels.[1,3] Indeed, Sm-C/IGF-I measurements find their major clinical utility in the diagnosis of disorders of growth hormone secretion. Serum Sm-C/IGF-I concentrations are elevated in patients with growth hormone excess (acromegaly) and the levels correlate

[16] R. I. Rosenfield, R. W. Furlanetto, and D. Bock, *J. Pediatr. (St. Louis)* **103**, 723 (1983).
[17] D. R. Clemmons, J. J. Van Wyk, E. C. Ridgway, B. Kliman, R. N. Kjelberg, and L. E. Underwood, *N. Engl. J. Med.* **301**, 1138 (1979).
[18] J. L. C. Borges, R. M. Blizzard, W. S. Evans, R. W. Furlanetto, A. D. Rogol, D. L. Kaiser, J. Rivier, W. Vale, and M. O. Thorner, *J. Clin. Endocrinol. Metab.* **59**, 1 (1984).

well with the activity of the disease.[17] The levels are low (usually less than 0.2 U/ml) in patients with growth hormone deficiency and they rise in response to growth hormone therapy.[1,3,18] However, low Sm-C/IGF-I levels also can occur for other reasons, making its measurement useful only as a screening test for growth hormone deficiency. If a low Sm-C/IGF-I is found, more definitive tests of growth hormone secretion must be performed. Other hormones which affect the serum Sm-C/IGF-I concentration include prolactin,[19] placental lactogen,[20] thyroid hormones, and sex steroids.[16,21] Prolactin and placental lactogen are weakly somatogenic and this correlates with their ability to bind to the growth hormone receptor. Thyroid hormone and sex steroids appear to modulate Sm-C/IGF-I levels indirectly through their effects on growth hormone secretion.[21]

Nutrition has important effects on the serum Sm-C/IGF-I concentration. Sm-C/IGF-I levels decline rapidly during caloric deprivation[22] and are a sensitive indicator of adequate nutritional balance.[23] This decline appears to be the result of a starvation-induced resistance to the somatogenic effects of growth hormone, since children with severe protein–calorie malnutrition have elevated growth hormone levels but very low Sm-C/IGF-I levels. The regulation of serum Sm-C/IGF-I levels has been the subject of a number of recent reviews.[24–26]

Acknowledgments

The authors would like to thank Patricia Cleary and Carole Wisehart for their technical assistance and Josephine Robinson for her help in preparing this manuscript. We would like to offer special thanks to the many physicians and physicians in training who assisted in collecting the blood samples used in these studies.

[19] D. R. Clemmons, L. E. Underwood, E. C. Ridgway, B. Kliman, and J. J. Van Wyk, *J. Clin. Endocrinol. Metab.* **52**, 731 (1981).

[20] R. W. Furlanetto, L. E. Underwood, J. J. Van Wyk, and S. Handwerger, *J. Clin. Endocrinol. Metab.* **47**, 695 (1978).

[21] R. I. Rosenfield and R. W. Furlanetto, *J. Pediatr. (St. Louis),* **107**, 415 (1985).

[22] A. Caufriez, J. Goldstein, P. Lebrun, A. Herchuelz, R. Furlanetto, and G. Copinschi, *Clin. Endocrinol.* **20**, 65 (1984).

[23] D. R. Clemmons, L. E. Underwood, R. N. Dickerson, R. O. Brown, L. J. Hak, R. D. MacPhee, and W. D. Heizer, *Am. J. Clin. Nutr.* **41**, 191 (1985).

[24] R. W. Furlanetto, *in* "Proceedings of the Symposium on Insulin-Like Factors/Somatomedins" (E. M. Spencer, ed.). de Gruyter, New York, 1983.

[25] R. W. Furlanetto and J. F. Cara, *Hormone Res.,* **24**, 177 (1986).

[26] J. J. Van Wyk, *in* "Hormonal Proteins and Peptides" (C. H. Li, ed.), p. 81. Academic Press, London, 1984.

[21] Estimation of Tissue Concentrations of Somatomedin C/Insulin-Like Growth Factor I

By A. Joseph D' Ercole and Louis E. Underwood

The somatomedins,[1] somatomedin C/insulin-like growth factor I (Sm-C/IGF-I) and insulin-like growth factor II (IGF-II), are thought to act on their cells of origin (autocrine action) or on cells near their origin (paracrine action).[2] Although traditionally assessed by estimation of serum or plasma concentrations, estimation of tissue concentrations may be a more appropriate indicator of the physiology of somatomedin. We have shown that when hypophysectomized rats are injected with a single dose of growth hormone (GH), Sm-C/IGF-I concentrations in tissues increase before a rise in blood is apparent.[3] A more precise understanding of somatomedin regulation might accrue, therefore, from studies of tissue concentrations of these growth factors.

Tissue Extraction

Principle

The somatomedins in blood are complexed to binding proteins, and they can be dissociated from these proteins under acid conditions (pH optimum 3.6).[4-6] Although less extensively studied, somatomedin-binding proteins also exist in cytosolic fractions of tissues.[7] The extraction method detailed below was designed on the premise that incubation at pH 3.6 will free somatomedin from its binding proteins. Although the dissociation of the Sm-C/IGF-I-binding protein complex in plasma is most rapid at 37°,[6]

[1] J. J. Van Wyk, in "Human Proteins and Peptides" (C. H. Li, ed.), p. 12. Academic Press, New York, 1984.

[2] M. B. Sporn and G. T. Todaro, N. Engl. J. Med. 303, 878 (1980).

[3] A. J. D'Ercole, A. D. Stiles, and L. E. Underwood, Proc. Natl. Acad. Sci. U.S.A. 81, 935 (1984).

[4] R. L. Hintz, in "Clinics in Endocrinology and Metabolism" (W. H. Daughaday, ed.), p. 31. Saunders, London, 1984.

[5] G. L. Smith, Mol. Cell. Endocrinol. 34, 83 (1984).

[6] P. G. Chatelain, J. J. Van Wyk, K. C. Copeland, S. L. Blethen, and L. E. Underwood, J. Clin. Endocrinol. Metab. 56, 376 (1983).

[7] H. J. Guyda and B. I. Posner, Endocrinology 106, 113 (1980).

tissues should be extracted at 4° for relatively short times, because kidney and other tissues contain proteases that degrade Sm-C/IGF-I.[8]

Procedure

The extraction of adult rat tissues for Sm-C/IGF-I presented here is a modification of our previously reported method.[3]

Procuring Tissues

1. Under ether anesthesia, rats are decapitated with a guillotine and whole blood is collected into disposable conical polystyrene centrifuge tubes. Blood is allowed to clot on ice, and the serum is separated by centrifugation and stored at −20°.

2. Dissection of tissues is delayed for approximately 30 sec to allow exsanguination.

3. Tissues are promptly blotted on absorbent toweling to remove adherent blood. Large organs, such as liver, lung, and heart, are cut into several pieces and blotted so as to promote further tissue exsanguination.

4. Tissues are then flash frozen by submersion in liquid nitrogen in styrofoam cups. With practice two individuals can secure 10 frozen tissues within 4 min of sacrifice of an animal.

5. Frozen tissues can be stored at −20° for at least 6 months without an apparent reduction in Sm-C/IGF-I content.

Extracting Tissues

1. Frozen tissues are placed in a porcelain 80-mm mortar, covered with liquid nitrogen, and pulverized with a pestle.

2. After the liquid nitrogen has evaporated, 0.1 to 0.5 g of the tissue powder is transferred to a tared disposable conical polystyrene centrifuge tube. The tube containing the tissue is weighed, the tissue weight is recorded, and the tissue is placed in a freezer or on ice until the next step.

3. To iced tubes containing tissues, 1.0 *M* acetic acid is added in a ratio of 5 ml for each gram of tissue (resultant pH is 3.5–3.9) and tubes are shaken vigorously in a mechanical test tube shaker for 15 sec.

4. Samples are incubated on ice for 2 hr.

5. Tubes are then centrifuged at 1000 *g* for 10 min in a refrigerated centrifuge.

6. A measured volume of the supernatant, generally 0.5 ml, is then removed for the Sm-C/IGF-I assay, and a second aliquot is saved for determination of hemoglobin (Hb).

7. The Sm-C/IGF-I aliquot, but not the Hb aliquot, is then neutralized with 5 *M* NaOH (about 12 µl of NaOH/0.1 ml of extract will result in a final pH between 6 and 8), and stored at −20° until the time of assay. The

[8] A. J. D'Ercole, C. J. Decedue, R. W. Furlanetto, L. E. Underwood, and J. J. Van Wyk, *Endocrinology* **101**, 247 (1977).

Hb aliquot is assayed without prior neutralization, because Hgb is precipitated by these conditions of neutralization.

Assays and Calculations

Sm-C/IGF-I Radioimmunoassay (RIA)

Extracts and sera are immunoassayed by a previously described non-equilibrium technique.[9,10] Because the cross-reactivity of the antibody raised against purified human Sm-C/IGF-I with the rat homolog is not known with certainty, we have utilized pools of serum as standards (either human or rat serum). These standards are aribitrarily defined as having 1 U of activity/ml. Under the conditions of the assay, $1-10$ μl from extract of normal rat tissues produces $20-80\%$ displacement of ^{125}I-labeled Sm-C/IGF-I. Generally, $5-10$ times more extract of tissues from hypophysectomized animals or fetuses is needed.

Because binding proteins in the tissue extracts could compete with the antibody for binding of Sm-C/IGF-I and artifactually influence quantification, we have performed a number of validation experiments: (1) serial dilution curves of extracts from each tissue were observed to be parallel to those of serum standards and purified Sm-C/IGF-I; (2) known amounts of pure Sm-C/IGF-I added to extracts of each tissue could be quantitatively recovered (accurately measured) in the RIA (range of recovery = $82-114\%$); (3) assay of extract fractions corresponding to free Sm-C/IGF-I from acid gel column chromatography resulted in Sm-C/IGF-I values that are nearly identical to those of acid-exposed extracts; and finally, (4) when extracts and ^{125}I-labeled Sm-C/IGF-I were reacted with the chemical cross-linking agent, disuccinimidyl suberate, and subjected to sodium dodecyl sulfate–polyacrylamide gel electrophoresis and autoradiography,[11] no ^{125}I-labeled Sm-C/IGF-I binding proteins complexes were visualized.

We have reported that the amounts of immunoreactive Sm-C/IGF-I measured in adult male rat sera are not different among fresh sera, acid-treated sera, and acid-chromatographed sera. We subsequently have found that in fetal, neonatal, young and pregnant rat sera, as well as in some adult rat sera, acidification effects an increase in immunoassayable Sm-C/IGF-I, as it does in human sera.[6] Therefore, we now expose all sera to acid when tissue concentrations of Sm-C/IGF-I are to be measured. A 1:2 dilution of serum with glycine–glycine · HCl buffer, pH 2.0, ionic strength 0.1,

[9] R. W. Furlanetto, L. E. Underwood, J. J. Van Wyk, and A. J. D'Ercole, *J. Clin. Invest.* **60**, 648 (1977).

[10] K. C. Copeland, L. E. Underwood, and J. J. Van Wyk, *J. Clin. Endocrinol. Metab.* **50**, 690 (1980).

[11] J. R. Wilkins and A. J. D'Ercole, *J. Clin. Invest.*, **75**, 1350 (1985).

achieves a final pH of 3.6–3.8.[6] Acid incubation of serum can be performed under conditions identical to that for tissues (2 hr at 4°), or at 37° for 24 hr in order to maximize the measureable serum Sm-C/IGF-I. Even with the latter, it is unusual for serum-derived Sm-C/IGF-I to account for more than 25% of the tissue-extractable Sm-C/IGF-I.

Hemoglobin Assay

Hb is assayed on extracts that have not been subjected to neutralization, using the modified benzidene method of Crosby and Furth,[12] and a purified rat Hb standard (Sigma). Aliquots for Hb should be stored at −20°. If stored at 4°, they should be assayed within 48 hr in order to limit proteolysis and spuriously low determinations. The extract Hb, together with blood Hb and hematocrit (these data can be directly determined or obtained from a variety of references), allow calculation of the volume of serum which may contaminate each extract. The volume (in milliliters) of serum in each milliliter of extract is calculated as follows:

$$\text{ml serum/ml extract} = ([\text{Hb}]_{ex}/0.1495)0.54 \qquad (1)$$

where $[\text{Hb}]_{ex}$ is grams of Hb in 1 ml of extract, 0.1495 is the mean Hb concentration (g/ml) of adult rat blood, and 0.54 is the portion of rat blood that is serum (mean hematocrit is 46%). Because the extraction volume and tissue wet weight are known, the volume of serum entrapped in a gram of tissue can be calculated as follows:

$$\frac{(\text{ml serum/ml extract})(\text{ml extract})}{\text{wet wt tissue extracted (g)}} = \text{ml serum/g tissue} \qquad (2)$$

With careful blotting and efforts to exsanguinate the tissues (see above), entrapment of serum (μl/g) in rat tissues should be within the following ranges: 4–20 for liver, 5–30 for lung, 1–10 for kidney, 5–25 for heart, 5–20 for muscle, 1–10 for brain, and <5 for cartilage, fat, and pancreas.

Calculations of Tissue Sm-C/IGF-I Concentrations

To obtain tissue Sm-C/IGF-I concentrations (U/g) the following calculations are performed.

1. Extract Sm-C/IGF-I concentration, $[\text{Sm-C/IGF-I}]_{ex}$, is calculated as units per milliliter from the RIA.
2. The Sm-C/IGF-I concentration in the extract is corrected for the

---

[12] W. H. Crosby and F. W. Furth, *Blood* **11**, 380 (1956).

dilution caused by neutralization:

$$[\text{Sm-C/IGF-I}]_{ex}\left[\frac{\text{aliquot volume (ml)} + \text{neutralization volume (ml)}}{\text{aliquot volume (ml)}}\right]$$
$$= \text{adjusted } [\text{Sm-C/IGF-I}]_{ex} \qquad (3)$$

3. To determine the total Sm-C/IGF-I extracted,

$$(\text{Adjusted } [\text{Sm-C/IGF-I}]_{ex})[\text{total extraction volume (ml)}]$$
$$= \text{total } [\text{Sm-C/IGF-I}]_{ex} \qquad (4)$$

Note that the total extraction volume is calculated by adding the tissue weight (g) to the recorded extraction volume, because the extracted Sm-C/IGF-I is distributed in the tissue as well. This assumes that a gram of tissue has a volume of 1 ml.

4. The total extracted Sm-C/IGF-I is then standardized for the tissue weight extracted:

$$\frac{\text{total Sm-C/IGF-I}_{ex}}{\text{tissue wt (g)}} = \text{total } [\text{Sm-C/IGF-I (U/g)}] \qquad (5)$$

5. To determine the possible contribution of serum-derived Sm-C/IGF-I to the tissue concentration (U/g):

$$(\text{ml serum/g tissue})(\text{serum } [\text{Sm-C/IGF-I (U/ml)}])$$
$$= \text{serum-derived } [\text{Sm-C/IGF-I (U/g)}] \qquad (6)$$

6. Then,

$$\text{Total } [\text{Sm-C/IGF-I (U/g)}] - \text{serum-derived } [\text{Sm-C/IGF-I (U/g)}]$$
$$= \text{tissue } [\text{Sm-C/IGF-I (U/g)}] \qquad (7)$$

These tedious calculations can be done easily using an inexpensive programmable calulator.

Validation of Method

To validate the modified extraction procedure reported here, we have extracted a variety of tissues from normal and hypophysectomized adult female rats (Table I). Using this procedure we extract more Sm-C/IGF-I than was possible using our previous method, presumably because significant quantities of Sm-C/IGF-I could not be solubilized from the precipitate formed in the original procedure after lyophilization.[3] Despite the differences in quantification, the two procedures result in a similar relationship between Sm-C/IGF-I concentration in normal and hypophysectomized rat tissues (Table I).

Additional evidence for the validity of this procedure is the finding that

TABLE I
EXTRACTABLE TISSUE Sm-C/IGF I CONCENTRATIONS IN FEMALE RATS

| | | Sm-C/IGF I (U/g)[a] | |
| | | Hypophysectomized | |
Tissue	Normal (mean ± SD)	Mean ± SD	Percentage normal
Liver	0.176 ± 0.037	0.017 ± 0.009[b]	9.4
Lung	0.095 ± 0.020	0.015 ± 0.006[b]	15.8
Kidney	0.641 ± 0.194	0.097 ± 0.028[b]	15.1
Heart	0.130 ± 0.018	0.042 ± 0.007[b]	32.3
Muscle	0.050 ± 0.016	<0.005[b]	<10.0
Brain	0.032 ± 0.006	0.025 ± 0.008	78.1
Cartilage (sternum)	<0.054>	<0.008>	14.8
Fat pad (perirenal)	0.176 ± 0.067	0.017 ± 0.003[b]	9.7
Pancreas	0.108 ± 0.015	0.038 ± 0.007[b]	35.2

[a] All values are expressed in rat units, that is the amount of immuno-reactive Sm-C/IGF-I in 1 ml of serum obtained from a pool of 10 adult male rats. Sera from these female rats contained 0.78 ± 0.08 U/ml (mean ± SD) in normals and 0.025 ± 0.003 in hypophysectomized (3.2% of normal). $N = 4$ for all values except cartilage, which was a pool of 4.

[b] Differences between Sm-C/IGF-I concentrations of normal and hypophysectomized rats are significant ($p < 0.01$).

in response to injection of ovine GH (oGH), most tissues from hypophysectomized rats have a marked increase in Sm-C/IGF-I concentration (Table II). In addition to GH, thyroxine appears to increase Sm-C/IGF-I in some tissues, and to have an additive effect on tissue concentrations (Table II). The latter results are, however, quite preliminary.

In the pancreas, it seemed possible that some of the measurable immunoreactive Sm-C/IGF-I was the result of cross-reactivity with high insulin concentrations, although the cross-reactivity of insulin in the Sm-C/IGF-I RIA is minimal (<0.001%). To test this possibility, pancreatic extracts were eluted on a cyanogen bromide-activated Sepharose 4B column (Pharmacia) that had been linked to a guinea pig anti-porcine insulin antibody. The eluate from the column contained only 7.3% of the original immunoassayable insulin, and 109.1% of the original Sm-C/IGF-I. It appears, therefore, that the pancreas synthesizes Sm-C/IGF-I.

Comments

We have utilized this extraction procedure in fetal rat and human tissues. In both, serial dilution of extracts produces curves that are parallel

TABLE II
Sm-C/IGF-I RESPONSE IN HYPOPHYSECTOMIZED FEMALE RATS TO THYROXINE
(T_4), OVINE GROWTH HORMONE (oGH), OR BOTH[a]

Tissue	Sm-C/IGF-I, percentage increase (mean ± SD)		
	T_4	oGH	T_4 + oGH
Serum	41.8 ± 5.4	81.8 ± 36.4	140.0 ± 27.3
Liver	38.9 ± 22.2	236.1 ± 27.8	283.3 ± 55.6
Lung	124.2 ± 145.4	193.9 ± 6.1	209.1 ± 27.3
Kidney	NI[b]	278.5 ± 97.1	419.6 ± 59.8
Heart	NI	74.7 ± 57.1	134.0 ± 64.8
Muscle	NI	NI	>680 ± 390
Brain	NI	NI	NI
Cartilage	100.0	127.0	138.0
Fat	NI	13.5	75.7 ± 64.9
Pancreas	80.5 ± 18.3	118.3 ± 39.0	63.4 ± 34.1

[a] Rats were given 2 μg T_4, 500 μg oGH, or both subcutaneously at 2000 hr for 3 consecutive days, and sacrificed 12 hr after the last dose of hormone.
[b] NI, No increase.

to those of sera or pure peptide; addition of pure peptide is quantitatively recovered in the RIA; and extracts and free peptide fractions of chromatographed extracts yield similar results. It seems likely that this procedure will help in the study of somatomedin physiology in other species. We have adapted our human plasma RIA for measurement of Sm-C/IGF-I in the mouse,[13] sheep,[14] baboon,[15] and rabbit,[16] as well as the rat, by modifying the time and temperature of acid incubation. Such techniques also should be useful for tissue extraction.

Acknowledgments

We thank Brian M. Sumner and Deborah K. Bell for technical assistance and Kimberly T. Shlanta for preparing this manuscript. This work was supported by a NICHHD grant (HD-08299) and a March of Dimes Basic Research Grant (1-758). A.J.D. is the recipient of a USPHS Research Career Development Award (HD-00435).

[13] A. J. D'ercole and L. E. Underwood, *Dev. Biol.* **79**, 33 (1980).
[14] L. E. Underwood, A. J. D'Ercole, K. C. Copeland, J. J. Van Wyk, T. Hurley, and S. Handwerger, *J. Endocrinol.* **93**, 31 (1982).
[15] K. C. Copeland, D. M. Johnson, L. E. Underwood, and J. J. Van Wyk, *Am. J. Primatol.* **5**, 161 (1983).
[16] A. J. D'Ercole, C. L. Bose, L. E. Underwood, and E. E. Lawson, *Diabetes* **33**, 590 (1984).

[22] Somatomedin C/Insulin-Like Growth Factor I Receptors on Human Mononuclear Cells and the IM-9 Lymphoid Cell Line

By RAYMOND L. HINTZ

Introduction

Over the last two decades, evidence has accumulated that hormone action is obligately linked to the binding of hormone to specific receptor sites on or within the cell, and that this interaction initiates the chain of events in the target cell that leads to the hormone action. Thus, the study of receptor sites allows quantification of a crucial early step in hormone action. It has become apparent that in many circumstances the concentration of receptor sites on or in the cell is just as important to hormone action as the circulating concentration of the hormone itself.

Somatomedin C (Sm-C) is a basic hormonal peptide that is under growth hormone (GH) control, and is responsible for many, if not all, of the growth promoting actions of GH.[1] It is now clear that Sm-C, and the insulin-like growth factor I (IGF-I) first sequenced by Rinderknecht and Humbel, are the same substance.[2,3] Sm-C/IGF-I shows a nearly 50% structural homology to insulin, and belongs to a family of related hormonal peptides that includes insulin, Sm-C/IGF-I, IGF-II, nerve growth factor, and relaxin. All the members of this peptide family are involved in the control of cell growth and/or differentiation. As is true of other hormonal peptides, the receptors for Sm-C/IGF-I are mainly on the outside of the cellular membrane. The measurement of these receptors allows insight into a crucial early step in the action of the Sm-C/IGF-I peptides, and thus the control of cellular growth. The number of receptors on circulating human blood cells, and in particular the mononuclear cells, changes with the circulating concentration of Sm-C/IGF-I, and reflects the dynamics of changes of Sm-C/IGF-I receptors throughout the body. This technique will be presented in detail later in this chapter. In addition, the closely related techniques for determining Sm-C/IGF-I receptors on a cultured human lymphoid cell line (IM-9) that has been useful for the *in vitro* study of receptor dynamics will also be discussed.

[1] R. L. Hintz, *Adv. Pediatr.* **28,** 293 (1980).
[2] E. Rinderknecht and R. E. Humbel, *Proc. Natl. Acad. Sci. U.S.A.* **73,** 2365 (1976).
[3] D. G. Klapper, M. E. Svoboda, and J. J. Van Wyk, *Endocrinology* **112,** 2215 (1983).

METHODS IN ENZYMOLOGY, VOL. 146

Receptors on Human Mononuclear Cells

Peptides

Until recently, there was no commercial source of Sm-C/IGF-I so that any investigators who wanted to measure Sm-C/IGF-I receptors had to either laboriously purify the peptide in their own laboratory, or obtain a supply as a gift from another investigator. Recently, a synthetic preparation of IGF-I has been manufactured by AMGEN Corporation (Thousand Oaks, CA; Amershaw International). This biosynthetic IGF-I has exactly the same structure as natural Sm-C/IGF-I purified from human plasma with the exception of a threonine substituted for methionine at position 59. This preparation has been used in my laboratory, and appears to be largely equivalent to the native hormone for receptor studies.

Iodination

The Sm-C/IGF-I, either native or synthetic, is iodinated by a modification of the chloramine-T method[4] using a limited amount of chloramine-T and without the use of sodium metabisulfite. The protocol we have used with the synthetic IGF-I is a follows:

IGF-I (100 μg/ml in 0.01 N HCl), 10 μl
NaPO$_4$ buffer (0.2 M, pH 7.4), 20 μl
Na^{125}I (100 mCi/ml, Amersham IMS.30a), 10 μl
Chloramine-T (4 mg/ml in H$_2$O), 20 μl

The reagents are mixed sequentially in a 400-μl polypropylene tube. After the chloramine-T is added and mixed, and a 30- to 40-sec incubation period, 50 μl of 0.1% BSA is added to quench the reaction. The reaction products are separated by either cellulose chromatography, or by Sephadex G-50 chromatography.

Isolation of Cells

The mononuclear cells are isolated by the method of Boyum.[5] Heparinized whole blood (25 ml minimum) is centrifuged for 10 min at room temperature. The top 70% of the plasma is removed. The remaining plasma along with the buffy coat plus the red cells immediately adjacent to the buffy coat are pipetted into a beaker and diluted 1:1 with phosphate-

---

[4] W. M. Hunter and F. C. Greenwood, *Nature (London)* **194,** 495 (1962).
[5] A. Boyum, *Scand. J. Clin. Lab. Invest. (Suppl. 97)* **21,** 77 (1968).

buffered saline. After mixing by gentle swirling, it is carefully layered over 10 ml Ficoll-Hypaque (Pharmacia) in 50-ml conical glass centrifuge tubes. The tubes are centrifuged at 1460 rpm for 25 min at room temperature using a Beckman J-6B centrifuge with JS-4.2 rotor. Four layers form in the tubes: (1) a top layer of PBS–plasma mix, (2) a thin, whitish layer containing the mononuclear cells (monocytes and lymphocytes), (3) a clear area of Ficoll-Hypaque, and (4) the bottom layer containing the red cells plus granulocytes. Using a disposable pipet, the whitish layer containing the mononuclear cells is removed and added to a clean 50-ml centrifuge tube. One should be very careful not to take any Ficoll-Hypaque solution, but some of the plasma–PBS layer may be included if necessary. The mononuclear cells are diluted with an equal volume of lymphocyte buffer (Na, 135 mM, K, 5 mM, Mg, 1.2 mM, Cl, 125 mM, bicarbonate, 15 mM, SO_4, 1.2 mM, EDTA, 1 mM, dextrose, 10 mM, HEPES, 100 mM, BSA, 1%, pH 7.4) and centrifuged at 1200 rpm for 10 min at 4°. The supernatant is aspirated, and the cells are washed once by resuspending them in 15 ml lymphocyte buffer and centrifuging again. Finally, the cell pellet is resuspended in 3 ml lymphocyte buffer, and 50 μl of the cell suspension is removed for determination of the cell concentration. The cells are then diluted to a concentration of 25 million cells/ml before use.

Assay of Receptor Sites

The assay for Sm-C/IGF-I binding sites is done by incubating the following mixture:

50 μl ^{125}I-labeled IGF-I (10,000–20,000 cpm)
50 μl buffer or peptide
400 μl cell suspension (add last)

The peptide standard curve covers the range of concentrations from 0 to 100 ng/ml Sm-C/IGF-I, with an additional nonspecific binding point of 250 ng/ml or the equivalent amount of partially purified Sm/IGFα peptide. A typical experiment in our laboratory is shown in Table I.

The assay is incubated for 2 hr at 15° in a shaking water bath. At the end of the incubation, duplicate 200-μl aliquots of the cell suspension are layered over 200 μl of ice-cold buffer in 500 μl Microfuge tubes, and the cells are sedimented by Microfuge (Beckman) centrifugation. The supernatant is aspirated, and the tip of the tube, containing the cell pellet, is cut off and counted for [125]Ia.

Results and Calculations

In our laboratory, the specific binding of ^{125}I-labeled IGF-I to adult mononuclear cells averages 4 to 6%/50 million mononuclear cells.[6,7] Half-

TABLE I
TYPICAL SM-C/IGF-I RECEPTOR ASSAY

Tube number	Peptide concentration (ng/ml)
1, 2	0
3, 4	Nonspecific (250 ng/ml)
5, 6	1
7, 8	2
9, 10	5
11, 12	10
13, 14	20
15, 16	50
17, 18	100

maximum displacement is typically found between 5 and 15 ng/ml pure IGF-I/Sm-C. The receptor on human mononuclear cells appears to be a typical type I receptor site, with a 10-fold higher specificity for IGF-I when compared with IGF-II, and insulin cross-reacts at approximately 200 times the concentration of IGF-I. Further analysis of the data can be done by Scatchard analysis,[8] in which the bound to free *(B/F)* ratio is plotted on the abscissa against the molar concentration of Sm-C/IGF-I bound (both radioactive and cold) on the ordinate. In this transformation, the slope of the line is equal to the K_d, and the ordinate intercept is equivalent to the receptor site concentration. The number of cells per milliliter is known, and the apparent concentration of receptor sites per cell can be calculated. Since the Scatchard plot is curvilinear, it can be analyzed either by resolution into two straight lines of high and low affinities ("two-site model"), or by other models.

Interpretation

Changes in receptor site number for ^{125}I-labeled IGF-I have been shown with development, with higher numbers of receptor sites found in the newborn,[9] and with changes in the circulating concentration of Sm-C/IGF-I peptide.[10] In general, higher levels of Sm-C/IGF-I in the plasma lead to a decrease in receptor site number ("down-regulation").

[6] A. V. Thorsson and R. L. Hintz, *Biochem. Biophys. Res. Commun.* **74**, 1566 (1977).

[7] R. G. Rosenfeld and R. L. Hintz, *Horm. Metab. Res.* **12**, 47 (1980).

[8] G. Scatchard, *Ann. N.Y. Acad. Sci.* **51**, 660 (1949).

[9] R. G. Rosenfeld, A. V. Thorsson, and R. L. Hintz, *J. Clin. Endocrinol. Metab.* **48**, 456 (1979).

[10] R. G. Rosenfeld, S. F. Kemp, S. Gaspich, and R. L. Hintz, *J. Clin. Endocrinol. Metab.* **52**, 759 (1981).

Receptors on Cultured Human Lymphoid Cells (IM-9)

Source of Cells

The IM-9 cell line is a human lymphoid cell in continous culture that has been demonstrated to have receptor sites for many hormones. These receptors include ones for insulin and glucocorticoids as well as IGF-I[11] and IGF-II.[12] They are available through the American Type Culture Collection (Rockville, MD), and through numerous investigators.

Culture of the Cells

The cells are cultured in HEPES-buffered RPMI 1640 (Gibco, Grand Island, NY) supplemented with 10% fetal calf serum, 4 mM 1-glutamine, 100 U/ml penicillin, and 100 μg/ml streptomycin. The cells may be split 1:4 with fresh medium every 3 to 4 days. The cell cultures attain a density of approximately 1 million cells/ml at stationary phase, and can be used for experiments 3 to 7 days after splitting.

Peptides and Iodination

The peptides and iodination procedure are exactly the same for the mononuclear cells.

Preparation of Cells

A 50 μl aliquot of the cells in removed and used to determine the cell concentration. The cells are harvested by sedimenting them from the medium at 1200 rpm for 10 min. They are then resuspended in lymphocyte buffer so that the final cell concentration is 25 million cells/ml.

Assay of Receptor Sites

The protocol for binding site assay is the same as for the mononuclear cells, with the exception that the incubation time is for 100 min at 15°.

Results and Calculations

The specific binding of ^{125}I-labeled IGF-I to IM-9 cells averages 20 to 25%/20 million cells at 15°. Like the receptor on the circulating mononuclear cells, the receptor on IM-9 cells for Sm-C/IGF-I is a type I receptor as

[11] R. G. Rosenfeld and R. L. Hintz, *Endocrinology* **107,** 1841 (1980).
[12] R. L. Hintz, A. V. Thorsson, G. Enberg, and K. Hall, *Biochem. Biophys. Res. Commun.* **118,** 774 (1984).

judged by affinity and specificity data. Scatchard analysis of the IM-9 data typically shows less curvilinearity than the mononuclear cell data.

Interpretation

Changes in receptor site concentration have been induced by increases of the medium concentration of both Sm-C/IGF-I[11] and insulin.[13] None of the hormones outside the insulin family has shown any effect on receptor site concentration. In addition, this model can be used for the study of intracellular receptor site synthesis, transport, and recycling.

Acknowledgments

Supported in part by Grants AM-24085, AM33190, and AG-01312 from the National Institutes of Health.

[13] R. G. Rosenfeld, R. L. Hintz, and L. A. Dollar, *Diabetes* **31**, 375 (1982).

[23] Human Insulin-Like Growth Factor I and II Messenger RNA: Isolation of Complementary DNA and Analysis of Expression

By LESLIE B. RALL, JAMES SCOTT, and GRAEME I. BELL

The insulin-like growth factors (IGF) are a family of polypeptides which are mitogenic for many cultured cells and, *in vivo,* have growth-promoting and anabolic properties.[1,2] Human IGF-I and IGF-II are single-chain serum proteins of 70 and 67 amino acids, respectively.[3,4] The sequences of these proteins possess 62% sequence identity with one another, and 47% identity with insulin. Both IGF-I and IGF-II are derived from precursors by proteolytic processing.[5-7] PreproIGF-I has 130 amino acids

[1] K. Hall and V. R. Sara, *in* "Vitamins and Hormones" (D. B. McCormick, ed.), Vol. 40, p. 175. Academic Press, New York, 1983.

[2] D. R. Clemmons and J. J. Van Wyk, *in* "Tissue Growth Factors" (R. Baserga, ed.), p. 161. Springer, Berlin, 1981.

[3] E. Rinderknecht and R. E. Humbel, *J. Biol. Chem.* **253**, 2769 (1978).

[4] E. Rinderknecht and R. E. Humbel, *FEBS Lett.* **89**, 283 (1978).

[5] M. Jansen, F. M. A. van Schaik, A. T. Ricker, B. Bullock, D. E. Woods, K. H. Gabbay, A. L. Nussbaum, J. S. Sussenbach, and J. L. Van den Brande, *Nature (London)* **306**, 609 (1984).

[6] G. I. Bell, J. P. Merryweather, R. Sanchez-Pescador, M. M. Stempien, L. Priestley, J. Scott, and L. B. Rall, *Nature (London)* **310**, 775 (1984).

and has the following structure: signal peptide (25 amino acids), IGF-I (70 amino acids), COOH-terminal or E-domain (35 amino acids). PreproIGF-II has 180 amino acids and its structure is signal peptide (24 amino acids), IGF-II (67 amino acids), E-domain (89 amino acids). In order to analyze the roles of these proteins and their precursors in normal growth and development as well as in tumorigenesis, cDNAs encoding the precursors to IGF-I and IGF-II have been isolated (Fig. 1). The cDNAs have been used as probes to identify tissues and cells which synthesize preproIGF-I and preproIGF-II mRNA and to measure changes in mRNA levels during development or in altered physiological states. In addition they can be used to program the synthesis of proIGF-I and proIGF-II in heterologous cells to determine if the progrowth factors have unique biological activities in addition to being the precursors to IGF-I and IGF-II.

Sources of mRNA Encoding PreproIGF-I and PreproIGF-II

The cDNAs encoding IGF-I and IGF-II were isolated from an adult human liver cDNA library.[5-8] Recently, we have used these cDNAs to identify other tissues[9] and cultured cells which contain IGF-I and IGF-II mRNA. We have found low levels of both mRNAs in all the tissues we have examined (Table I). However, the amount of each varied. Liver has the highest levels of IGF-I mRNA; the other tissues contained 3–20% as much. The highest concentration of IGF-II mRNA was also present in the liver; slightly lower values were observed in the heart, kidney, spleen, and several areas of the brain. Hybridization of ^{32}P-labeled IGF-I and IGF-II cDNA probes to Northern blots[10] of adult human liver poly(A)$^+$ RNA has indicated that the IGF-I and IGF-II mRNAs are extremely large.[11] There are multiple IGF-I transcripts; the major one is 7700 bases and there are minor species of about 5300 and 900 bases. IGF-II mRNA is about 5300 bases in length. Comparison of the sizes of the prominent transcripts with those of the cDNAs which have been isolated to date indicates that the cDNAs do not contain the entire sequence of the mRNAs. We have also identified two human cell lines which have high levels of IGF-II mRNA;

[7] M. Jansen, F. M. A. van Schaik, H. Van Tol, J. L. Van den Brande, and J. S. Sussenbach, *FEBS Lett.* **179**, 243 (1985).

[8] D. E. Woods, A. F. Markham, A. T. Ricker, G. Goldberger, and H. R. Colten, *Proc. Natl. Acad. Sci. U.S.A.* **79**, 5661 (1982).

[9] J. Scott, J. Cowell, M. E. Robertson, L. M. Priestley, R. Wadey, J. Pritchard, G. I. Bell, L. B. Rall, C. F. Graham, and T. J. Knott, *Nature (London)* **317**, 260 (1985).

[10] P. S. Thomas, *Proc. Natl. Acad. Sci. U.S.A.* **77**, 5201 (1980).

[11] G. I. Bell, D. S. Gerhard, N. L. Fong, R. Sanchez-Pescador, and L. B. Rall, *Proc. Natl. Acad. Sci. U.S.A.* **82**, 6450 (1985).

A

```
CTTCAGAAGCAATGGAAAAATCAGCAGTCTTCCAACCAATTATTTAAGTGCTGCTTTTTGTGATTTCTTGAAGGTGAAG

                                                          -25        -20
                                                  Met His Thr Met Ser Ser His Leu
                                                  ATG CAC ACC ATG TCC TCG CAT CTC

        -10                                   1                     10
Phe Tyr Leu Ala Leu Cys Leu Leu Thr Phe Thr Ser Ser Ala Thr Ala |Gly Pro Glu Thr Leu Cys Gly Ala Glu Leu Val Asp Ala Leu
TTC TAC CTG GCG CTG TGC CTC CTC ACC TTC ACC AGC TCT GCC ACG GCT GGA CCG GAG ACG CTC TGC GGG GCT GAG CTG GTG GAT GCT CTT

          20                            30                          40
Gln Phe Val Cys Gly Asp Arg Gly Phe Tyr Phe Asn Lys Pro Thr Gly Tyr Gly Ser Ser Ser Arg Arg Ala Pro Gln Thr Gly Ile Val
CAG TTC GTG TGT GGA GAC AGG GGC TTT TAT TTC AAC AAG CCC ACA GGG TAT GGC TCC AGC AGT CGG AGG GCG CCT CAG ACA GGC ATC GTG

          50                            60                          70
Asp Glu Cys Cys Phe Arg Ser Cys Asp Leu Arg Arg Leu Glu Met Tyr Cys Ala Pro Leu Lys Pro Ala Lys Ser Ala|Arg Ser Val Arg
GAT GAG TGC TGC TTC CGG AGC TGT GAT CTA AGG AGG CTG GAG ATG TAT TGC GCA CCC CTC AAG CCT GCC AAG TCA GCT CGC TCT GTC CGT

          80                            90                          100
Ala Gln Arg His Thr Asp Met Pro Lys Thr Gln Lys Glu Val His Leu Lys Asn Ala Ser Arg Gly Ser Ala Gly Asn Lys Asn Tyr Arg
GCC CAG CGC CAC ACC GAC ATG CCC AAG ACC CAG AAG GAA GTA CAT TTG AAG AAC GCA AGT AGA GGG AGT GCA GGA AAC AAG AAC TAC AGG

105
Met AM
ATG TAG GAAGACCCTCCTGAGGAGTGAAGAGTGACATGCCACCGCAGGATCGTTTCCGAGTGCTGCAGCCCCAGGGCTCTGAATCGCTGGGATTTGATAACATTAA

AAGATGGGCGTTTCCCCAATGAAATACACAAGTAAACATTCCAACATTGTCTTTAGGAGTGATTTGCACCTTGCAACTTGGAGTTGGTAGATTGCTGTTGATCTTTTATCAA

TAATGTTCTATAGAAAGAAA
```

B

```
CAGGGGCCGAAGAGTCACCACCGAGCTTGTGTGGGAGGAGGTGGATTCCAGCCCCCAGGGCTCTGAATCGCTGCCCCAGCCTGCCCCAGCCTGAGC

CCCAGCAGGCCAGAGAGCCCAGTCCTGAGGTGAGCTGCTGTGGCCTGTGGCCCCCAGGGCCCTGTCTGTGGCTCGGACGAACTGAGGCTCCCAGGACTGAGAGCCCCCAGCCTCAGCCCCAACTGCGAGC

          -24          -20                    -10                         20                     1
AGAGAGACACCA    Met Gly Ile Pro Met Gly Lys Ser Met Leu Val Leu Leu Thr Phe Leu Ala Phe Ala Ser Cys Cys Ile Ala|Ala Tyr
                ATG GGA ATC CCA ATG GGG AAG TCG ATG CTG GTG CTT CTC ACC TTC TTG GCC TTC GCC TCG TGC TGC ATT GCT GCT TAC

          10                            20                          30
Arg Pro Ser Glu Thr Leu Cys Gly Gly Glu Leu Val Asp Thr Leu Gln Phe Val Cys Gly Asp Arg Gly Phe Tyr Phe Ser Arg Pro Ala
CGC CCC AGT GAG ACC CTG TGC GGC GGG GAG CTG GTG GAC ACC CTC CAG TTC GTC TGT GGG GAC CGC GGC TTC TAC TTC AGC AGG CCC GCA

          40                            50                          60
Ser Arg Val Ser Arg Arg Ser Arg Gly Ile Val Glu Glu Cys Cys Phe Arg Ser Cys Asp Leu Ala Leu Leu Glu Thr Tyr Cys Ala Thr
AGC CGT GTG AGC CGT CGC AGC CGT GGC ATC GTG GAG GAG TGC TGC TTC CGC AGC TGT GAC CTG GCC CTG GAG ACG TAC TGT GCT ACC

          70                            80                          90
Pro Ala Lys Ser Glu|Arg Asp Val Ser Thr Pro Pro Thr Val Leu Pro Asp Asn Phe Pro Arg Tyr Pro Val Gly Lys Phe Phe Gln Tyr
CCC GCC AAG TCC GAG AGG GAC GTG TCC ACC CCT CCG ACC GTG CTT CCC GAC AAC TTC CCC AGA TAC CCC GTG GGC AAG TTC TTC CAA TAT

          100                           110                         120
Asp Thr Trp Lys Gln Ser Thr Gln Arg Leu Arg Arg Gly Leu Pro Ala Leu Leu Arg Ala Arg Arg Gly His Val Leu Ala Lys Glu Leu
GAC ACC TGG AAG CAG AGC ACC CAG CGG CTG CGC AGG GGC CTG CCT GCC CTC CTG CGT GCC AGG CGT GGC CAC GTG CTG GCC AAG GAG CTC

          130                           140                         150
Glu Ala Phe Arg Glu Ala Lys Arg His Arg Pro Leu Ile Ala Leu Pro Thr Gln Asp Pro Ala His Gly Gly Ala Pro Pro Glu Met Ala
GAG GCG TTC AGG GAG GCC AAA CGT CAC CGT CCC CTG ATT GCT CTA CCC ACC CAA GAC CCC GCC CAC GGG GGC GCC CCA CCA GAG ATG GCC

          156  OP
Ser Asn Arg Lys
AGC AAT CGG AAG TGA GCAAAACTGCCGCAAGTCTGCAGCCGGCAGCGGCCACCATCTGCACCTCCTCTGACCACGGACGTTTCCATCAGGTTCCATCCGAAAATCTCTCGGTTCCA

CGTCCCCCTGGGGCTTCTTCTGACCCAGTCCCGTGCCCCGCCTCTCCCGAAACAGGCTACTCTCCTGCCGCCCCTCCATCGGGCTGAGGAAGCACAGCAGCATCTTCAAACATGTACAA

ATCGATTGGCTTTAAACACCTTCACATACCT
```

FIG. 1. The sequence of cDNAs encoding preproinsulin-like growth factors I (A) and II (B). The IGF moiety in each sequence is boxed. The first residue of the progrowth factor is 1; the residues of the signal peptide have negative numbers. These cDNAs were isolated from an adult human liver cDNA library.[8]

TABLE I

RELATIVE ABUNDANCE OF INSULIN-LIKE GROWTH FACTOR
mRNA IN FEMALE RHESUS MONKEY TISSUES

Tissue	IGF-I mRNA	IGF-II mRNA	IGF-I/IGF-II mRNA ratio
Liver	1.0	1.0	2.3
Heart	0.1	0.7	0.5
Lung	0.1	0.1	1.2
Kidney	0.1	0.6	0.3
Pancreas	0.03	0.05	1.5
Spleen	0.2	0.4	1.2
Small intestine	0.05	0.05	2.5
Colon	0.2	0.05	7.7
Brain			
Brainstem	0.1	0.5	0.4
Frontal lobe	0.1	0.2	1.0
Occipital lobe	0.1	0.3	0.7
Cerebellum	0.05	0.4	0.3

HepG2 and HepG3 cells[12] contain 10–100 times more IGF-II mRNA than does human liver. Because of the ease of isolation of undegraded RNA from cultured cells, these cell lines represent a good source of RNA for construction of cDNA libraries enriched in sequences coding for human IGF-II.

Methods

Isolation of RNA from Tissues

The method that we use for the isolation of RNA from tissues is similar to that described by Cathala et al.[13] It utilizes guanidinium thiocyanate as a chaotropic agent during tissue homogenization,[14] direct precipitation of RNA from the guanidinum solution by LiCl,[15] followed by purification of the RNA from residual DNA and protein by successive urea–LiCl washes.

[12] B. B. Knowles, C. C. Howe, and D. P. Aden, *Science* **209**, 497 (1980).
[13] G. Cathala, J.-F. Savouret, B. Mendez, B. L. West, M. Karin, J. A. Martial, and J. D. Baxter, *DNA* **2**, 329 (1983).
[14] J. M. Chirgwin, A. E. Przybyla, R. J. MacDonald, and W. J. Rutter, *Biochemistry* **18**, 5294 (1979).
[15] C. Auffray and F. Rougeon, *Eur. J. Biochem.* **107**, 303 (1980).

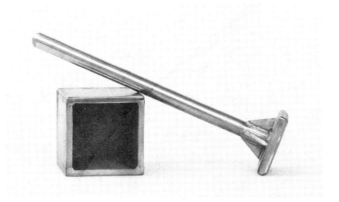

FIG. 2. Stainless steel box for pulverizing frozen tissue. The box is $3 \times 3 \times 3\frac{1}{2}$ as measured from the outside and the plates are $\frac{3}{8}$ in. thick. The shaft of the plunger is 10 in. long and $\frac{5}{8}$ in. wide, and the base is $2\frac{1}{4} \times 2\frac{1}{4} \times \frac{1}{4}$ in.

Messenger RNAs of greater than 9.5 kb have been isolated using this procedure.[16]

Human tissue should be procured as soon after death as possible, cut into 1- to 5-g pieces and stored in liquid N_2 or at $-80°$. The tissue is pulverized in a stainless steel box with a loose-fitting square plunger whose shaft is struck with a hammer (Fig. 2). The box and plunger is chilled in crushed dry ice before use. The pulverized tissue is added to the guanidinium thiocyanate solution and solubilized with a Tissumizer as described below. All solutions are prepared with distilled water which has been treated with 0.1% diethyl pyrocarbonate (DEP) and then autoclaved. Solid and liquid reagents are dispensed from previously unopened containers. All solutions are filtered through 0.2-μm filters (Nalgene) and stored in sterile disposable plastic containers. With these precautions, nonspecific degradation of the RNA is minimized. When it is necessary to use glassware, it is acid washed and heat treated at $175°$ before use.

One gram of pulverized tissue is homogenized using a Tekmar Tissumizer for 1 min at room temperature in 10 ml of lysis buffer containing 5 M guanidinium thiocyanate, 100 mM Tris–HCl (pH 7.5), 10 mM EDTA (pH 8.0), and 14% (v/v) 2-mercaptoethanol. After dissolution, the homogenate is clarified by centrifugation (10,000 g, $4°$, 10 min). The RNA is precipitated (16–24 hr, $4°$) by adding 5.5 vol of 4 M LiCl. The RNA and some protein and DNA is then pelleted by centrifugation (10,000 g, $4°$, 30 min) using two 30-ml Corex tubes. Each pellet is mechanically resus-

[16] L. B. Rall, G. I. Bell, D. Caput, M. A. Truett, F. R. Masiarz, R. C. Najarian, P. Valenzuela, H. D. Anderson, N. Din, and B. Hansen, *Lancet* **1**, 44 (1985).

pended in 5 ml of a solution of 3 M LiCl containing 4 M urea by using a Pasteur pipet sealed at the end, and then vortexed vigorously for about 1 min. The volume of the suspension is adjusted to 30 ml and the precipitate collected by centrifugation as described above. This procedure is repeated two to three times or until the supernatant is clear. Each RNA pellet is finally dissolved in 2.5 ml of 1% sodium dodecyl sulfate (SDS). One volume of phenol [equilibrated with 50 mM Tris–HCl (pH 8.0), 1 mM EDTA, and 0.1% 2-mercaptoethanol] is added and the mixture vortexed vigorously for about 1 min. An equivalent volume of chloroform:isoamyl alcohol (24:1) is added and the solution vortexed for 1 min. The mixture is centrifuged (4000 g, 4°, 20 min); the aqueous phase is extracted with chloroform:isoamyl alcohol and centrifuged (4000 g, 4°, 10 min). The RNA is precipitated (16–24 hr, −20°) from the aqueous phase by adding 1/10 vol of 2 M potassium acetate (pH 5.0) and 3 vol of absolute ethanol (Goldshield). The precipitated RNA is collected by centrifugation (10,000 g, 4°, 30 min) and dissolved in a minimal volume (1 ml/g of tissue used) of DEP-treated water and precipitated as described above. The RNA–ethanol slurry can be stored for years at −20° without apparent degradation occurring. For subsequent analyses including the preparation of poly(A)$^+$ RNA,[17] the RNA is centrifuged, washed in 70% ethanol (−20°), and dissolved in the appropriate solution.

Isolation of RNA from Cultured Cells

Cells are grown to confluency in 100-mm tissue culture dishes, placed on ice, and the medium removed. In our experience, 40 plates can be conveniently handled at 1 time. Each monolayer of cells is rinsed with 5 ml of phosphate-buffered saline (PBS) (4°) and then scraped into 1 ml of PBS using a polyethylene policeman. The cells are collected by centrifugation (150 g, 5 min) and vortexed for 1 min in 0.5 ml/100-mm plate of a solution containing 10 mM Tris–HCl (pH 8.0), 150 mM NaCl, 10 mM EDTA, 1% Nonidet-P40, and 1000 U/ml of RNasin (Promega Biotec). The nonionic detergent and ribonuclease inhibitor are added to the lysis buffer just prior to use. The nuclei are pelleted by centrifugation (2000 g, 5 min, 4°) and 1/20 volume of 10% SDS is added to the supernatant which contains the cytoplasmic RNA. This solution is extracted with an equal volume of phenol:chloroform:isoamyl alcohol (25:24:1, respectively). Following centrifugation, the aqueous phase is extracted with an equal volume of chloroform:isoamyl alcohol. The RNA is precipitated by adding 3 vol of ethanol and is stored at −20° for 16–24 hr.

---

[17] H. M. Goodman and R. J. MacDonald, this series, Vol. 68, p. 75.

Residual DNA from lysed nuclei is removed by selective LiCl precipitation. (In many instances this step is unnecessary.) The RNA is pelleted by centrifugation (10,000 g, 30 min, 4°), dissolved in DEP-treated water (1 ml/40 plates), and the concentration adjusted to about 1 mg/ml. LiCl (12 M stock) is added to a final concentration of 3 M and, after 16–24 hr at 4°, the RNA is centrifuged as described above. The pellet is suspended by vortexing 1 min in an equivalent volume of 3 M LiCl, centrifuged, and finally dissolved in 1 ml of DEP-treated water. The RNA is precipitated by adding 1/10 vol of 2 M potassium acetate (pH 5.0) and 3 vol of ethanol, and the solution is stored at −20°. The yield of RNA is about 0.25 mg/10^8 cells; the number of cells per 10-cm plate varies from 0.5–1.0 × 10^7 depending on the characteristics of the cells. This procedure is effective for cells that have a low content of RNase. The guanidinium thiocyanate method described above is used for those cells that have high levels of RNase.

Identification of IGF cDNAs

Human IGF-I and IGF-II cDNAs have been isolated from an adult human liver cDNA library[5-8] which was prepared from poly(A)[+] RNA using standard procedures.[17,18] Double-stranded cDNA was synthesized, treated with S_1 nuclease to cleave the hairpin loop, and inserted into the PstI site of the plasmid vector using the dG:dC tailing technique. Due to the low abundance of IGF-I and IGF-II mRNA in most tissues, the complexity of the cDNA library must be at least 10^4–10^5. Gubler and Hoffman[19] have recently described a procedure for synthesizing double-stranded cDNA which simplifies the construction of cDNA libraries and hence the cloning of low-abundance mRNAs. It combines the classical first-strand synthesis with RNase H–DNA polymerase I-mediated second strand synthesis. The double-stranded cDNA is then tailed and cloned without further manipulation. Large numbers of recombinants are screened by using high-density screening procedures.[20-22]

The initial IGF-I and IGF-II cDNAs were identified using a pool of oligonucleotides since the sequence of the mRNAs were unknown.[5,6] These pools also hybridized to cDNAs other than those coding for the IGFs and therefore, it was necessary to sequence each clone. As the sequences of both human IGF-I and IGF-II cDNAs have now been determined (Fig. 1),

---

[18] H. Land, M. Grez, H. Hauser, W. Lindenmaier, and G. Schütz, this series, Vol. 100, p. 285.

[19] U. Gubler and B. J. Hoffman, Gene 25, 263 (1983).

[20] M. Grunstein and J. Wallis, this series, Vol. 68, p. 379.

[21] D. Hanahan and M. Meselson, this series, Vol. 100, p. 333.

[22] D. Ish-Horowicz and J. F. Burke, Nucleic Acids Res. 9, 2989 (1981).

it is possible to synthesize a probe of 20 bases or longer based upon this information. We usually use an oligonucleotide of 45 bases to avoid nonspecific hybridization. The oligonucleotides can be synthesized manually or with one of the commercial nucleotide synthesizers. It is purified away from failure sequences by electrophoresis in a denaturing polyacrylamide gel as described by Urdea.[23]

The 5' end of the oligonucleotide is labeled with ^{32}P using [γ-^{32}P]ATP and polynucleotide kinase.[24] One hundred picomoles of oligonucleotide is labeled in a volume of 50 μl containing 1.4 mCi (~ 200 pmol) of [γ-^{32}P]ATP (ICN, crude, 7000 Ci/millimole), 50 mM Tris–HCl (pH 7.8), 10 mM MgCl$_2$, 10 mM dithiothreitol, and 25 U of polynucleotide kinase. The reaction is incubated at 37° for 60 min and then terminated by heating at 70° for 10 min. The ^{32}P-labeled oligonucleotide is separated from [γ-^{32}P]ATP and other reactants on a Sep-Pak C$_{18}$ cartridge (Waters Associates). The inactivated reaction is diluted with 3 ml of 0.5 M NaCl, 0.10 M Tris–HCl (pH 8.0), 5 mM EDTA, and 5 μg of poly(dA) and then slowly applied to a Sep-Pak cartridge which has been washed with 5 ml of methanol and then 10 ml of distilled water. (These steps are conveniently done using disposable syringes; air and fluid should move in only one direction.) ATP, ADP, and P$_i$ are removed by washing with 10 ml of distilled water. The labeled oligonucleotide is eluted with 1.5–3 ml of 1:1 solution of 0.1 M ammonium acetate (adjusted to pH 7.5 with NH$_4$OH) and methanol. This oligonucleotide solution can be used for hybridization directly, or evaporated to dryness in a Savant Speed-Vac and dissolved in 100 μl of distilled water.

Recombinant bacteria containing plasmids encoding IGF-I and IGF-II precursors are identified by hybridization. Duplicate filters are screened to exclude nonspecific hybridization. Nitrocellulose filters containing immobilized plasmid or phage DNA are soaked at 37° for 60 min in a solution of 4 × SSC (SSC is 0.15 M NaCl and 0.015 M sodium citrate), 40 mM sodium phosphate (pH 6.8), 2 × Denhardt's solution (1 × Denhardt's is 0.02% bovine serum albumin, 0.02% Ficoll 400 and 0.02% poly(vinylpyrrolidone)360, 0.1% SDS, and 300 μg/ml of sonicated and heat-denatured salmon testes DNA. The filters are then transferred to another container of the same solution that also has 1–2 × 10^6 cpm/ml of the labeled oligonucleotide. Hybridization is at 37° for 16–20 hr, although 4–6 hr may be sufficient. To remove the unhybridized probe, the filters are washed in 2 × SSC containing 0.1% SDS at room temperature for 30 min and then at 50° for 30 min in the same solution for an oligonucleotide of 20 bases, or 1 ×

[23] M. Urdea, this volume [3].
[24] A. M. Maxam and W. Gilbert, this series, Vol. 65, p. 499.

SSC and 0.1% SDS for an oligonucleotide of 45 bases. The filters are blotted dry, wrapped in Saran Wrap, and autoradiographed with a screen overnight. Duplicate regions of hybridization indicate a clone encoding IGF-I or IGF-II.

Although the IGF-I and IGF-II cDNAs do not cross-hybridize under stringent hybridization conditions, cross-hybridization is observed under conditions of reduced stringency. Consequently, if a clone encoding IGF-I is isolated, it can be used as a probe to identify those encoding IGF-II and vice versa. Such a strategy was used by Jansen et al.[7] to isolate IGF-II cDNAs.

Analysis of RNA Preparations for IGF-I and IGF-II mRNA

The dot–blot technique described below is a combination of procedures described by Kafatos et al.[25] and Thomas.[26] It has been used to determine the abundance of IGF-I and IGF-II mRNA in a variety of Rhesus monkey tissues.[9] RNA (up to 10 μg) is precipitated just prior to dot–blot analysis and the dried pellets are dissolved in 10 μl of DEP-treated water. Thirty-five microliters of a solution containing 50 mM morpholinopropanesulfonic acid (MOPS) (pH 7.0), 13 mM sodium acetate, 1.3 mM EDTA (pH 8.0), 2.8 M formaldehyde, and 65% (v/v) deionized formamide is added and the RNA is denatured at 65° for 10 min.[27] To determine the amount of hybridization to contaminating DNA present in the RNA preparation, the RNA is hydrolyzed with NaOH. The RNA solution in 10 μl of DEP-treated water is adjusted to 0.4 M NaOH (2 M stock) and heated at 65° for 20 min and then neutralized with an equal volume of 2 M ammonium acetate (4°). The solutions of denatured or hydrolyzed RNA are diluted to 200 μl with 20× SSC. (The large volume and high salt concentration are critical for the effective binding of the RNA.) The samples are spotted onto a nitrocellulose filter presoaked in 20× SSC using a dot–blot apparatus (Bethesda Research Laboratories). The filter is dried in vacuo at 80° for 45 min, treated with 20 mM Tris–HCl (pH 8.0) at 100° for 1 min, and then hybridized with nick-translated cDNA inserts (specific activity of 2–4 × 10⁸ cpm/μg) as described previously.[28] The intensity of the hybridization signals is determined by densitometry. The data can be normalized to the amount of mRNA present in a specific tissue, for example liver in the case of IGF-I

[25] F. C. Kafatos, C. W. Jones, and A. Efstratiadis, Nucleic Acids Res. 7, 1541 (1979).
[26] P. S. Thomas, this series, Vol. 100, p. 255.
[27] T. Maniatis, E. F. Fritsch, and J. Sambrook, "Molecular Cloning: A Laboratory Manual," p. 202. Cold Spring Harbor Lab., Cold Spring Harbor, New York, 1982.
[28] G. I. Bell, J. H. Karam, and W. J. Rutter, Proc. Natl. Acad. Sci. U.S.A. 78, 5759 (1981).

and IGF-II mRNA (Table I). In addition, a quantitative estimate of the amount of an mRNA sequence can be obtained by comparing the hybridization signal of the RNA samples with that observed for variable amounts of cDNA. Since only one strand of the DNA is transcribed, yet both strands hybridize with a nick-translated probe, it is necessary to adjust the values obtained by a factor of two when comparing hybridization between the cDNA standards and RNA samples. The amount of a specific mRNA can be expressed as DNA equivalents per 10 μg of RNA. In addition, the ratio of one mRNA to another, for example IGF-I to IGF-II (Table I), can be determined after taking into account the variation in hybridization due to the difference in the sizes of the cDNA probes. The cDNA samples, either insert or plasmid DNA, are prepared by denaturing (100°, 2 min) 1 μg of DNA in 25 μl of 0.3 M NaOH. This solution is neutralized with an equal volume of 2 M ammonium acetate and diluted with 20 × SSC such that the final concentration ranges from 0.5 to 10 pg/100 μl.

This procedure has not been compared with the results using RNA denatured by glyoxal as described in detail by Thomas.[26] However, it is at least 10-fold more sensitive than the NaI technique refined by Bresser *et al.*[29] for the selective retention of mRNA. We estimate that as little as 1 pg of a specific mRNA is detectable in the formamide/formaldehyde system following autoradiography for a period of 24 hr.

Acknowlegments

We would like to thank Diana M. Wyles for careful preparation of the manuscript.

[29] J. Bresser, J. Doering, and D. Gillespie, *DNA* 2, 243 (1983).

[24] Radioligand Assays for Insulin-Like Growth Factor II

By WILLIAM H. DAUGHADAY

The work of Rinderknecht and Humbel established that there are two principal somatomedins in human serum which they named insulin-like growth factors I and II (IGF-I and IGF-II).[1,2] Two structurally similar IGFs have been isolated from rat serum.[3] There is assay evidence that chicken

[1] E. Rinderknecht and R. E. Humbel, *J. Biol. Chem.* 253, 2769 (1978).
[2] E. Rinderknecht and R. E. Humbel, *FEBS Lett.* 89, 283 (1978).
[3] J. S. Rubin, I. Mariz, J. W. Jacobs, W. H. Daughaday, and R. A. Bradshaw, *Endocrinology* 110, 734 (1982).

serum and turtle serum also contain these two IGFs but toad and trout serum only seem to contain an IGF-I-like peptide.[4] Because of the presence of these two IGFs, efforts have been made in several laboratories to develop methods which specifically measure these peptides. This chapter will focus on methods of measurement of IGF-II.

Iodination of IGF Peptides

The various radioligand displacement assays depend on the availability of highly purified peptides for iodination. Limited quantities of hIGF-II have been laboriously isolated from serum in small quantities in only a few laboratories and microgram quantities for iodination have been generously supplied to others by some investigators. Rat IGF-II (MSA-III 2) has been isolated from conditioned medium of Buffalo rat hepatocytes. Most multiplication-stimulating activity (MSA) preparations include a high proportion of partially processed precursors. (See Greenstein *et al.* [25], this volume, for a further description of IGF-II (MSA) properties.)

Iodination of IGF-II has been achieved with various modifications of the chloramine-T and lactoperoxidase methods (Table I). In the author's laboratory a carefully controlled lactoperoxidase method which was developed for nerve growth factor by Tait *et al.*[5] has been most successful in providing unaggregated labeled IGF which retains immune and receptor binding properties.

The reagents are prepared as follows:

1. Lactoperoxidase, 145 μg/ml (1.86 μM): Dissolve 1 mg of partially purified bovine milk lactoperoxidase (No. L 2005, Sigma Chemical Co., St. Louis, MO) in 2 ml distilled H_2O. On the basis of Σ_{412} of 114 nM^{-1} cm^{-1} adjust the OD to 0.212 to obtain a 1.86 μM solution.

2. Hydrogen peroxide, 0.5 nM: The 30% solution is approximately 10 M. On the basis of Σ_{230}=72.4 M^{-1} cm^{-1}, to obtain a 0.5 mM solution, dilute stock to OD 0.036 before each assay.

3. Quenching solution: 1 M NaCl, 0.1 M sodium iodide, 50 mM sodium phosphate, pH 7.5, 0.2% BSA, 1 mM sodium azide, 0.1% protamine sulfate.

The iodination is carried out in 12 × 75 polystyrene tubes with the following additions:

1. IGF-II, 2 μg in 20 μl of 0.2 M sodium phosphate buffer, pH 7.4, is prepared in advance and stored at −70°.

[4] W. H. Daughaday, M. Kapadia, C. E. Yanow, K. Fabrick, and I. K. Mariz, *Gen. Comp. Endocrinol.*, in press (1985).

[5] J. F. Tait, S. A. Weinman, and R. A. Bradshaw, *J. Biol. Chem.* **256**, 11086 (1981).

TABLE I
TECHNIQUES OF IODINATION OF IGF-II

Reference	IGF-II	^{125}I technique	Purification
Zapf et al.[a]	hIGF-II	Chloramine-T SA—40–80 μCi/μg	Sephadex G-50 0.1 M ammonium acetate, 0.2% BSA
Moses et al.[b]	rIGF-II (MSA III-2)	Chloramine-T SA—50–190 μCi/μg	Sephadex G-50 Fine 1 M acetic acid, 0.1% BSA
Hintz and Liu[c]	Tyr-IGF-II (33-40) (Tyr)Ser-Arg-Val-Ser-Arg-Arg-Ser-Arg	Chloramine-T SA—300–400 μCi/μg	BioGel P-2 in 0.04 M sodium phosphate, pH 7.4, 0.5% BSA, 0.15 M NaCl, 0.1% sodium azide
Daughaday et al.[d]	hIGF II	Lactoperoxidase[e] SA—50–100 μCi/μg	Sephadex G-50 0.025 M Tris buffer, pH 7.4

[a] J. Zapf, H. Walter, and E. R. Froesch, *J. Clin. Invest.* **68,** 1321 (1981).
[b] A. C. Moses, S. P. Nissley, P. A. Short, and M. M. Rechler, *Eur. J. Biochem.* **103,** 401 (1980).
[c] R. L. Hintz and F. Liu, *J. Clin. Endocrinol. Metab.* **54,** 442 (1982).
[d] W. H. Daughaday, B. Trivedi, and M. Kapadia, *J. Clin. Endocrinol. Metab.* **53,** 282 (1981).
[e] J. F. Tait, S. A. Weinman, and R. A. Bradshaw, *J. Biol. Chem.* **256,** 11086 (1981).

2. Twenty microliters sodium phosphate buffer (0.2 M, pH 7.5).
3. Ten microliters containing 1 mCi ^{125}I. Mix on Vortex.
4. Lactoperoxidase solution (2.5 μl). Mix on Vortex.
5. Two microliters of hydrogen peroxide, 0.5 mM. Mix on Vortex.
6. After 3 min, add 2 μl of hydrogen peroxide. Mix on Vortex.
7. After 3 min, add 150 μl of quenching solution.
8. Transfer to Sephadex G-50 column (0.9 × 30 cm) and elute with 0.03 M sodium phosphate buffer, pH 7.4, 0.25% BSA. Collect 1-ml fractions.
9. Determine radioactivity in 5-μl aliquots from each tube. Three peaks are usually seen: (1) aggregate, (2) monomer, (3) unreacted ^{125}I. Use monomer peak for radioligand assay.

The ^{125}I-labeled IGF-II is stored at 4° and remains usable for most ligand assays for 1 to 3 months. Degradation products may be removed by repeated Sephadex G-50 gel filtration or by octyl-Sepharose CL-4B columns according to the method of Baxter and Brown.[6]

Antibodies to IGF-II

Polyclonal antibodies against IGF-II have been raised to preparations identified as Sm-A, IGF-II, IGF-II (C peptide), and rat IGF-II (MSA) (Table II). Most authors have selected the rabbit as the recipient of antigen.

[6] R. C. Baxter and A. S. Brown, *Clin. Chem.* **28,** 485 (1982).

TABLE II
ANTIBODIES AGAINST IGF-II (MSA)

Reference	Immunogen	Animal	$B/B_0 50^a$	Cross-reactivity	
Moses et al.[b]	rIGF-II (MSA) coupled to ovalbumin, Freund's complete adjuvant	Rabbit	1–2 ng/ml	MSA-III-2 MSA-II hIGF-I hIGF-II	100% 33% 1.3% 3%
Zapf et al.[c]	hIGF-II, NaHCO$_3$–KAl (SO$_4$)$_2$, Freund's complete adjuvant	Rabbit	1 ng/ml	hIGF-II IGF-I Sm-C Sm-A	100% 10% 10% 10%
Enberg and Hall[d]	Sm-A (10–40% pure), Freund's complete adjuvant	Rabbit	1.5 ng/ml	hIGF-II hIGF-I MSA-II	100% 10% <1%
Hintz and Liu[e]	IGF-II (C peptide) conjugated to bovine thyroglobulin	Rabbit	80 ng/ml	IGF-II hIGF-I, other hormonal peptides	100% unreactive

[a] Concentration of IGF-II required for 50% displacement of ^{125}I-labeled IGF-II from antibody.
[b] A. C. Moses, S. P. Nissley, P. A. Short, and M. M. Rechler, *Eur. J. Biochem.* **103**, 401 (1980).
[c] J. Zapf, H. Walter, and E. R. Froesch, *J. Clin. Invest.* **68**, 1321 (1981).
[d] G. Enberg and K. Hall, *Acta Endocrinol. (Copenhagen)* **107**, 164 (1984).
[e] R. L. Hintz and F. Liu, *J. Clin. Endocrinol. Metab.* **54**, 442 (1982).

Moses *et al.* complexed rIGF-II with ovalbumin.[7] Hintz and Liu conjugated IGF-II C domain peptide to thyroglobulin.[8] Conjugation may not be required because satisfactory antibodies have been raised with hIGF-II suspended in a mixture of NaHCO$_3$ and KAl(SO$_4$)$_2$ and Enberg and Hall merely dispersed an impure mixture of neutral somatomedins (Sm-A) in Freund's adjuvant.[9,10] Multiple site injections are regularly employed.

The polyclonal antibodies against hIGF-II have exhibited 10% cross-reactivity with IGF-I. The polyclonal antibody raised against IGF-II C peptide is highly specific with negligible cross-reactivity with IGF-I. None of the antibodies reacts significantly with insulin or other hormonal peptides.

Recently a monoclonal antibody was raised against an IGF-II preparation isolated from conditioned medium of a human pituitary cell line

[7] A. C. Moses, S. P. Nissley, P. A. Short, and M. M. Rechler, *Eur. J. Biochem.* **103**, 401 (1980).
[8] R. L. Hintz and F. Liu, *J. Clin. Endocrinol. Metab.* **54**, 442 (1982).
[9] J. Zapf, H. Walter, and E. R. Froesch, *J. Clin. Invest.* **68**, 1321 (1981).
[10] G. Enberg and K. Hall, *Acta Endocrinol. (Copenhagen)* **107**, 164 (1984).

(18-54 SF) (Tanaka et al.[11] This antibody produced and distributed by the Amano Pharmaceutical Co., Ltd., of Nagoya, Japan, has proved to have excellent properties for radioimmunoassay employing a conventional double antibody separation method. Although the company reports an B/B_0 50 of 5 ng/ml using their IGF-II for labeling and standards, we have found the B/B_0 50 to be about 300 pg/ml using highly purified bovine IGF-II prepared in our laboratory. Moreover, the company reports a cross-reactivity with human IGF-I of 10%, but under our conditions of assay the cross reactivity is less than 3% with (Thr_{59}) hIGF-I or partially purified human serum IGF-I free of IGF-II. Another valuable property of this monoclonal antibody is wide cross-reactivity with IGF-II of other animal sera. We have found specific IGF-II immunoreactivity in the acid ethanol extracts of rat, guinea pig, porcine, bovine, and chicken serum. We have no information concerning the relative immunoreactivities of these various animal sera.

Avoidance of Interference from Binding Proteins

The IGFs circulate tightly bound to one or more high-affinity binding proteins (reviewed in Ref. 12). The effect of these binding proteins on IGF radioligand assays is complex. In native adult human serum the large-molecular-weight IGF complex (150K) equilibrates poorly with added ^{125}I-labeled IGF-II and is hindered in its reactivity with immuno and receptor binding sites.[13] It was originally proposed by Furlanetto et al. that prolonged preincubation for 72 hr would obviate interference from binding protein in the Sm-C RIA and render displacement curves for serum and standards parallel.[14] Possible perturbations of direct serum assays by protamine, heparin, and EDTA have been investigated.[15] (Also, see Furlanetto and Marino, [20], this volume.)

Even brief acidification of serum followed by neutralization converts the large binding protein complex (150K) to a smaller IGF binding protein complex (50 K). This complex can equilibrate with tracer. Bala and Bhau-

[11] H. Tanaka, K., Nishikawa, Y. Yamada, T. Hayano, O. Asami, Y. Nakanishi, and K. Yamashita, in "Proceedings of the International Symposium on Growth and Differentiation of Cells in Defined Environment held in Fukuoka, Japan on September 2–6, 1984" (H. Murakami, I. Yamane, D. W. Barnes, J. P. Mather, I. Hayashi, and G. H. Sato, eds.), p. 345. Springer-Verlag, Berlin, 1985.

[12] G. L. Smith, Mol. Cell. Endocrinol. 34, 83 (1984).

[13] W. H. Daughaday, I. K. Mariz, and S. L. Blethen, J. Clin. Endocrinol. Metab. 51, 781 (1980).

[14] R. W. Furlanetto, L. Underwood, J. J. Van Wyk, and A. J. D'Ercole, J. Clin. Invest. 60, 648 (1977).

[15] J. R. Clemmons, L. E. Underwood, and P. G. Chatelain, J. Clin. Endocrinol. Metab. 56, 384 (1983).

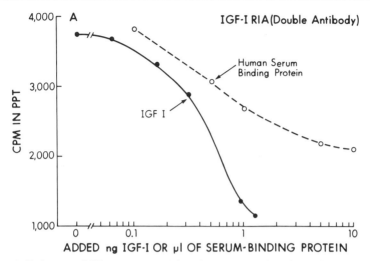

FIG. 1. Endogenous IGFs were removed from human serum by acid gel filtration. Different amounts of the "stripped" serum, expressed as microliters of original serum, were introduced in the IGF-I RIA (double antibody). On the abscissa cpm of ^{125}I-labeled IGF-I in the immunoprecipitate is given. Reprinted by permission from W. H. Daughaday, M. Kapadia, and I. K. Mariz, *J. Lab. Clin. Med.*, in press, 1987.

mick have advocated addition of formic acid to serum in a final concentration of 1%.[16] After 4 hr at 4°, the acid is removed by lyophilization. The authors claim that "the Sm binding equilibrium may be markedly shifted in the direction of the high affinity Sm antibodies, enabling the Sm RIA to measure total immunoreactive somatomedin in serum." The empiric usefulness of this procedure has been established in extensive clinical studies.

The notion that the affinity of the antibodies for IGFs is sufficiently higher than that of the binding proteins so that the latter do not affect binding equilibrium is not borne out by direct study. When serum is stripped of endogenous IGF-II by acid Sephadex gel filtration and then added back to the double-antibody IGF-I RIA, additions of less than 1 μl of human serum caused substantial displacement of label (Fig. 1).[17] Because rat serum contains higher concentrations of binding protein with somewhat higher binding affinity, the displacement of more than 50% of the label was observed with 0.1 μl of stripped serum (Fig. 2) and the displacement curve was nearly parallel to IGF-I standards.

To avoid possible interference from binding proteins in assays, a number of strategies have been utilized. We have preferred the ease and conve-

---

[16] R. M. Bala and B. Bhaumick, *J. Clin. Endocrinol. Metab.* **49,** 770 (1979).
[17] W. H. Daughaday, M. Kapadia, and I. K. Mariz, *J. Lab. Clin. Med.* **109,** in press (1987).

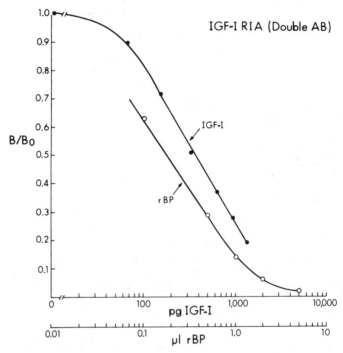

FIG. 2. Endogenous IGFs were removed from normal adult rat serum by acid gel filtration. Different amounts of the stripped serum were introduced in the IGF-I RIA (double antibody). On the abscissa cpm of ^{125}I-labeled IGF-I in the immunoprecipitate is given. Reprinted by permission from W. H. Daughaday, M. Kapadia, and I. K. Mariz, *J. Lab. Clin. Med.,* in press (1987).

nience of an acid–ethanol extraction method Table III).[13] In this method 0.8 ml of 87.5% ethanol, 12.5% 2 *M* HCl (v/v) is added to 0.2 ml of serum. After thorough mixing and standing 30 min, the tube is centrifuged 1800 *g* for 30 min. Supernatant (0.5 ml) is removed and neutralized with 0.2 ml 0.855 *M* Tris base. Aliquots are removed for IGF-I and IGF-II assays.

A modification of this method may be used in which 2 *M* formic acid is substituted for HCl. Aliquots of the supernatant are lyophilized directly in the Speed Vac. The acid–ethanol extraction procedure has worked effectively in the author's laboratory with human serum but the removal of binding protein from rat serum and fibroblast conditioned medium is incomplete (Fig. 3).[17]

As an alternative procedure, the absorption of IGFs from acidified serum can be achieved with a C_{18} cartridge (Sep Pak). The Sep Pak cartridges, held in a Multisample Rack (Waters, Milford, MA), are conditioned by washing with 5 ml isopropanol, 5 ml methanol, and 10 ml of 4%

TABLE III
PROPERTIES OF IGF-II RECEPTOR AND SERUM PROTEIN BINDING ASSAYS

Reference	Binding site	B/B_0 50[a]	Specificity	
Zapf et al.[b,c]	Serum binding protein: human serum was gel filtered through Sephadex G-50 in 1 M HAc; bed volume 30–50% was pooled, dialyzed	8 ng/ml	IGF-II IGF-I Insulin	100% 7.7% <1%
Rechler et al.[d]	Rat liver membranes prepared according to method of Neville[e]	1 ng/ml	IGF-II IGF-I Insulin	100% 3–10% <1%
Daughaday et al.[f]	Rat placentas from days 19 to 22 of pregnancy. Membranes prepared by a modification of the method of Marshal et al.[g]	6 ng/ml	IGF-II IGF-I Insulin	100% <1% <<1%

[a] Concentration of IGF-II required for 50% displacement of ^{125}I-labeled IGF-II from binding site.
[b] J. Zapf, U. Kaufman, J. Eigenmann, and E. R. Froesch, *Clin. Chem.* **23**, 677 (1977).
[c] J. Zapf, E. Schoenle, and E. R. Froesch, *Eur. J. Biochem.* **98**, 285 (1978).
[d] M. M. Rechler, J. Zapf, S. P. Nissley, E. R. Froesch, A. C. Moses, J. M. Podskalny, E. E. Schilling, and R. E. Humbel, *Endocrinology* **107**, 1451 (1980).
[e] D. M. Neville, *Biochim. Biophys. Acta* **154**, 540 (1968).
[f] W. H. Daughaday, B. Trivedi, and M. Kapadia, *J. Clin. Endocrinol. Metab.* **53**, 289 (1981).
[g] R. N. Marshall, L. E. Underwood, S. J. Voina, D. B. Foushee, and J. J. Van Wyk, *J. Clin. Endocrinol. Metab.* **39**, 283 (1974).

acetic acid. Serum, 0.2 ml (the cartridge will accept up to 3 ml), is acidified by adding an equal volume of 0.5 M HCl and passing through the cartridge. The cartridge is then washed with 10 ml of 4% acetic acid and the IGFs are eluted with 7 ml of 50% acetonitrile in water (v/v). Aliquots for assay are lyophilized in a Speed Vac. Sixty to 70% of added IGFs can be recovered.

The gold standard for methods of removing binding protein from IGFs is the acid gel filtration column. This has been the standard procedure in the IGF-II RIA of Zapf et al.[9] and of Hintz and Liu.[8] It has been routinely used in the rIGF-II (MSA) RIA of Moses et al.[7] Satisfactory separation of binding protein has been achieved with Sephadex G-50 and G-75 and with Bio-Rad P100. Elution has generally been with 0.5 to 1.0 M acetic acid. Addition of 100 mM NaCl or ammonium acetate will tighten up the binding protein elution. This is the recommended procedure for assays on rat serum.

Incubation and Separation Conditions

IGF-II RIAs have been conducted in Tris–saline or phosphosaline buffers of pH 7.4. To minimize nonspecific binding of IGF-II to glass or plastic tubes, 0.2 to 0.5% human serum albumin is added. The albumin

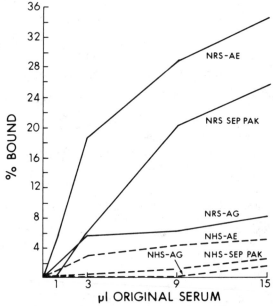

Fig. 3. Normal human (NHS) and normal rat (NRS) serum were extracted by the acid–ethanol (AE), acid gel filtration (AG), and Sep Pak methods. Residual binding activity was determined by incubation with ^{125}I-labeled IGF-II overnight at 4° and separation of unbound ^{125}I-labeled IGF-II by charcoal. The abscissa give microliters of original serum represented. Reprinted by permission from W. H. Daughaday, M. Kapadia, and I. K. Mariz, *J. Lab. Clin. Med.*, in press (1987).

should be of high purity and free of contained IGF-I, IGF-II, and of IGF binding protein.

When there is no binding protein in the reactive mixture, equilibrium is reached after 16 to 24 hr at 4° and even more rapidly at room temperature. Because of the high concentrations of IGF-II in normal human serum, there is no need for disequilibrium assays to increase sensitivity.

At the conclusion of the incubation, separation of antibody bound from free ^{125}I-labeled IGF-II may be accomplished by double antibody, Sepharose-bound antibody, or polyethylene glycol separation methods.[8-10] The author prefers a final concentration of 12.5% polyethylene glycol to separate free and antibody-bound IGFs because of speed and economy.

Receptor Binding Assays

Three IGF-II competitive binding assays have been described which depend on membrane receptors or specific IGF binding sites of a serum carrier protein. These methods were widely employed prior to the development of RIAs and remain useful procedures.

The serum binding protein assay is easily performed.[18] Human or rat serum is stripped of endogenous IGFs by gel filtration through Sephadex G-50 or G-75 in 0.5 M acetic acid. The serum protein fractions are pooled and the acid removed by dialysis against a neutral phosphosaline buffer.

As done in the author's laboratory, the reaction mixture contains binding protein from 5 to 10 μl of serum, ^{125}I-labeled IGF-II (8000– 12,000 cpm), IGF-II standards or aliquots from extracted serum in 0.5 ml of neutral phosphosaline buffer, 0.25% (fatty acid free) BSA (Sigma Chemical Co., St. Louis, MO). Blank tubes without binding protein are also prepared. Binding equilibrium is reached within 2 hr at room temperature or overnight at 4°. Separation of bound ^{125}I-labeled IGF-II is accomplished by adding 0.6 ml of 0.5% charcoal (activated, neutralized, Sigma Chemical Co., St. Louis, MO). The tubes were vortexed, centrifuged, and the supernatant removed and counted. Greater than 80% of the ^{125}I-labeled IGF-II in the tube lacking binding protein should be bindable by charcoal.

If care is taken to exclude serum binding protein from the reaction and the conductivity and pH of the samples to be measured approximate that of the assay buffer, the method is quite sensitive and specific. Seven percent cross-reactivity with IGF-I compares favorably with the described hIGF-II polyclonal antibodies.[19] There is no cross-reactivity with insulin. Residual ampholytes from isoelectric focusing experiments increase the binding of ^{125}I-labeled IGF-II by binding protein (W. H. Daughaday, unpublished).

Two IGF-II membrane receptor assays have been described. The first uses the type II (IGF-II) receptor on rat liver membranes and the second uses the type II receptor of rat placental membranes. These membranes are enriched in the type II receptors and relatively deficient in the type I receptors.

The rat liver membrane assay was originally developed by Megyesi et al. for use with ^{125}I-labeled NSILAs (a mixture of IGF-I and IGF-II).[20] The properties of the assay were later characterized with ^{125}I-labeled IGF-I and IGF-II by Rechler et al.[21] With ^{125}I-labeled IGF-II as a ligand, IGF-I is only 3 to 10% as potent as IGF-II in displacing ^{125}I-labeled IGF-II from the membrane receptor.

Rat placental membranes have been preferred by the author for mea-

[18] J. Zapf, U. Kaufman, J. Eigenmann, and E. R. Froesch, *Clin. Chem.* **23**, 677 (1977).
[19] J. Zapf, E. Schoenle, and E. R. Froesch, *Eur. J. Biochem.* **87**, 285 (1978).
[20] K. Megyesi, C. R. Kahn, J. Roth, D. M. Neville, Jr., S. P. Nissley, R. E. Humbel, and E. R. Froesch, *J. Biol. Chem.* **250**, 8990 (1975).
[21] M. M. Rechler, J. Zapf, S. P. Nissley, E. R. Froesch, A. C. Moses, J. M. Podskalny, E. E. Schilling, and R. E. Humbel, *Endocrinology* **107**, 1451 (1980).

surement of IGF-II.[22,23] In preparing membranes placentas are removed from 30 to 40 rats in the last 3 days of pregnancy. Extraneous membranes are removed and the placentas are weighed. If desired, the placentas may be stored at $-17°$ for several months before processing. The placentas are minced finely with scissors into $0.25 M$ sucrose, 1 ml/g placenta, and homogenized with a Polytron (Brinkman Instruments, Westbury, NY) with a PT10 probe head at a rheostat setting of 5 for 30 sec and 7 for 30 sec.

The homogenate is then centrifuged at 770 g to remove gross debris and the supernatant recentrifuged at 14,000 g. NaCl, 580 mg, and $MgSO_4$, 12 mg, are added per 100 ml of the membrane mixture and stirred at room temperature for 15 min.

The partially purified membranes are recovered by centrifugation at 40,000 g for 40 min. The pellet is resuspended in 50 ml of $0.05 M$ Tris–HCl, pH 7.8, and recentrifuged at 100,000 g for 90 min.

The final membrane pellet is resuspended in 20 ml of $0.05 M$ Tris–HCl buffer and protein content determined with the Lowry method.

The assay is conducted in 1.5-ml Microfuge tubes. The assay reaction mixture contains 200 μg of placental membrane, IGF-II as standards or unknown sample, ^{125}I-labeled IGF-II, 10,000 cpm, in a total volume of 0.5 ml of $0.1 M$ Tris, $0.25 M$ NaCl, 3% BSA. The mixture is incubated overnight at $4°$ and centrifuged at 8700 g for 10 min. The supernatant is carefully removed by aspiration and the radioactivity of the pellet measured.

We have repeatedly determined that this assay is highly specific for IGF-II with less than 1% cross-reactivity with hIGF-I, biosynthetic hIGF-I (Amgen Biological, Thousant Oaks, CA) and rIGF-I. In contrast, Pilistine *et al.* find that IGF-I is from 5 to 20% as effective as IGF-II in displacement of ^{125}I-labeled IGF-II from its receptor on rat placental membranes.[24] The reason for this discrepancy is not established but may reflect differences of membrane storage or assay buffer.

Binding proteins must be removed before assaying IGF-II in serum and conditioned medium. The assay has been employed in studies of human, rat, guinea pig, bovine, chicken, and turtle serum and so appears to have wide species applicability.[25] As in all membrane assays, care must be taken to avoid differences in salt content and pH of the samples.

[22] W. H. Daughaday, I. K. Mariz, and B. Trivedi, *J. Clin. Endocrinol. Metab.* **53,** 282 (1981).
[23] W. H. Daughaday, B. Trivedi, and M. Kapadia, *J. Clin. Endocrinol. Metab.* **53,** 282 (1981).
[24] S. J. Pilistine, A. C. Moses, and H. Munro, *Endocrinology* **115,** 1060 (1984).
[25] W. H. Daughaday, M. Kapadia, C. E. Yanow, K. Fabrick, and I. Mariz, *Gen. Comp. Endocrinol.,* **59,** 316 (1985).

Standards for IGF-II Assays

Because highly purified rat IGF-II (MSA peak III) and human IGF-II are not readily available, partially purified preparations of hIGF-II and rIGF-II (MSA peak II) have been used to define the displacement in IGF-II RIAs and RRAs. The working standard can be standardized against the purified IGF-II peptides. It is also helpful to include assays on pooled serum from young adults in each assay. We have found that the IGF-II content in such serum is very stable over long periods of time while the more purified working standards may lose potency. Until a universal IGF-II standard preparation is available, some investigators prefer to express assay results in terms of pooled normal adult human serum whose potency is assigned as 1 U/ml.

Acknowledgments

Supported by Research Grant RO1 AM05105, Diabetes Research and Training Center Grant P60 AM20579, National Institute of Arthritis, Diabetes, Digestive and Kidney Diseases, and General Clinical Research Center Grant RR 00036, National Institutes of Health, Division of Research Resources, Bethesda, MD.

[25] Purification of Rat Insulin-Like Growth Factor II

By Lawrence A. Greenstein, Lynne A. Gaynes, Joyce A. Romanus, Lilly Lee, Matthew M. Rechler, and S. Peter Nissley

Dulak and Temin[1,2] first described the purification of multiplication-stimulating activity (MSA) or rat IGF-II (rIGF-II) from serum-free medium conditioned by the BRL-3A rat liver cell line. There are three size classes of rIGF-II found in the BRL-3A conditioned medium: 16,300, 8700, and 7100 Da.[3] In addition, there is microheterogeneity within the two smaller size classes.[3] Marquardt et al.[4] purified one of the smaller species and demonstrated extensive amino acid sequence homology with human IGF-II. Results of biosynthetic labeling experiments suggest that

[1] N. C. Dulak and H. M. Temin, J. Cell. Physiol. **81**, 153 (1973).
[2] N. C. Dulak and H. M. Temin, J. Cell. Physiol. **81**, 161 (1973).
[3] A. C. Moses, S. P. Nissley, P. A. Short, M. M. Rechler, and J. M. Podskalny, Eur. J. Biochem. **103**, 387 (1980).
[4] H. Marquardt, G. J. Todaro, L. E. Henderson, and S. Oroszlan, J. Biol. Chem. **256**, 6859 (1981).

the various size rIGF-II species are derived from a common pro-rIGF-II (~20 kDa).[5,6] The smallest rIGF-II species (MSA-III-2) has the highest specific activity in radioreceptor assays and bioassays and is most easily purified to homogeneity. However, greater quantities of the 8700-Da species (MSA-II) are produced by the BRL-3A cells and a mixture of MSA-II species can be obtained following a two-step purification procedure.

Culture of the BRL-3A Rat Liver Cell Line

Coon[7] used low initial plating density and serial subcloning to establish parenchymal cell lines from normal rat liver. The BRL-3A line was cloned from the liver of a 5-week-old female Buffalo rat. It is available from American Type Culture Collection (Rockville, MD, ATCC CRL 1442).

The BRL-3A cells have a cobblestone appearance with variable cell size upon reaching confluency, but the cells continue to multiply, forming ridges of multiple cell layers.[8] The doubling time in medium containing 5% serum is approximately 24 hr and only slightly longer in serum-free medium.[8]

Cultivation of Stock Cultures. We do not passage cells from dense cultures from which conditioned media collections have been made. Instead, production cultures are started from frozen cell stocks or from small-scale cultures that have been grown in serum-free Ham's F12 or Dulbecco's modified Eagle's medium (DMEM). Cells can be detached from the plastic surface by exposure to Coon's CTC formulation (2% chicken serum, 0.075% trypsin, 7 U/ml collagenase) or 0.125% trypsin in TD buffer (NaCl, 8 g/liter; KCl, 0.038 g/liter; Na_2HPO_4, 0.1 g/liter; dextrose, 1 g/liter; Tris, 3 g/liter, pH 7.4). Cells are plated in medium containing 5% serum and then switched to serum-free medium at the time of the first feeding. Cells are passaged (1:5 split) upon reaching confluency; dense cultures are difficult to free from the plastic as single cells.

Cultivation of Cells for the Collection of Serum-Free Conditioned Medium. BRL-3A cells have been grown in Temin's modified Eagle's medium with 20% tryptose phosphate broth,[1,2] DMEM,[9] and Waymouth's medium[4] (DR-1027, Meloy Laboratories, Springfield, VA). The cells are maintained at 37° in an atmosphere of 95% air–5% CO_2. IGF-II produc-

[5] Y. W.-H. Yang, J. A. Romanus, T.-Y. Liu, S. P. Nissley, and M. M. Rechler, *J. Biol. Chem.* **260,** 2570 (1985).

[6] Y. W.-H. Yang, M. M. Rechler, S. P. Nissley, and J. E. Coligan, *J. Biol. Chem.* **260,** 2578 (1985).

[7] H. G. Coon, *Year Book—Carnegie Inst. Washington* **67,** 419 (1968).

[8] S. P. Nissley, P. A. Short, M. M. Rechler, and J. M. Podskalny, *Cell* **11,** 441 (1977).

[9] N. C. Dulak and Y. W. Shing, *J. Cell. Physiol.* **90,** 127 (1976).

tion has not been quantitatively compared for cells grown in these different media; we currently use DMEM. Dishes, flasks, and roller bottles have been used; roller bottles are convenient for large-scale harvesting of conditioned medium.

Cells from stock cultures are plated in medium containing 5% calf serum. At the time of the first feeding serum-free medium is utilized. Medium collections are not begun until the cells become confluent and the first serum-free medium collection following switching from serum-containing medium is discarded to avoid contamination with serum proteins.

Medium is harvested every 3 days for DMEM and every 2 days for Waymouth's medium. Because there is a lag in the appearance of biologically active IGF-II in the medium following medium change, and IGF-II accumulates linearly with time,[10] there is no advantage to making more frequent medium collections. The harvested medium is centrifuged at low speed to remove cells and then stored at −20°.

Measurement of rIGF-II during Purification

Competitive binding assays and bioassays have been utilized to measure rIGF-II during purification. A convenient radioreceptor assay which we describe here utilizes rat placental membranes and radiolabeled rIGF-II (MSA-III-2).[11] (See also Daughaday [24], this volume.) An alternative method which does not depend upon the availability of rIGF-II tracer is measurement of [^{3}H]thymidine incorporation into DNA in chick embryo fibroblasts.[12]

Rat Placental Membrane Preparation. The placentas from Sprague-Dawley rats (20 days gestation), trimmed of amnion and chorion, are harvested in 0.25 M sucrose on ice, and then washed three times in 0.25 M sucrose. Placentas (10 placentas/25 ml in a 50-ml polypropylene tube) are homogenized on ice with a Polytron (Brinkmann Instruments, Westbury, NY) at setting 5 for 1 min. The homogenate is centrifuged at 12,000 g (4°) for 30 min; the pellets are discarded. The supernatant is adjusted to 0.1 M NaCl and 0.2 mM MgSO$_4$, then centrifuged at 40,000 g for 40 min. The pellets are suspended in 0.05 M Tris–HCl at pH 7.4 with a Polytron at setting 1 to 2 for 15 sec and then centrifuged at 40,000 g for 40 min. The

[10] S. P. Nissley, M. M. Rechler, A. C. Moses, H. J. Eisen, O. Z. Higa, P. A. Short, I. Fennoy, C. B. Bruni, and R. M. White, *Cold Spring Harbor Conf. Cell Proliferation* **6,** 79 (1979).
[11] W. H. Daughaday, I. K. Mariz, and B. Trivedi, *J. Clin, Endocrinol. Metab.* **53,** 282 (1981).
[12] L. A. Greenstein, S. P. Nissley, A. C. Moses, P. A. Short, Y. W.-H. Yang, L. Lee, and M. M. Rechler, *in* "Methods for Preparation of Media, Supplements, and Substrata for Serum-Free Animal Culture" (D. W. Barnes, D. A. Sirbasku, and G. H. Sato, eds.), pp. 111–138. Liss, New York, 1984.

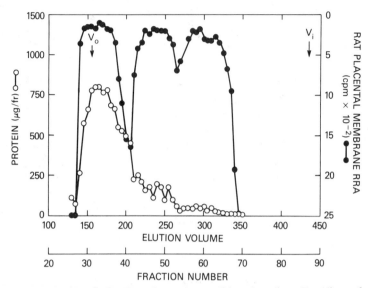

FIG. 1. BioGel P-10 gel filtration of BRL-3A conditioned medium. Four liters of conditioned medium that had been dialyzed and lyophilized was taken up in 15 ml of 1 M acetic acid, centrifuged, and the supernatant was applied to a 90 × 2.6 cm column of BioGel P-10 in 1 M acetic acid. Five-milliliter fractions were collected. Aliquots (25, 50, or 100 μl) were dried onto Immulon-1 flat 96-well microtiter plates (Dynatech Industries, Inc., Alexandria, VA) and protein was determined using the Bio-Rad protein assay kit. Following development of color the plate was read on a Titertek Multiscan (Dynatech Industries, Inc.) with a 595-nm filter. Protein is expressed as micrograms/fraction. Aliquots (25 μl) from column fractions were taken to dryness and assayed for IGF-II in the rat placental membrane radioreceptor assay as described in the text. Results are expressed as cpm ^{125}I-labeled IGF-II present in the membrane pellet and are plotted on an inverted scale.

last procedure is repeated two times. Membrane protein is determined by the Bradford method[13] and the aliquots of preparation are stored at −20°.

Rat Placental Membrane Radioreceptor Assay. Samples of rIGF-II standard are suspended in 200 μl of assay buffer [0.1 M Tris–HCl, 0.25 M NaCl, pH 7.4, 3% bovine serum albumin (Sigma Chemicals, St. Louis, MO, catalog #A7638)], then combined with ^{125}I-labeled rIGF-II (10,000 cpm/200 μl assay buffer) and finally rat placental membranes (20 to 40 μg/200 μl assay buffer) in 1.5-ml polypropylene microfuge tubes. After vortexing, the assay tubes are incubated at 4° for 18 hr. The tubes are centrifuged at 10,000 rpm for 10 min (Beckman microfuge, Beckman Instruments, Palo Alto, CA), after which the supernatants are aspirated, and the tube tips are cut and counted in a gamma counter.

[13] M. M. Bradford, *Anal. Biochem.* **72**, 248 (1976).

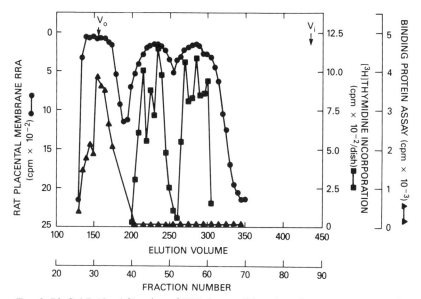

FIG. 2. BioGel P-10 gel filtration of BRL-3A conditioned medium; measurement of IGF carrier protein and rIGF-II. Four liters of BRL-3A conditioned medium was processed as in Fig. 1. To measure IGF binding protein, 50-μl aliquots were dried and ^{125}I-labeled rIGF-II (MSA-III-2) binding was measured using albumin-treated charcoal to separate bound from free tracer at the end of the incubation.[15] Protein-bound ^{125}I-labeled rIGF-II is plotted (▲). IGF-II was measured by the rat placental membrane radioreceptor assay (●) as in Fig. 1. In addition, 10-μl aliquots were dried and used to simulate [3H]thymidine incorporation into DNA in chick embryo fibroblasts (■) as described in detail elsewhere.[12]

Rat IGF-II Purification

Earlier purification schemes utilized ion-exchange chromatography, gel filtration in acid, and preparative polyacrylamide gel electrophoresis.[3] Marquardt et al.[4] described a simplified method which used gel filtration and reversed-phase high-performance liquid chromatography (HPLC). It is this method which we describe here.

The methods of culture of BRL-3A rat liver cells and collection of conditioned media are described above. Four liters of conditioned medium is thawed and dialyzed for 60 hr at 4° against 16 liters of 0.1 M acetic acid (Spectrapor 6 dialysis tubing, cut-off = 3500, Spectrum Medical Industries, Inc., Los Angeles, CA). The 0.1 M acetic acid is changed twice daily. The dialysate including any precipitate is lyophilized in 2-liter bottles. The sides of the lyophilization bottles are rinsed twice with 25 ml of 1 M acetic acid and lyophilization is repeated.

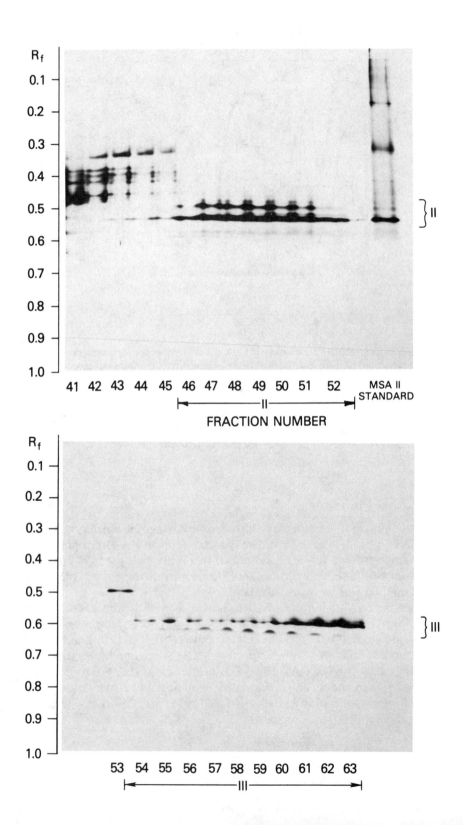

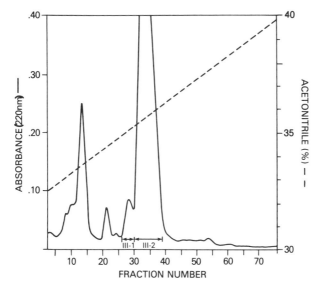

FIG. 4. HPLC of MSA-III purified by BioGel gel filtration. Fractions comprising the MSA-III region from BioGel gel filtration (Figs. 1–3) were pooled, lyophilized, and dissolved in 3.0 ml of 1 M acetic acid and 1 ml was injected onto the HPLC column (see text). The absorbance at 220 nm (solid line) is shown. The acetonitrile concentration is shown by the dotted line. IGF-II was measured by the rat placental membrane radioreceptor assay (not shown). Fractions were pooled as indicated by the arrows for repeat HPLC (Fig. 5), using 50 μl aliquots from each column fraction.

BioGel P-10 Gel Filtration. The lyophilized conditioned medium is resuspended in 15 to 20 ml of 1 M acetic acid and centrifuged in a 50-ml conical propylene tube at 2000 g for 20 min at 4°. The supernatant is applied to a 90 × 2.6 cm column packed with BioGel P-10 (Bio-Rad Laboratories, Richmond, CA) equilibrated with 1 M acetic acid. The column is run at room temperature at a flow rate of 40 ml/hr; 5-ml fractions are collected.

Aliquots from each of the BioGel P-10 column fractions are dried (Speed Vac Concentrator, Savant Instruments, Hicksville, NY) and assayed for protein using the Bradford method[13] and for rIGF-II using the rat

FIG. 3. Polyacrylamide gel electrophoresis of fractions from BioGel P-10 gel filtration of BRL-3A conditioned medium. Aliquots (10 μg of protein) were dried and analyzed by polyacrylamide gel electrophoresis in an acid–urea system as described in detail elsewhere.[12] The gels were stained for protein using a modified silver staining procedure.[12] MSA-II and MSA-III species are indicated. Upper panel: Fraction: 41–52, Fig. 1. Lower panel: Fractions 53–63 from BioGel P-10 gel filtration (Fig. 1).

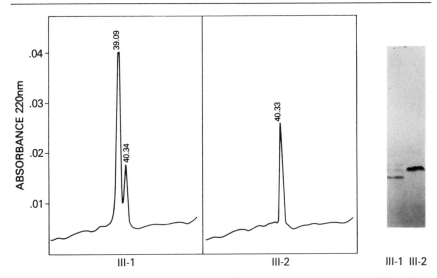

FIG. 5. Repeat HPLC of MSA-III-1 and -III-2. MSA-III-1 and MSA-III-2 pools from Fig. 4 were applied to the reversed-phase HPLC column using the program given in the text. The elution time in minutes is indicated above each peak. The MSA-III-1 and MSA-III-2 pools were also analyzed by gel electrophoresis as in Fig 3.

placental membrane radioreceptor assay described above (Fig. 1), or by the [³H]thymidine incorporation assay in chick embryo fibroblasts[12] (Fig. 2).[15] There are three peaks of activity by radioreceptor assay of the P-10 column fractions. The first peak near the exclusion volume of the column is accounted for, at least in part, by IGF carrier protein[14] (Fig. 2). The middle peak of activity measured by the radioreceptor assay corresponds to rIGF-II, 8700-Da species (MSA-II), while the third peak is rIGF-II, 7100-Da species (MSA-III). While in Fig. 2 it appears that the bioassay provides better resolution of the MSA-II and MSA-III species than does the radioreceptor assay, this is only because larger than optimal aliquots were tested in the radioreceptor assay in this example.

To confirm the distribution of the MSA-II and MSA-III species among the BioGel P-10 column fractions, aliquots are analyzed in a polyacrylam-

---

[14] In conditioned medium rIGF-II is found bound to a carrier protein (35 kDa) and as free rIGF-II. The rIGF-II is dissociated from the carrier protein by 1 M acetic acid and the carrier protein is resolved from rIGF-II by gel filtration in acid. The carrier protein behaves as apparent activity in the radioreceptor assay by competing with receptor for rIGF-II tracer; apparently, the complex formed by the acid-stripped carrier protein and rIGF-II tracer is not fully recognized by the receptor.

[15] A. C. Moses, S. P. Nissley, J. Passamani, R. M. White, and M. M. Rechler, *Endocrinology* **104,** 536 (1979).

ide gel electrophoresis system described in detail elsewhere[12] (Fig. 3).
Although pools of fractions containing MSA-II or MSA-III could be made
simply on the basis of the BioGel P-10 radioreceptor assay profile, analysis
by gel electrophoresis helps to eliminate from these pools fractions with
contaminating proteins. As many as four MSA-II (8700 Da) species have
been detected by gel electrophoresis in some preparations.[3] The relative
amounts of these different MSA-II species are variable; in the preparations
shown in Fig. 3 two MSA-II species predominate. Similarly, the relative
abundance of MSA-III-1 and MSA-III-2 species[3] is variable; in the preparation shown in Fig. 3 MSA-III-2 (upper band) predominates over MSA-
III-1 (lower band).

MSA-II pools such as shown in Fig. 3 can be utilized without further
purification for preparation of IGF-II affinity columns, determination of
nonspecific binding levels in competitive binding assays, and certain biologic studies requiring large amounts of IGF such as the growth of cells in
chemically defined medium. The yield of MSA-II from 4 liters of conditioned medium usually ranges from 300 to 500 μg, with higher yields
occasionally seen.

Purification of MSA-III-2 by Reversed-Phase HPLC. MSA-III-2 from
BioGel P-10 (Fig. 3) is further resolved from MSA-III-1 and other contaminating proteins by reversed-phase HPLC using a Waters system (Waters

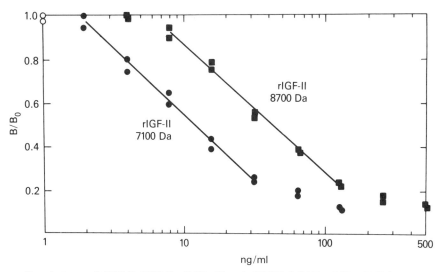

FIG. 6. Assay of rIGF-II, 8700 Da (MSA-II) and rIGF-II, 7100 Da (MSA-III-2) in the rat
placental membrane radioreceptor assay. The radioreceptor assay is described in the text.
B, cpm of ^{125}I-labeled rIGF-II bound to membranes at a given concentration of competing
unlabeled ligand; B_0, cpm bound without unlabeled ligand.

Associates, Inc., Milford, MA) and a Synchropak RP-P column (250 × 4.1 mm, Synchrome, Linden, IN). The solvents are 0.1% trifluoroacetic acid (TFA) (Pierce Chemicals, Rockford, IL) in water (pH 1.9 to 2.1) as the aqueous solvent and 0.075% TFA in acetonitrile as the organic solvent. Water is filtered through a Milli-Q Reagent Grade Water System (Millipore Corp., Bedford, MA). Solvents are filtered with a solvent clarification kit (Waters Associates) through HAWP and FHUP filters (Millipore) for aqueous and organic solvents, respectively. The solvents are degassed under vacuum before use.

The column is operated at a flow rate of 1 ml/min at room temperature. The column is initially equilibrated with 20% organic solvent. After injection of the sample (MSA-III pool in up to 2 ml of 1 M acetic acid), the concentration of the organic solvent is kept at 20% for 10 min. A linear gradient (20 to 40% organic solvent) is programmed from minutes 10 to 50 (Fig. 4). After a second linear gradient to 80% organic solvent from min-

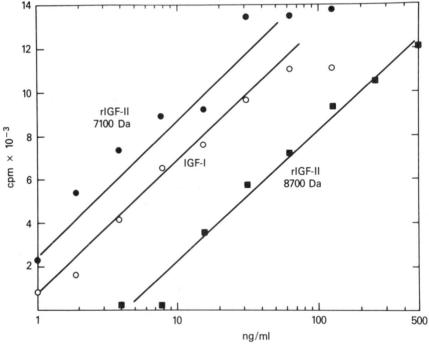

FIG. 7. Assay of rIGF-II, 8700 Da (MSA-II), rIGF-II, 7100 Da (MSA-III-2), and IGF-I in the [³H]thymidine incorporation assay using chick embryo fibroblasts.[12] IGF-I was purchased from AmGen (Thousand Oaks, CA). Radioactivity incorporated into DNA minus basal incorporation (4990 cpm) is plotted versus concentration of peptide.

utes 50 to 60, the column is maintained at 80% organic solvent for an additional 10 min before being reequilibrated to initial conditions (20% organic solvent) for 30 min. The column is rinsed and stored in 50:50 water:methanol (Burdick and Jackson Laboratories, Inc., Muskegon, MI).

Fractions (0.2 min) are collected in 12×75 mm polypropylene tubes during the 20 to 40% acetonitrile gradient, taken to dryness with a Speed-Vac concentrator, and dissolved in 200 μl 1 M acetic acid. Aliquots are taken to dryness and assayed in the rat placental membrane radioreceptor assay (not shown). Fractions comprising the active peaks are pooled. Figure 5 shows that MSA-III-2 has been resolved from MSA-III-1.

The amount of MSA-III-2 in different lots of conditioned media is invariably less than the amount of MSA-II. The amount of MSA-III-2 obtained from 4 liters is usually in the range of 150 to 250 μg, although the yield from some lots of conditioned media is much less than this.

Evaluation of rIGF-II Preparations

While analysis by gel electrophoresis and reversed-phase HPLC can be used to monitor purity, it is desirable to assess the specific biologic activity of MSA-II and MSA-III-2 preparations. In the rat placental membrane radioreceptor assay (Fig. 6) rIGF-II, 8700 Da (MSA-II) is approximately 3.5-fold less potent than rIGF-II, 7100 Da (MSA-III-2). Although 10 ng/ml of MSA-III-2 causes half-maximal competition in the assay shown in Fig. 6, this value will vary somewhat depending on the amount of receptor used in the assay and cannot be used as a strict criterion of biologic potency.

At the present time the only commercially available preparation of highly purified IGF is IGF-I from AmGen (Thousand Oaks, CA). In the chick embryo fibroblast [3H]thymidine incorporation assay, IGF-I and IGF-II are approximately equipotent. Consequently, MSA-III-2 preparations can be compared with IGF-I in this assay to assess biologic potency (Fig. 7).

[26] Purification of Multiplication-Stimulating Activity Carrier Protein

By GARY L. SMITH, RUSSETTE M. LYONS, RICHARD N. HARKINS, and DANIEL J. KNAUER

Multiplication-stimulating activity (MSA) is the designation originally given by Dulak and Temin[1] to a family of related polypeptides produced by a cell line (BRL-3A) derived from normal rat liver. MSA is a representative of the somatomedin hormone family which is characterized by its growth hormone dependency, its weak insulin-like activity resulting from an amino acid sequence homology to proinsulin, and its mitogenic activity for various cell types.

Terminology within this hormone family currently recognizes two major forms of somatomedin. The first is represented by somatomedin C (Sm-C) and insulin-like growth factor I (IGF-I), which have been shown by amino acid sequence data to be identical.[2] This basic peptide exhibits greater growth hormone dependency and is a more potent mitogen than other members of the family. The second form of somatomedin includes insulin-like growth factor II (IGF-II) and MSA. These neutral peptides are more insulin-like and less growth hormone dependent than SM-C/IGF-I. The amino acid sequence of MSA has been shown to differ from that of human IGF-II at only five amino acid residues and the term, rat IGF-II, has often been used in place of the previous descriptor.[3]

The somatomedins are unique among the polypeptide hormones in that they circulate in the blood in association with high-molecular-weight carrier proteins. Various forms of somatomedin–carrier protein complexes have been described.[4-6] The major endogenous complex found in serum has a molecular weight of about 150,000 and its presence is dependent on growth hormone. A second complex which is not growth hormone dependent, and which predominates in hypopituitary serum, has a molecular weight of 40,000 to 50,000. While the role of these carrier proteins in the physiology of the somatomedins remains speculative, it is clear that an

[1] N. C. Dulak and H. M. Temin, *J. Cell. Physiol.* **81**, 153 (1973).
[2] D. G. Klapper, M. E. Svoboda, and J. J. Van Wyk, *Endocrinology* **112**, 2215 (1983).
[3] H. Marquardt, G. J. Todaro, L. E. Henderson, and S. Orosylan, *J. Biol. Chem.* **256**, 6859 (1981).
[4] J. Zapf, M. Waldvogel, and E. R. Froesch, *Arch. Biochem. Biophys.* **168**, 638 (1975).
[5] R. L. Hintz and F. Liu, *J. Clin. Endocrinol. Metab.* **45**, 988 (1977).
[6] W. H. Daughaday, A. P. Ward, A. C. Goldberg, B. Trivedi, and M. Kapadia, *J. Clin. Endocrinol. Metab.* **55**, 916 (1982).

understanding of their function can best be achieved following purification and characterization. The present chapter presents a simple, three-step method for the purification of the MSA carrier protein (MCP) from serum-free BRL-3A cell conditioned medium using affinity chromatography as the major purification step.

Assay Methods

Various methods are available for monitoring the presence of somatomedin carrier proteins based on their ability to bind the polypeptide hormone. However, it is important to recognize that carrier protein complexes present in serum of BRL-3A conditioned medium must first be dissociated and the carrier protein separated from the somatomedin prior to any assay which measures binding activity. Complexes in serum and conditioned medium are largely saturated with somatomedin[7] and any assay based on binding activity would grossly underestimate the amount of carrier protein present. The association between the somatomedin polypeptides and the carrier proteins can be disrupted by exposure to acidic conditions. Separation is achieved by chromatography on Sephadex G-50 or G-75 equilibrated in 1.0 M acetic acid. Under these conditions, the acid-stable carrier proteins elute near the void volume region of the column effectively separated from the free somatomedin polypeptides. Such carrier protein preparations from serum, conditioned medium, or subsequent purification steps can be assayed for somatomedin binding activity in various ways.

Bioassay. While bioassays are expensive in terms of time, labor, and quantities of expended materials, they may be found suitable to accommodate the initial needs of some investigators. Acid-dissociated carrier protein preparations as well as purified MSA carrier protein (MCP) have been shown to inhibit the biological activity of the somatomedins in various cell systems.[8-10] For example, purified MCP will inhibit the stimulation of DNA synthesis in quiescent chicken embryo fibroblasts by MSA or IGF-I. This can be easily monitored in a standard [^{3}H]thymidine incorporation assay.

Radioassays. A convenient and sensitive assay for the detection of carrier protein is the competitive protein binding assay. Purified MSA can be radioactively labeled to high specific activity using the chloramine-T

[7] P. G. Chatelain, J. J. Van Wyk, K. C. Copland, S. L. Blethen, and L. E. Underwood, *J. Clin, Endocrinol. Metab.* **56**, 376 (1983).

[8] D. J. Knauer and G. L. Smith, *Proc. Natl. Acad. Sci. U.S.A.* **77**, 7252 (1980).

[9] J. Zapf, E. Schoenle, G. Jagors, I. Sand, J. Grunwald, and E. R. Froesch, *J. Clin. Invest.* **63**, 1077 (1979).

[10] C. Meuli, J. Zapf, and E. R. Froesch, *Diabetologia* **14**, 255 (1978).

procedure[11] or the iodogen method.[12] Following incubation of the labeled ligand with carrier protein preparations, BSA-activated charcoal is used to separate carrier protein bound ^{125}I-labeled MSA from free ^{125}I-labeled MSA.[13]

In addition, a receptor assay can also be utilized with facility based on the observation that carrier proteins compete for labeled somatomedin with cell surface receptors. Such an assay has been used in the purification of a human somatomedin binding protein from amniotic fluid.[14]

Starting Material

The MSA carrier protein (MCP) present in BRL-3A cell conditioned medium has been purified using, as the source, a previously discarded fraction from the purification scheme of MSA itself. The initial steps in the isolation procedure which are common to the purification of both MSA and MCP have been adequately described elsewhere and will be addressed only briefly here.[15,16] (see also Greenstein et al., this volume [25]). Serum-free conditioned medium is prepared from roller bottle cultures of BRL-3A rat liver cells and is subjected to ion exchange chromatography using Dowex 50W-X8 resin in the sodium form. At this stage, the MSA–MCP complex adsorbs to the resin under neutral conditions and the column is washed with 0.1 M sodium bicarbonate (pH 9). The complex is then eluted using 0.1 M ammonium hydroxide (pH 11). The eluant is dialyzed exhaustively against 1.0 M acetic acid, lyophilized, and the residue is resuspended in 1.0 M acetic acid. This preparation is referred to as Dowex-50 MSA. This fraction (50 mg of protein) is then applied to a 2.5 × 100 cm column containing Sephadex G-75 equilibrated in 1.0 M acetic acid at 4°. Under these strongly dissociating conditions the carrier protein can be effectively separated from the low-molecular-weight MSA polypeptides which elute in the retarded fractions of the column. The MCP elutes with the majority of the protein near the void volume (Fig. 1). We have found that fractions comprising the region eluting immediately after the majority of the void volume proteins are greatly enriched in MCP relative to total void volume

[11] M. M. Rechler, J. M. Podskalny, and S. P. Nissley, J. Biol. Chem. 252, 3989 (1977).
[12] B. R. Smith and R. Hall, this series, Vol. 74, p. 405.
[13] A. C. Moses, S. P. Nissley, P. A. Short, M. M. Rechler, and J. M. Podskalny, Eur. J. Biochem. 103, 387 (1980).
[14] G. Povoa, G. Enberg, H. Jornvall, and K. Hall, Eur. J. Biochem. 144, 199 (1984).
[15] A. C. Moses, S. P. Nissley, P. A. Short, M. M. Rechler, and J. M. Podskalny, Eur. J. Biochem. 103, 387 (1980).
[16] D. J. Knauer, F. W. Wagner, and G. L. Smith, J. Supramol. Struct. Cell. Biochem. 15, 177 (1981).

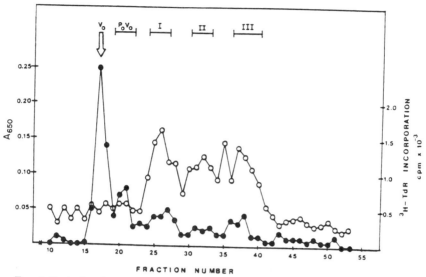

FIG. 1. Sephadex G-75 chromatography of Dowex-50 MSA. Dowex-50 MSA prepared from 12 liters of BRL-3A cell conditioned medium was fractionated on a column of Sephadex G-75 (fine) resin in 1.0 M acetic acid. Fractions of 10 ml each were collected and assayed for protein content (●) and for the ability to stimulate DNA synthesis as measured by an [³H]thymidine incorporation assay in quiescent cultures of chicken embryo fibroblasts (○).

proteins. These fractions, designated post-void (P_0V_0), are pooled for subsequent purification of MCP by affinity chromatography.

Affinity Chromatography

In this major purification step, the MCP-containing fractions from the Sephadex G-75 column are pooled, lyophilized, and resuspended in phosphate-buffered saline (PBS), pH 7.4, to a final concentration of 500 μg protein/ml. This preparation is then applied to an affinity column consisting of a mixture of MSA polypeptides covalently linked to an immobile support matrix. Initially we prepared the activated N-hydroxysuccinimide ester of Sepharose 4B as described by Parikh et al.[17] For convenience, we have more recently been using Affi-Gel 10 (Bio-Rad Laboratories, Richmond, CA), a cross-linked agarose bead matrix with the same functional group.

The MSA used in the coupling reaction is a mixture of the last two peaks of activity eluting from the Sephadex G-75 column, since it is not feasible to purify the necessary quantity of a single species of MSA for this

[17] I. Parikh, S. March, and P. Cuatrecasas, this series, Vol. 34, p. 77.

purpose. MSA pools from several Sephadex G-75 runs are lyophilized and resuspended in PBS at a concentration of 10–12 mg/ml.

To prepare the affinity column, the Affi-Gel 10 resin (1.5 ml) is washed with cold deionized water on a Büchner funnel and the moist gel cake is transferred to a small flask. The MSA solution (1.0 ml) is added and the mixture is incubated with continuous rocking for 1 hr at 4°. Following incubation with MSA, any remaining active ester groups are blocked by the addition of 0.1 ml of 1.0 M glycine/ml of gel and the incubation is continued for an additional hour. The gel is then transferred to a small column and extensively washed with water to remove free reactants, prior to equilibration in PBS.

For the affinity purification of MCP, 5 ml (2.5 mg protein) of the Sephadex G-75 post–void preparation described above is applied to the column at 22° at a flow rate of 5 ml/hr. The protein content of the effluent is continuously monitored by absorbance at 280 nm (Fig. 2). After sample application, the column is washed with PBS until the effluent is free of unadsorbed protein. Bound protein is then eluted from the affinity column with 0.1 M acetic acid (pH 3.5), at a flow rate of 15 ml/hr. The eluted protein peak is pooled, dialyzed against 1.0 M acetic acid, and concen-

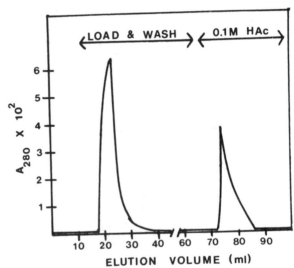

FIG. 2. Affinity chromatographic purification of MCP. Five milliliters of the Sephadex G-75 post–void pool, prepared at 500 μg/ml in PBS, was applied to a column of Affi-Gel 10 resin containing covalently immobilized MSA. Following extensive washing to remove unadsorbed protein, the bound fraction was eluted with 0.1 M acetic acid. The protein content of the eluant was monitored by measuring absorbance at 280 nm.

trated by lyophilization. The protein is then resuspended in 1.0 ml of 1.0 M acetic acid and chromatographed on a 1 × 50 cm column of Sephadex G-75 equilibrated in 1.0 M acetic acid. This final step in the purification of MCP is to ensure that the MCP is free of any residual MSA polypeptides which may have been dissociated from the affinity column. The yield of highly purified MCP using this isolation procedure is approximately 50 – 70 μg/liter of conditioned medium.

Characteristics of Purified MCP

Purified MCP is a single-chain polypeptide with a molecular weight of 34,000 when fully reduced (Fig. 3). Under nonreducing conditions MCP exhibits a molecular weight of 31,500 after electrophoresis on an SDS–polyacrylamide gel, indicating the presence of intrachain disulfide bridges

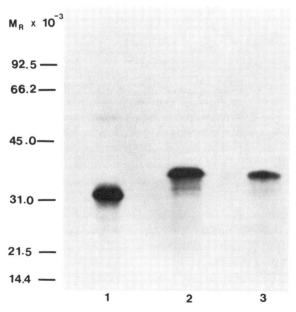

FIG. 3. Characterization of purified MCP. Radioiodinated MCP (1 × 10⁵ cpm) was subjected to various treatments prior to electrophoresis on a 7.5 – 15% linear gradient acrylamide gel with a 4% acrylamide stacking gel using the procedure of U. K. Laemmli [*Nature (London)* **227**, 680 (1970)]. Lane 1: nonreduced ^{125}I-labeled MCP. Lane 2: ^{125}I-labeled MCP reduced in the presence of 2-mercaptoethanol. Lane 3: ^{125}I-labeled MCP reduced with 2-mercaptoethanol and carboxymethylated with iodoacetamide. Radioactive bands were visualized by autoradiography. The positions of a mixture of low-molecular-weight markers (Bio-Rad Laboratories) are given.

TABLE I
AMINO ACID COMPOSITION OF MCP[a]

Amino acid	Residues/100 residues
Asx	9.02
Thr[b]	3.98
Ser[b]	4.19
Glx	16.02
Pro	11.60
Gly	9.86
Ala	7.10
Cys[c]	3.59
Val	4.40
Met[c]	0.25
Ile	2.18
Leu	8.88
Tyr	2.27
Phe	2.25
His	3.35
Lys	3.85
Arg	7.27
Trp[d]	ND

[a] All figures are average values of two 24-hr, two 48-hr, and two 72-hr hydrolyses.

[b] The average values for threonine and serine were obtained by extrapolation to zero time. The 72-hr values for valine and isoleucine were used.

[c] Cysteine and methionine were underivatized prior to analysis.

[d] Tryptophan content was not determined.

which serve to compact the structure. The results of an amino acid analysis demonstrate that MCP is a highly charged molecule containing relatively large amounts of glutamic and aspartic acids as well as an unusually high proline content (Table I).

Purified MCP will specifically bind radiolabeled MSA, forming a complex which is similar in size to the low-molecular-weight somatomedin–carrier protein complex seen in serum. When MCP is covalently cross-linked to radioiodinated MSA by disuccinimidyl suberate and then analyzed by SDS–polyacrylamide gel electrophoresis and autoradiography, the molecular weight of the MCP–MSA complex is approximately 42,000.[18] As discussed earlier, the biological activity of MSA is inhibited when in association with the carrier protein, MCP.

[18] R. M. Lyons and G. L. Smith, *Mol. Cell. Endocrinol.* **45,** 263 (1986).

The use of somatomedin affinity columns, as described here, should prove useful for the isolation of carrier proteins from acid-dissociated preparations following separation of the carrier protein from the low-molecular-weight polypeptide hormone. However, the isolation of intact complexes from biological fluids cannot be achieved in this manner since the endogenous complexes are, to a large extent, saturated with hormone. Antibodies prepared against MCP, purified as described, may facilitate the isolation of endogenous complexes by affinity methods and further our understanding of somatomedin–carrier protein complexes.

Acknowledgments

This work was supported by USPHS Research Grant CA17620, The University of Nebraska–Lincoln Research Council, and NIH Biomedical Research Support Grant No. RR-07055.

Section IV

Bone and Cartilage Growth Factors

[27] Purification of Bovine Skeletal Growth Factor

By JOHN C. JENNINGS, SUBBURAMAN MOHAN, and DAVID J. BAYLINK

Introduction

Unlike most other organ systems, the specific cellular components of bone — osteoblasts, osteocytes, and osteoclasts — account for only a minor portion of tissue (5–10%) weight. The major component of bone is matrix (90–95%), which consists of a mineral phase (60–65% of noncellular bone weight) and a protein phase (35–40%). The largest protein component (90% of matrix protein) is type I collagen. The noncollagenous proteins (10%) are heterogeneous in origin; some appear to be produced by bone cells while others are incorporated from or are concentrated from serum. The noncollagenous proteins are laid down into bone matrix either by binding to the mineral phase or to collagen or by binding to other matrix proteins which do attach to bone mineral or collagen.

The identification and functions of bone noncollagenous proteins (bone derived and serum derived) are only now beginning to be explored in depth. Two proteins, albumin and α_2HS-glycoprotein, are major serum-derived components of bone matrix.[1,2] Although both appear to be concentrated in matrix their functions (if any) are not known. Several additional noncollagenous proteins have now been isolated which are apparently unique to bone matrix. Osteocalcin (also known as bone GLA protein or BGP), is a 6-kDa protein whose synthesis by osteoblasts is vitamin D dependent and which undergoes a posttranslational vitamin K-dependent amino acid modification (Glu → Gla).[3,4] Osteonectin, a 30-kDa protein, is also produced by osteoblasts.[5,6] From their *in vitro* actions on crystallization of hydroxylapatite both of these bone-derived, bone-specific proteins have been proposed to alter bone mineralization.[3,5,7]

[1] B. A. Ashton, H. J. Hohling, and J. T. Trifitt, *Calcif. Tissue Res.* **22**, 27 (1976).

[2] J. D. Termine, A. B. Belcourt, K. M. Conn, and H. K. Kleinman, *J. Biol. Chem.* **256**, 10403 (1981).

[3] P. A. Price, R. S. Otsuka, J. W. Poser, J. Kristaponis, and N. Raman, *Proc. Natl. Acad. Sci. U.S.A.* **73**, 1447 (1976).

[4] P. A. Price and S. A. Baukol, *J. Biol. Chem.* **255**, 11660 (1980).

[5] J. D. Termine, H. K. Kleinman, S. W. Whitson, K. M. Conn, M. L. McGarvey, and G. R. Martin, *Cell* **26**, 99 (1981).

[6] S. W. Whitson, W. Harrison, M. K. Dunlap, D. E. Bowers, Jr., L. W. Fisher, P. G. Robery, and J. D. Termine, *J. Cell Biol.* **99**, 607 (1984).

[7] R. W. Romberg, P. G. Werness, P. Lollar, B. L. Riggs, and K. G. Mann, *J. Biol. Chem.* **260**, 2728 (1985).

Other bone matrix noncollagenous proteins have been identified, but these are less well characterized than osteocalcin or osteonectin.[2]

Calciotrophic hormones (parathyroid hormone and vitamin D metabolites) are major determinants of bone formation and of bone resorption. Nonetheless, local rates of bone formation and bone resorption appear to be regulated independently of systemic concentrations of those calciotrophic hormones in certain pathologic situations (for example, during fracture repair or in Paget's disease of bone). Such observations raise the possibility that some type of local regulatory factors produced or stored within bone could be responsible for modulating local rates of bone formation (i.e., paracrine or autocrine regulation). It has been inferred for a substantial period of time that bone matrix contains biologically active factors. In a series of reports beginning in 1965 Urist and collaborators have demonstrated that implantation of demineralized bone matrix at ectopic sites induces a complex series of events similar to endochondral bone formation.[8] (Also, see Urist et al. [28], this volume.) These investigators have now isolated a protein, which they have named bone morphogenetic protein (BMP), which promotes these processes in soft tissue.[8] Reddi and colleagues have confirmed many of these findings.[9] An important concept inherent in the results of these studies is that a component of bone matrix proteins may possess biologic activity.

That bone matrix may contain, and bone cells produce, proteins which may have biological actions on bone and other cell types has now been amply confirmed. Canalis and co-workers have partially isolated a protein, bone-derived growth factor (BDGF), which is released into media from cultured rat calvariae, and which enhances osteoblast proliferation and stimulates collagen synthesis in vitro.[10] Howard et al. have shown that a substance which stimulates DNA synthesis in cultured bone cells is apparently released from cultured chick embryo long bones during the dissolution of bone matrix.[11] This latter observation suggested that a growth factor(s) might be present within bone matrix and that such a growth factor could be released during the process of bone resorption (i.e., dissolution of the mineral and protein phases of bone matrix). Because of accumulating evidence for local control of bone formation and because a factor (or factors) which stimulated DNA and matrix synthesis in cultured bone tissue/bone cells in vitro was apparently released by cultured bone tissue during bone resorption Farley and Baylink[12] reasoned that the proteins

[8] M. R. Urist, R. J. Lange, and G. A. M. Finerman, *Science* **220**, 680 (1983).

[9] T. K. Sampath, O. P. DeSimone, and A. H. Reddi, *Exp. Cell Res.* **142**, 460 (1982).

[10] E. Canalis, W. A. Peck, and L. G. Raisz, *Science* **210**, 1021 (1980).

[11] G. A. Howard, B. L. Bottemiller, R. T. Turner, J. I. Rader, and D. J. Baylink, *Proc. Natl. Acad. Sci. U.S.A.* **78**, 3204 (1981).

[12] J. R. Farley and D. J. Baylink, *Biochemistry* **21**, 3502 (1982).

released during demineralization of the bone with EDTA might contain that active substance. Subsequently they purified a factor from the mixture of proteins released during EDTA demineralization of human bone. This factor, an 83-kDa protein, which they have named human skeletal growth factor (hSGF), enhances DNA synthesis in cultured bone cells and stimulates collagen synthesis in chicken embryo long bones in organ culture.[12] A similar factor released from cultured chick embryo bones has been partially characterized.[13]

In order to obtain greater quantities of SGF, we have used bovine bone since large quantities of fresh bone are readily available for extraction of matrix proteins. We have also substantially modified the initial purification procedures used to isolate the 83-kDa hSGF. With these modified techniques, to be described below, we have purified a small molecular weight (11–14 kDa) polypeptide, bovine SGF, with growth factor activity similar to, but more potent than, the activity of the 83-kDa hSGF described by Farley and Baylink.[12]

Because a sensitive assay procedure is needed for purification, we will first describe the chick embryo calvarial cell bioassay used to monitor purification. Subsequently we will describe the method of matrix protein extraction and finally the techniques used for purification of the small-molecular-weight bSGF.

Bioassay Technique

The bone cell isolation and culture procedure has been modified from that originally described by Puzas et al.[14] The frontal and parietal bones are removed from 13-day-old chick embryo calvariae, cleaned of adherent tissue, and incubated with 1.5 mg/ml crude collagenase (Worthington) for 1 hr at 37°. Released cells are collected from the incubation medium by centrifugation at 100 g for 5 min at 4°, resuspended in medium, and stored on ice. Calvariae are resuspended in fresh collagenase for an additional 1 hr at 37° and the released cells are again collected by centrifugation and then combined with the cells obtained in the first isolation. Twenty-five to 50% of the cells freed from matrix with collagenase stained positively for alkaline phosphatase. Upon long-term culture these cells grow in multilayers typical for bone and cartilage cells and not for fibroblasts. This finding, in conjunction with the alkaline phosphatase staining, suggests that the cultured cells are predominantly osteoblasts and osteoblast precursors.

[13] R. H. Drivdahl, G. A. Howard, and D. J. Baylink, *Biochim. Biophys. Acta* **714**, 26 (1982).
[14] J. E. Puzas, R. H. Drivdahl, G. A. Howard, and D. J. Baylink, *Proc. Soc. Exp. Biol. Med* **166**, 113 (1981).

The bone cell suspension obtained as described above is diluted to 70,000 cells/ml in DMEM, and plated into 24-well (1.0 ml/well) or 48-well (0.5 ml/well) tissue culture plates. After 24 hr in culture at 37° in 5% CO_2 atmosphere, buffer or test fractions (50 μl maximum volume) are added. After a 16-hr incubation 1.5 μCi of [^3H]thymidine in 20 μl of medium is added to each well. Two hours later the medium is aspirated, the attached cells washed twice with cold phosphate-buffered saline, and the wells aspirated and frozen. Radioactivity incorporated into acid-precipitable products is determined by swabbing the plates with cotton pledgets saturated with 12.5% cold trichloroacetic acid (TCA). The cotton swabs are washed with two changes of cold 12.5% TCA, washed with 95% ethanol, and then air dried. Pledgets are then placed into vials with 5 ml of scintillation fluid (ScintiVerse E, Fisher). This method for isolating and counting TCA-precipitable material was described in detail by Gospodarowicz et al.[15] Thymidine incorporation assays may not always reflect DNA synthesis or cell proliferation (see Dickson et al. [31], this volume). Thus, the results of this assay are confirmed by cell counts using conditions identical to those described above: after 24- and 48-hr exposures to control or experimental fractions, cells are released from the wells with trypsin–EDTA, and cell numbers determined in a Coulter counter.

Bioassay Design

All test factors are diluted in DMEM containing 1 mg/ml bovine serum albumin to prevent adsorption of active factors to exposed surfaces and added in a volume of 50 μl (24-well plates) per well. BSA containing wells (maximum stimulation at 50 μg/well) give [^3H]thymidine incorporation approximately 10–50% above wells with protein-free medium. In order to ensure that stimulation of [^3H]thymidine incorporation by experimental fractions is not a nonspecific protein effect due to BSA in the diluted fractions, a 50-μl aliquot of DMEM containing 1 mg/ml BSA is added to each control well. In each experiment we have approximately six control wells (one lane, four lanes/plate) for each four plates. In addition to the BSA (nonspecific protein) controls, each experiment contains a minimum of six wells each with bovine or fetal calf serum [0.5% (v/v) final] and with crude bone extract (10 μg/ml final). These latter two controls are included to determine (1) the maximal cell response (no further stimulation is seen with serum concentrations > 0.5%) and (2) the maximal cell response to bone matrix proteins (BBE). There is little or no further stimulation of DNA synthesis with concentrations of BBE proteins > 10 μg/ml and this

[15] D. Gospodarowicz, H. Bialeck, and G. Greenburg, *J. Biol. Chem.* **253**, 3736 (1978).

TABLE I
PURIFICATION AND RECOVERY OF bSGF

Fraction	Protein (mg)	Protein (percentage of initial)	Units/ microgram	Purifica- tion	Total units ($\times 10^{-8}$)	Percentage recovery activity
Bovine bone extract	3,000	100	111	—	3.33	100
Hydroxylapatite 1	1,143	38	422	3.8	4.82	145
Hydroxylapatite 2	170	5.7	1,539	13.9	2.62	79
Sephacryl S-200	15	0.5	14,125	127	2.12	64
Reversed phase	0.53	0.018	260,114	2,343	1.38	41

stimulation with BBE is usually equal to maximal serum stimulation.[16] In *all* assays, experimental fractions or fraction pools are tested at equal final protein (see protein estimation method below) concentrations.

For calculation of results the mean counts per minute of all wells with BSA only is determined and used as the control. The mean counts of test fractions (minumum of six replicate wells for each concentration) are determined. Results are then expressed either as mean ($\pm$ SD) cpm/well or as mean percentage of BSA controls. Units of activity (SA) are calculated as follows: test percentage of control $-$ 100/protein concentration in test fraction. In Table I, only protein concentrations with results falling on the linear portion of the log dose – response curve were used for the determination of specific activity.

Protein Assay

The protein concentrations of the extracts and pools or individual fractions from purification are determined by the dye-binding method described by Bradford[17] using bovine γ-globulin as a standard. In some instances protein concentrations are also determined by the Lowry method.[18] The two methods give similar results both with mixtures of proteins and with the more purified fractions of bSGF. However, independent analysis – absorbance at 280 nm and band density on SDS gels suggest that these quantitative measures of protein *overestimate* the protein concentration in the more highly purified preparations of bSGF.

[16] J. Jennings and D. J. Baylink, *in* "The Chemistry and Biology of Mineralized Tissues" (W. T. Butler, ed.), p. 48. EBSCO Media, Birmingham, AL, 1985.
[17] M. Bradford, *Anal. Biochem.* **72**, 248 (1976).
[18] O. H. Lowry, N. J. Rosebrough, A. L. Farr, and R. J. Randall, *J. Biol. Chem.* **193**, 265 (1951).

Source of Bone. Fresh bovine femurs, tibias, and, in some instances, pelvic bones are obtained from local abbatoirs and kept frozen at −20° until use.

Preparation of Bovine Bone for Extraction

Adherent tissue is removed by mechanical scraping. Both ends are removed and the shafts, free of joint and cartilage, are cut into ~1 × 1 in. pieces with a band saw fitted with a special blade. Marrow is removed from the marrow cavity. Bone pieces are washed briefly in water and frozen at −20°. Bone pieces are cooled in liquid nitrogen, mixed with crushed dry ice, and ground in a Wiley mill (precooled with dry ice) into ~2 mm³ pieces.

The ground bone (mixed with dry ice) is extensively washed first with tap water (~40°) to remove fat and then with deionized water to lyse cellular elements and remove as much of the remaining serum as possible.

Extraction of Matrix Proteins: Method

The washed bone powder is packed into inner chambers of modified pipet washers. Matrix proteins are removed from the bone particles by continuous recirculation of 0.5 M EDTA (adjusted to pH 7.0) containing 0.005% sodium azide, and the following protease inhibitors: phenylmethylsulfonyl fluoride (PMSF, 0.001 M), benzamidine–HCl (0.005 M), and ε-aminocaproic acid (0.05 M). Extraction is carried out at 25° and the extract collected at 7 days. Fresh extraction buffer is added and extraction is carried out for an additional 7 days. The bone powder is discarded after the second extraction. The extract is clarified by filtration. The resulting filtrate is concentrated and dialyzed against 10 mM sodium phosphate (pH 7.0) in a hollow fiber concentrator/dialyzer (Erika). After dialysis the extract is filtered on glass fiber filters, cellulose acetate filters (0.45 μm), and either frozen for later analysis or immediately added to hydroxylapatite (see below).

Results of Extraction

With two 7-day extraction periods we obtain approximately 2 g of protein for every 1 kg (wet weight) of ground bone. With additional extraction the yield of protein decreases. While the specific activity in these later extracts is similar to that in the initial two extracts the total activity recovered is not sufficient to justify continued extraction beyond 2 weeks. The crude extract stimulates [³H]thymidine incorporation into DNA in the chick calvarial cells. Maximum stimulation is usually seen with 10 μg/

ml of BBE protein. The extract also stimulates an increase in proliferation of these cells as assessed by cell number. When the extracted proteins are separated by molecular sieve chromatography on BioGel A-0.5 m in 0.03 M Tris–acetate (pH 7.4) containing 0.15 M NaCl the majority of activity is recovered in large-molecular-weight (>100 kDa) fractions.[16]

Purification of Bovine SGF

The degree of purification is given in Table I. Details of each purification step are given below.

Adsorption and Elution from Hydroxylapatite Using Native Conditions

Five to 6g bone extract proteins (previously dialyzed against 0.01 M sodium phosphate, pH 7.0) is loaded onto a 2.5 × 16 cm column of hydroxylapatite (Sigma, type I). This HA column (approximately 400 ml of packed HA) is made by pouring suspended HA onto a medium qualitative filter in a 10 × 16 cm Büchner funnel which is suspended over a 2-liter vacuum flask. The HA is then covered with another filter, the sample added, and the unbound proteins removed by washing (under vacuum with regulator) with 0.01 M sodium phosphate. Bound proteins are then eluted sequentially with 0.15 M sodium phosphate followed by 0.4 M sodium phosphate.

Results and Discussion. Virtually all protein and mitogenic activity are recovered at this step. Based on the mean values from six experiments we find that 10% of the protein is unbound, 51% is eluted with 0.15 M and 39% is eluted with 0.4 M sodium phosphate. The specific activity of the material eluted with 0.4 M phosphate is approximately 3-fold greater than the crude extract and is used for further purification. In early experiments aliquots of HA after 0.4 M phosphate elution were solubilized by suspension in dialysis bags and dialysis against 20% EDTA with protease inhibitors. The specific activity of the resultant solution is approximately twice that of the 0.4 M sodium phosphate-eluted fraction. This step was abandoned due to the long period of dialysis required to recover the small amount of protein (0–2%).

We have also used sodium phosphate gradient elution of a similar HA preparation in small columns. However, in these small columns the protein binding capacity is low; the maximum flow rates are insufficient to allow rapid separation; and the resolution of active and inactive proteins is insufficient to justify this modification. While we have also used other preparations of HA for the HA adsorption step, the other preparations either have too little capacity to adequately batch process large quantities of bone matrix proteins or give inconsistent patterns of elution of activity using sodium phosphate step gradients.

Elution from Hydroxylapatite Using Dissociating Conditions

The proteins eluted from HA with 0.4 M sodium phosphate containing the active factor are dialyzed against 0.01 M sodium phosphate. Solid guanidine–HCl (Gu–HCl) is added so that the final concentration is 4 M and this is allowed to stand overnight at 4°. Approximately 2 g of protein from this fraction is then added to 4.4 × 25 cm column of hydroxylapatite [HA Fast Flow (HAFF), CalBiochem] previously equilibrated with 0.01 M sodium phosphate in 4 M Gu–HCl, pH 7.0. The column is first washed with equilibration-starting buffer. The bound proteins are then eluted with 0.4 M sodium phosphate in 4 M Gu–HCl.

Results and Discussion. Approximately 30% (33 ± 11%, $\bar{x}$ and SD, $N = 10$ experiments) of the protein does not bind to HAFF in 0.01 M sodium phosphate in the presence of 4 M Gu–HCl. All of the mitogenic activity is present in these unbound proteins. Another 30% of the protein elutes with 0.4 M sodium phosphate–4 M Gu–HCl. This fraction does not contain mitogenic activity when dialyzed and tested on the chick calvarial cells. The remainder of the protein remains bound to HA column. These "irreversibly bound" proteins require that the HA be replaced after approximately 20 g of cumulative loaded sample protein.

We have not systematically explored other types of hydroxylapatite for this procedure because Fast Flow performed well in that is has a large binding capacity, rapid flow rates, and gives consistent results from batch to batch.

Molecular Sieve Chromatography

The unbound, active fraction from the second HA procedure is concentrated in a 400-ml stirred cell (Amicon fitted with a YM5 (nominal molecular weight cutoff of 5K, Amicon) membrane. Protein samples (100–150 mg) (in 5 ml of 0.01 M sodium phosphate–4 M Gu–HCl) are loaded onto a 2.5 × 95 cm column of Sephacryl S-200 (Pharmacia) equilibrated with 0.03 M Tris–acetate, pH 7.4, containing 4 M Gu–HCl (Tris–Gu). Proteins are eluted with Tris–Gu at 20 ml/hr and 3.7-ml fractions collected. Aliquots of individual fractions are pooled according to protein peaks (or shoulders), dialyzed against 0.01 M sodium phosphate, protein concentration determined, and pools assayed for mitogenic activity.

Results and Discussion. A representative protein elution profile and the results of assay for mitogenic activity are shown in Fig. 1. All active material elutes in fractions after the 17.5-kDa marker.

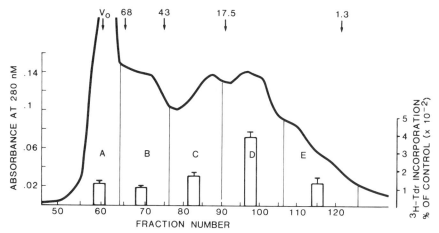

FIG. 1. Proteins (100 mg) eluted from the hydroxylapatite column with 0.01 M sodium phosphate in 4 M guanidine–HCl (Gu) were separated on a 2.5 × 95 cm column of Sephacryl S-200 (Pharmacia). The column was equilibrated with 0.03 M Tris–acetate (pH 7.4), in 4 M Gu and proteins eluted with the same buffer at a flow rate of 20 ml/hr. Fractions of 3.7 ml were collected and fractions were pooled as indicated. Each pool was concentrated, dialyzed against 0.01 M sodium phosphate, (pH 7.0), and tested for mitogenic activity at 20 ng/ml. The elution volumes of the following standard proteins were determined: thyroglobulin (V_0), bovine serum albumin (68 kDa), ovalbumin (43 kDa), myoglobin (17.5 kDa), and vitamin B_{12} (1.3 kDa). The majority of active proteins elute after the 17.5-kDa marker.

In addition to the Sephacryl S-200 we have also used HPLC with a 2.15 × 60 cm column of TSK-G 3000 SWG (LKB) equilibrated with Tris–Gu. All mitogenic activity elutes from the HPLC column in the small-molecular-weight (after the 17.5-kDa marker) fractions. However, the resolution in the small-molecular-weight range with TSK-G 3000 SWG is inferior to the resolution with Sephacryl S-200 when these separations are done in 4 M Gu–HCl.

Reversed-Phase HPLC

The fractions eluting from the Sephacryl column between K_{av} of approximately 0.3 and 0.5 are pooled and dialyzed against 0.01 M sodium phosphate (pH 7.0). The soluble proteins are separated by centrifugation, concentrated by lyophilization, and reconstituted in 25% acetonitrile containing 0.1% trifluoroacetic acid (TFA). Ten milligrams of protein in 2 ml (clarified by microfuge centrifugation) is loaded onto a 10 × 250 mm C_4 reversed-phase column (Bio-Rad, Hi-Pore RP 304). The column is washed for 5 min (flow rate 1.5 ml/min) with equilibration buffer. The bound proteins are eluted with a linear 25 to 60% acetonitrile gradient developed over 70 min. One-minute fractions (1.5 ml) are collected and concentrated by Speed Vac (Savant) centrifugation. Fractions are reconstituted in sterile deionized water, protein concentration determined, and individual fractions assayed or aliquots of fractions pooled and the pool assayed.

Results and Discussion. A representative reversed-phase (RP) protein elution profile and corresponding results of assay for mitogenic activity are shown in Fig. 2. In this example the greatest activity is seen in fraction 35 (35–36 min). The elution of the material with the greatest specific activity

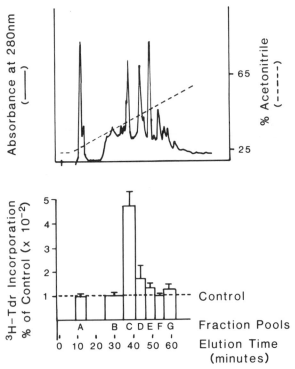

Fig. 2. Ten milligrams of active proteins from Sephacryl S-200 was concentrated by lyophilization, reconstituted in 2 ml of 25% acetonitrile containing 0.1% trifluoroacetic acid, and loaded onto a 10×250 mm C_4 reversed-phase column (Bio-Rad, Hi-Pore RP 304). The column was then washed for 5 min (at a flow rate of 1.5 ml/min with 25% acetonitrile). Following this was a 25–60% linear acetonitrile gradient was developed over 70 min. One-minute (1.5 ml) fractions were collected and concentrated by Speed Vac (Savant) centrifugation. Each fraction was reconstituted in distilled water, aliquots of individual fractions pooled, and each pool (A–G) was assayed for mitogenic activity at 2 ng/ml. In a separate assay, individual fractions (30–42 min) were tested for stimulation of DNA synthesis at 1 ng/ml (data not shown). In this preparation maximum activity was present in proteins which eluted between 34 and 35 min (fraction #35, ~40% acetonitrile).

is consistent (± 0.5 min) with multiple runs using the same S-200 pool. However, the elution of the peak of activity of different preparations of S-200 starting material varies between 32 and 38 min.

A precipitate often forms during dialysis (to remove Gu – HCl) of the active Sephacryl S-200 pool. This precipitate is soluble in 25% acetonitrile with 0.1% TFA. When run on the RP column with conditions identical to those described above, more than 90% of the protein elutes later than 45 min. Some activity is present in a broad area of the earlier eluting proteins (before 45 min). We have not yet attempted to determine whether this activity resides in proteins identical to bSGF.

Assessment of Purity. The principal method we have used to assess purity of SGF preparations is SDS – PAGE. Our method for gel electrophoresis is a modification of that described by Bethesda Research Laboratories and is done using the following conditions: (1) stacking gel of 3.5% polyacrylamide; (2) running gel of (a) 10–20% gradient of PA with 6 M urea, 0.1% SDS, and 0.1 M sodium phosphate (pH 7.2), or (b) 20% PA with urea, SDS, and phosphate as in (a); (3) running buffer of 0.1 M sodium

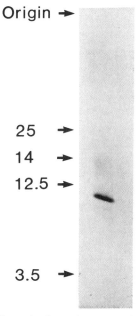

FIG. 3. Ten micrograms of protein from that reversed-phase fraction with maximum biological activity (bSGF) was examined by a 10–20% polyacrylamide gradient gel with SDS, urea, and 2-mercaptoethanol as described in the text. Molecular weight standards were α-chymotrypsinogen (25 kDa), lysozyme (14 kDa), cytochrome c (12.5 kDa), and insulin (3.5 kDa). This gel was stained with Coomassie Blue – Copper as given in the text.

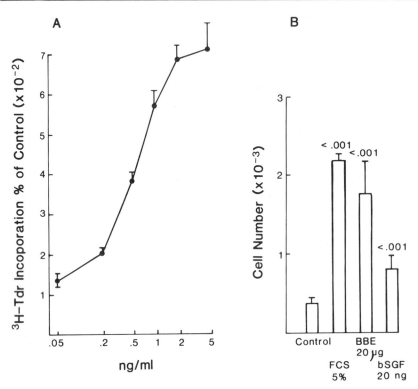

FIG. 4. The reversed-phase preparation of bSGF shown in Fig. 3 was assayed for stimulation of [^{3}H]thymidine incorporation (A) at final protein concentrations between 0.05 and 20 ng/ml of calvarial cell incubation media. There was log-linear stimulation of DNA synthesis between 0.2 and 2 ng/ml. There was no additional stimulation (or inhibition) at 10 and 20 ng/ml (data not shown). A different preparation of bSGF increased calvarial cell number (B) after a single addition of 20 ng/ml at time 0 and a 48-hr exposure of the calvarial cells to bSGF. Both fetal calf serum [FCS, 5% (v/v)] and the matrix protein extract (BBE, 20 μg/ml) also enhanced cell number.

phosphate (pH 7.2) and 0.1% SDS; (4) sample buffer of 0.01 M sodium phosphate (pH 7.2) with 7 M urea, 1% SDS, 1% 2-mercaptoethanol, and 0.01% bromphenol blue. Gels are stained either with Coomassie blue with 0.1% cupric sulfate or with the silver staining procedure of Marshall[19] Figure 3 shows bSGF on a 10–20% gradient gel stained with Coomassie blue–copper. In different RP preparations of bSGF there is some minor size heterogeneity (11–14 kDa). However, the biological properties, pH and temperature stability, and lability to treatment with DTT, in different preparations of bSGF are identical.

[19] T. Marshall, *Anal. Biochem.* **136**, 340 (1984).

Purified bSGF is biologically active at picomolar concentrations (Fig. 4) similar to other purified polypeptide growth factors. However, some RP preparations of bSGF are not homogeneous on SDS gels. In order to consistently purify bSGF to homogeneity we have rerun bSGF on C_4 reversed phase using the same conditions described above. The eluted proteins are not biologically active. Thus procedures other than this RP technique will have to be developed.

Conclusions

The overall technique with two batch procedures and two analytical procedures is quite simple and can be applied to large quantities of bone matrix proteins. With the exception of the slight variability in the time of elution of the SGFs from the C_4 reversed-phase column the above procedures have been quite reproducible in our hands.

However, consistent purification to complete homogeneity will most likely require an additional step after the described reversed-phase chromatography. In addition, other techniques/schemes will be necessary to isolate the native, large molecular weight form of bSGF, in order to compare the chemistry and biological actions with the small molecular weight form of bSGF described here. By the methods described here, sufficient quantities of nearly homogeneous bSGF can be purified to further explore its biologic actions and interactions with other hormones and growth factors both *in vitro* and *in vivo*.

Note Added in Proof. Since submission of this manuscript we and others have shown that the noncollagenous proteins isolated from bone matrix contain multiple growth factors.[20-23] Several of these factors do not retain their biological activity under the denaturing conditions described here for the purification of skeletal growth factor.[21,22] However, SGF and a type β transforming growth factor copurify through the Sephacryl S-200 step, and in some preparations the activities do not completely separate from one another on C_4 reverse phase. Using a shallower acetonitrile gradient (25 to 45% in 100 min rather than the gradient described above (25 to 60% in 70 min), βTGF and SGF are readily separated.

[20] T. A. Linkhart, J. C. Jennings, S. Mohan, G. K. Wakley, and D. J. Baylink, *Bone* **7**, 479 (1986).

[21] J. C. Jennings, S. Mohan, T. A. Linkhart, M. W. Lundy, and D. J. Baylink, *in* "Proceedings of the IXth International Conference on Calcium Regulating Hormones and Bone Metabolism" (D. V. Cohn, ed.), in press (1987).

[22] S. M. Seyedin, T. C. Thomas, A. Y. Thompson, S. M. Rosen, and K. A. Puz, *Proc. Natl. Acad. Sci. U.S.A.* **82**, 2267 (1985).

[23] P. V. Hauschka, A. E. Mavrakos, M. D. Iafrati, S. E. Doleman, and M. Klagsbrun, *J. Biol. Chem.* **261**, 12665 (1986).

Acknowledgments

Dr. Jennings is a Research Associate in the Veterans Administration Research and Development, Career Development Program. These studies were funded in part by research and development funds from the Veterans Administration and by NIH Grant AM31062-03.

The technical assistance of Rick Langer and George Evans and the secretarial help of Penny Nasabal are gratefully acknowledged.

[28] Preparation and Bioassay of Bone Morphogenetic Protein and Polypeptide Fragments

By Marshall R. Urist, J. J. Chang, A. Lietze, Y. K. Huo, A. G. Brownell, and R. J. DeLange

Among the many cellular and extracellular components of the organic matrix of dentin, bone, and various tumors is a bone morphogenetic protein (BMP).[1-5] The identity of BMP is based upon its capacity to induce differentiation of mesenchymal type perivascular connective tissue cells into bone. A tissue (living or killed), derivative of tissue, crude chemical extract of tissue, or sample of purified BMP is bioassayed by implantation in a muscle pouch, subcutaneous space, or diffusion chamber.[6-10] When the sample is positive for BMP activity, it induces differentiation of cartilage and woven bone within 10 days, lamellar bone within 20 days, and an ossicle containing bone marrow within 30 days. The quantity of new bone is proportional to the dose of implanted material. The BMP molecule is morphogenetic because it induces cell disaggregation, migration, reaggregation, hyaluronate accumulation, and corresponding increase

[1] M. R. Urist, *Science* **150**, 893 (1965).
[2] M. R. Urist, *Dev. Biol. (Suppl.)* **4**, 125 (1971).
[3] M. R. Urist, M. A. Conover, A. Lietze, J. T. Triffitt, and R. DeLange, in "Partial Purification and Characterization of Bone Morphogenetic Protein" (D. Cohn, R. Talmage, and J. L. Matthews, eds.), p. 307. Excerpta Medica, Amsterdam, 1981.
[4] M. R. Urist, *Perspect. Biol. Med.* **22**, S89 (1979).
[5] M. R. Urist, R. J. DeLange, and G. A. M. Finerman, *Science* **220**, 680 (1983).
[6] M. R. Urist, H. Iwata, P. W. L. Ceccotti, R. L. Dorfman, S. D. Boyd, R. M. McDowell, and C. Chien, *Proc. Natl. Acad. Sci. U.S.A.* **70**, 3511 (1973).
[7] M. R. Urist, A. Mikulski, and A. Lietze, *Proc. Natl. Acad. Sci. U.S.A.* **76**, 1828 (1979).
[8] M. R. Urist, in "New Bone Formation Induced in Postfetal Life by Bone Morphogenetic Protein" (R. Becker, ed.), p. 406. Charles C. Thomas, Springfield, IL, 1981.
[9] M. R. Urist, H. Mizutani, M. A. Conover, A. Lietze, and G. A. M. Finerman, in "Factors and Mechanisms Influencing Bone Growth" (A. Dixon and B. Sarnat, eds.), p. 61. Liss, New York, 1983.
[10] M. R. Urist, A. Lietze, H. Mizutani, K. Takagi, J. T. Triffitt, J. Amstutz, R. DeLange, J. Termine, and G. A. M. Finerman, *Clin. Orthop.* **162**, 219 (1982).

in hyaluronidase activity before cell proliferation and differentiation are established.[11] BMP is also morphogenetic because it induces development of an ossicle with three-dimensional form and shape, which is a permanent part of the structure of the body. The histophysiology of BMP-induced cartilage cell differentiation *in vitro* has been described in detail in previous publications.[12-16]

We outline here a standard method of preparing BMP from either bovine or human bone. We also present methods of cleavage of the molecular structure of BMP to produce the smallest unit of protein structure with osteoinductive activity, designated as bone morphogenetic polypeptide (BMP-p).

Procedure. The most available source of BMP is demineralized bovine or human cortical bone matrix. BMP is the least abundant of about 17–25 presently classified bone matrix proteins. The most direct method of preparation of BMP consists of chemical extraction of HCl-demineralized bone in a 4 M solution of GuHCl. This extract consists of a complex mixture of intra- and extracellular protein that is extremely difficult to fractionate. A slightly more manageable extract with a high level of BMP activity is obtainable from insoluble gelatinized bone matrix prepared by sequentially extracting the relatively soluble noncollagenous proteins,[6] as outlined in Table I. This removes about 20% of the non-BMP relatively soluble proteins from the system. The BMP is tightly bound to insoluble protein aggregates within the densely packed cross-linked structure of cortical bone matrix.

Table II presents 10 steps in the extraction of BMP with the aid of 4 M GuHCl, and differential precipitation of non-BMP proteins, M_r of 34K, 22K, and 17K and other lower M_r components, with 1.5 M GuHCl. Table II also includes a special step for removal of other non-BMP proteins which are soluble in Triton X-100 and have M_r values of 65K, 45K, 38K, 34K, 22K, 12K, and other lower values. The 1.5 M GuHCl-soluble and *Triton X-100-insoluble* components consist of a protein aggregate which includes BMP and has an SDS–PAGE pattern composed of components with M_r values of 22K, 21K–18K, 14K, 12K, and 5K, contaminated with trace amounts of lower and higher molecular weight material. The protein aggregate collected by this arduous procedure is characterized by about 1000× more osteoinductive activity than the original bovine or human

[11] M. R. Urist, Y. Yerashima, M. Nakagawa, and C. Stamos, *In Vitro* **14**, 697 (1978).
[12] H. Mizutani and M. R. Urist, *Clin. Orthop.* **171**, 213 (1982).
[13] K. Takagi and M. R. Urist, *Clin. Orthop.* **171**, 224 (1982).
[14] M. R. Urist and K. Takagi, *Ann. Surg.* **198**, 100 (1982).
[15] O. Nilsson and M. R. Urist, *Clin. Orthop.* **195**, 275 (1985).
[16] K. Sato and M. R. Urist, *Clin. Orthop.* **183**, 180 (1984).

TABLE I
DEMONSTRATION OF PROGRESSIVE REDUCTION OF DRY WEIGHT OF BONE MATRIX[a]

Sequentially used extraction procedure	Percentage reduction in dry fat-free weight of bone matrix	Milligrams of nondialyzable protein extracted per gram of dry weight	Induced bone, milligrams of bone ash per gram of preimplanted matrix
2 M CaCl$_2$, 2°	3.87 ± 0.91	2.52	499 ± 36
0.5 M EDTA, pH 7.4, 2°	2.32 ± 0.20	2.41	401 ± 54
8 M LiCl, 2°	2.39 ± 0.26	5.91	502 ± 47
H$_2$O, 55°	11.30 ± 1.36	17.16	409 ± 36
Total	19.88 ± 2.73	28.00	

[a] By removal in sequence of *soluble* noncollagenous proteins, and by converson of collagen to cold water insoluble gelatinized matrix, with retention of high osteoinductive activity.

demineralized bone matrix. The protein aggregate obtained by step 10 (Table II) is therefore excellent starting material for separation of BMP and for limited proteolysis preparation of BMP-p. Five years ago, we employed an inexpensive 0.5 M CaCl$_2$/6 M urea inorganic solvent mixture. We replaced it with the 4 M GuHCl solutions in an effort to remove the possibility of an isocyanate block at the N-terminal amino acid of the putative BMP molecules.[17,18] However, the nature of the block is still under investigation.

The weight of the total protein, components of SDS–PAGE patterns, and bioassay results obtained at each of the 10 steps listed in Table II are presented in Table III.

BMP is found in an aggregate of proteins that are soluble only in 6 to 8 M urea or 4 M GuHCl. The aggregate is insoluble in water (collected at steps VI and VII) and contains about 2000× more osteoinductive activity than bone matrix as measured by the level of ^{45}Ca incorporation into calcifying new bone. Under dissociative conditions in 4 M urea, ultrafiltration through a 10K hollow fiber cartridge separates matrix Gla protein.[17,19]

[17] M. R. Urist, Y. K. Huo, A. G. Brownell, W. M. Hohl, J. Buyske, A. Lietze, P. Tempst, H. Hunkapillar, and R. J. DeLange, *Proc. Natl. Acad. Sci. U.S.A.* **81,** 371 (1984).

[18] M. R. Urist, K. Sato, A. G. Brownell, T. I. Malinin, A. Lietze, Y. K. Huo, D. J. Prolo, S. Oklund, G. A. M. Finerman, and R. J. DeLange, *Proc. Soc. Exp. Biol. Med.* **173,** 194 (1983).

[19] P. A. Price, M. R. Urist, and Y. Otawara, *Biochem. Biophys. Res. Commun.* **117,** 765 (1983).

TABLE II

TEN STEPS IN PREPARATION OF A CRUDE EXTRACT OF BONE MATRIX GELATIN WITH
GuHCl AND WITH DIFFERENTIAL PRECIPITATION OF BMP

1. The starting material is gelatinized insoluble bone matrix prepared by sequential extraction as outlined in Table I.

2. Extract 10 kg of the gelatinized matrix with 4 M GuHCl/0.5 M CaCl$_2$ including 1.0 mM NEM, 0.1 mM benzamidine HCl/liter. Stir at room temperature for 24 hr.

3. Filter extract through cheesecloth and reextract residual matrix with fresh solution. Filter again through Whatman paper #1 and then ultrafilter through a 10K hollow fiber ultrafilter (Amicon). Concentrate to approximately 2 liters, to allow for more efficient dialysis.

4. Dialyze vs water for 3 days at 4° (three changes of water each day). Heat retentate in the dialysis bags to 35° each time gelling occurs.

5. When the retentate has been dialyzed 3X, heat to 35° and centrifuge out the precipitate. Wash the precipitate with cold water and redissolve in 3 liters of 4 M GuHCl/0.05 M CaCl$_2$.

6. To remove gelatin peptides, centrifuge and dialyze the supernatant in 11 vol of 0.25 M sodium citrate – citric acid buffer (pH 3.1). Following dialysis, heat the retentate to 40°, centrifuge out the precipitate (at 30–35°), and wash three times with cold water. To remove the tightly bound lipoprotein and lipid, extract the washed precipitate thoroughly with 2 liters of 1 : 1 chloroform – methanol. Filter the solvent through a Büchner funnel and let the product dry at room temperature.

7. Dissolve defatted 4 M GuHCl-soluble proteins in 100 vol of GuHCl. Centrifuge out the insoluble material and save for bioassay. Dialyze the supernatant vs a volume of cold water that will decrease the concentration from 4 to 1.5 M GuHCl. This separates the insoluble 24K protein, which has no BMP activity, from the 1.5 M GuHCl-soluble proteins, including BMP, by differential precipitation. Centrifuge the solution at 9000 rmp for 30 min at 0°.

8. Wash the precipitate three times with cold water and lyophilize. Dialyze the 1.5 M GuHCl-soluble proteins against water to produce a precipitate which includes BMP. Centrifuge and wash the 1.5 M GuHCl-soluble, water-insoluble proteins three times with cold water.

9. Redissolve the wet protein in 1 liter of 4 M GuHCl/0.5 M CaCl$_2$, and dialyze vs an equal volume of 4 M GuHCl containing 0.10 M Trizma – HCl, pH 7.2, and 0.2% Triton X-100 (0.2 g/100 ml). Add the buffer to approximately 300 ml of water and adjust the pH with HCl; then add the GuHCl and Triton X-100. Most of the 34K protein is soluble in Triton X-100.

10. Centrifuge the Triton X-100-insoluble precipitate, which includes BMP, and is now separated from the bulk of the proteins with M_r values of 6K, 24K, and 34K proteins. When tested individually and in combination with other proteins, these proteins have no osteoinductive activity in the mouse thigh assay system. Wash the BMP and associated water-insoluble noncollagenous proteins three times with cold water and lyophilize.

TABLE III
BMP Activity in GuHCl Extracts of Bovine Bone[a]

Step Number		Weight	M_r of the major electrophoretic components ($\times 10^3$)	Induced bone formation[b]		
				Incidence (%)(N)[c]	Yield (mm^3/mg)	^{45}Ca uptake in CPM/ml ($\times 10$)
I:	Wet undermineralized bone	10.0 kg	—	0 (10)	0	0
II:	Dry fat-free HCl-demineralized matrix	3.0 kg	—	0 (30)	0	0
III:	Bone matrix gelatin	1.4 kg	—	8 (40)	0.5	520
IV:	4 M GuHCl, 0.5 M CaCl$_2$-soluble, water-soluble substances	1.0 kg	—	28 (50)	1.0	965
V:	Heat to 37°, centrifuged water-insoluble substances	100 g	—	45 (20)	0.6	901
VI:	4 M GuHCl-soluble proteins, dialyze vs citrate buffer, centrifuge, extract precipitate in chloroform methanol	32.0 g	18, 68, 45, 34, 17–15, 14, 12, 5	50 (50)	1.8	1605
VII:	4 M GuHCl-soluble, 1.5 M GuHCl-soluble proteins	12.23 g	45, 34, 24, 22 18–17, 14, 12	55 (60)	1.5	1005
VIII:	1.5 M GuHCl-insoluble protein	16.0 g	34, 22, 17	8 (40)	0.3	510
IX:	Triton X-soluble, 1.5 M GuHCl-soluble proteins	7.3 g	65, 45, 38, 34, 22, 12	0 (20)	0	120
X:	Triton X-insoluble protein	3.5 g	22, 21–18, 14, 12, 5	96 (160)	2.0	1960

[a] BMP activity = < 0.001% of wet bone; BMP activity = 0.003% of demineralized lyophilized fat-free bone matrix.
[b] Implants in thigh muscle pouch of Swiss Webster mice.
[c] N, Number of implants.

The 18.5K, 17.5K, and 17K M_r proteins are soluble in water, and then ready for purification by means of hydroxyapatite chromatography.

Triton X-100-Soluble Proteins. A Triton X-100-soluble, 34K protein (isolated by step IX) was one of the single most abundant noncollagenous proteins of the group. The identity of the component may be determined by cross reactions of antibodies to osteonectin.

Sodium Dodecyl Sulfate (SDS) Polyacrylamide Slab Gel Electrophoresis (PAGE). The above-described protein fractions are solubilized by incubation for 24 hr in 0.06 M Tris–HCl (pH 6.8) containing 2 M urea and 0.2% SDS. Five microliters (2.5 mg/ml) of each sample was applied to a 12.6% gel with a 3% stacking gel and electrophoresed at 25 mA. The gels are stained with 0.25 Coomassie Brilliant Blue R-250 in methanol/acetic acid/H_2O(5:1:5), and with a standard method with silver. The relative migration mass (M_r) of each component is estimated with the aid of standards (Pharmacia, Uppsala) with a range of 94K to 14.4K. The putative BMP with an M_r of 18.5K ± 0.5K is positive staining with Coomassie Blue and produces a negative image by repelling silver.

Preparative Flat-Bed Electrofocusing in a Granulated Gel. The protein aggregate prepared by step 10 (Table II) can be fractionated and analyzed by isoelectric focusing as follows. Two hundred-milligram samples are dissolved in 4 M urea, and added to 5 ml of ampholyte (Biolyte, Bio-Rad, Richmond, CA). After electrofocusing, each band is cut out of the gel separately with a spatula. Each of the gel samples are eluted with 5 ml of 6 M urea in 0.5 M $CaCl_2$ for a solvent. The eluate is collected in tubes and the pH is determined for each sample. Two fractions with isoelectric points (pI) of 4.9 and 5.3, collected by elution of the gel slices with 0.5 M $CaCl_2$ in 6 M deionized urea, consist of three electrophoretic components with M_r values of 22K, 18.5K ± 0.5K, and 14K. These fractions have about 3× higher specific activity of BMP than the original aggregate, i.e., 3.3 mg induced as much bone as 10 mg of the aggregate. Seven other fractions in the pI range below 4.3 and above 7.5 have no 18.5K component and no BMP activity. Fortuitously, osteoinductive segments of the BMP molecule are not deactivated by the SDS–PAGE above-described procedure.

Slices of stained and unstained tube gels, extracted with 6 M urea to isolate the individual proteins, and reelectrophoresed on slab gels show that the 24K, 17.5K, 17.0K, and 14K components migrated the same distances as in the unsliced gels and represented separate entities. The 18.5K also migrates the expected distance, and, when eluted with a buffered solution of SDS and urea, is recovered by dialysis. When assayed in tissue culture, this component displays chondrogenetic activity.

Hydroxyapatite (HA) Chromatography. The separation of the BMP with M_r of 18.5K from other bone matrix proteins is accomplished by HA chromatography.[17] HA columns of BioGel HTP (Bio-Rad Laboratories, CA) are prepared by adding 15 g of HA to 100 ml of 0.05 M phosphate buffer, pH 7.3, containing 6 M urea. After 10 min the solvent is decanted and the HA is resuspended in 100 ml to give a settled bed height of 9 cm (volume, approximately 55 ml). The column is equilibrated with 100 ml of starting buffer, using a gravity flow rate of 0.6 ml/min. A sample of 100 mg of the preparation obtained by step 10, or a 10- to 20-mg sample of proteins with M_r <30K collected by S-200 gel filtration, is dissolved in deionized 6 M urea, applied to the column, and fractionated as follows. Figure 1 illustrates the absorbance patterns of various protein fractions eluted by a gradient of 0.05 to 0.3 M phosphate buffer at pH 7.3. Three proteins in the range of M_r 17K to 18K are isolated from other electrophoretic components by elution at specified concentrations of phosphate ion. A 22K, 14K, and other proteins lacking BMP activity are recovered in solutions of phosphate ions ranging from 0.5 to 0.10 M in concentration. Fractions containing components in the range of 17K to 18K were generally eluted at concentrations of phosphate buffer at 0.18 to 0.2 M. By rechromatographing this fraction it was possible to purify each of the three with variable degrees of contamination with other proteins of a lower molecular weight. The protein with BMP activity was white in color and fluffy in texture and had an M_r of 18.5K ± 0.5K. The proteins with M_r valves, of 17.5K, 17.0K, and 14K do not induce bone formation *in vivo* or cartilage cell differentiation *in vitro* (Fig. 1).

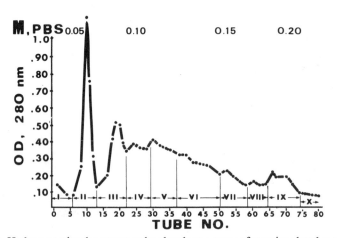

FIG. 1. Hydroxyapatite chromatography absorbance curve of proteins eluted at successive phosphate ion concentrations. The proteins with osteoinductive activity in the mouse's thigh muscle pouch assay (Fig. 4) were eluted at ion concentrations (fraction VIII) of 0.18 to 2.0 M, pH 7.3. Nine protein fractions with higher and lower molecular weights eluted at other phosphate ion concentrations did not induce bone formation in the mouse thigh.

Amino Acid Analysis. The individual proteins isolated by preparative gel electrophoresis were hydrolyzed at 110° for 24 hr in 6 M HCl in evacuated sealed tubes. Analysis for tryptophan was performed following hydrolysis at 110° for 24 hr in 2.5 M KOH, crystine, following performic acid oxidation; γ-carboxyglutamic acid (Gla) was determined by the method of Price.[19] Amino acid analysis was performed on an amino acid analyzer (Beckman 119 C) equipped with a Spectra Physics 4000 data reduction system. Bovine BMP has an M_r of 18.5K and the amino acid composition of an acidic polypeptide, and may include one residue of hydroxyproline but no γ-carboxyglutamic acid (Gla) (Table IV). Three Gla residues are found in the 14K protein while none was detected in the 18.5K, 22K, or 34K proteins.[17]

Hydrophobic Column High-Performance Liquid Chromatography (HPLC). Purification of the protein fraction isolated by HA chromatography is accomplished by HPLC using a hydrophobic column (Ultrapore, RPSC, Beckman Co.). For hydrophobic chromatography, 1.0-mg samples are dissolved in the following solutions: 1% phosphoric acid, 0.1 M sodium perchlorate, 5% trichloracetic acid, 5% acetic acid. The solution is applied to the column and eluted with a gradient of (1) 100% water and (2) 70%

TABLE IV
AMINO ACID ANALYSES

Amino Acid	M_r, 22K (mol%)	Residues (per 200 residues)	M_r, 18K ± 0.5K (mol%)	Residues (per 180 residues)	M_r, 14K (mol%)	Residues (per 130 residues)
Asx	12.29	26	16.8	17	11.40	15
Thr	3.64	7	7.1	7	2.57	3
Ser	5.43	11	17.6	18	6.58	9
Glx	14.54	29	23.4	23	14.84	19
Pro	6.58	13	9.0	9	5.08	7
Gly	8.71	17	12.4	12	6.14	8
Ala	5.99	12	14.4	14	8.56	11
Val	4.51	9–10	11.0	11	3.88	5
Met	2.21	4–5	2.75	3	1.30	2
Ile	2.67	5–6	5.9	6	4.09	5–6
Leu	6.27	14	15.0	15	7.54	10
Tyr	8.30	17	7.8	8	6.59	9
Phe	4.10	8	7.2	7	3.98	5
Lys	2.83	6	9.6	10	4.46	6
His	2.35	5	3.4	3	1.70	2
Arg	6.04	12	16.6	17	9.23	12
½Cys	3.74	7–8	4.2	4	2.06	3
Gla		0		0		3

acetonitrile in water containing 0.1% phosphoric acid/0.1 M sodium perchlorate, 1.0 ml/min in 100 min. This procedure further purifies the BMP with M_r of 18.5K and isolates the component with M_r of 17.5K. Amino acid analysis suggests that the 17.5K component has an amino acid sequence comparable to histone 2B. The 18.5K component isolated as outlined above has a blocked N-terminal amino acid, and is presently under investigation by means of peptide mapping.

Characteristics and Definitions of Units of BMP Activity. Table V is a tentative list of characteristics of BMP. The by-products, obtained at various steps in the production of BMP, are shown in Table VI. A review of the literature on bone morphogenetic and growth factors suggests that morphogens may initiate, while growth factors stimulate, bone development.[5] Knowledge of local hormones or growth factors associated with bone is just beginning to emerge. The chemistry and physiology of cartilage and bone growth factors are reviewed in detail by Seyedin *et al.,*[20] Linkhart *et al.,*[21] Nathanson,[22] and Muthukumaran and Reddi.[23]

Bone Morphogenetic Polypeptides (BMP-p). The smallest unit of protein structure with osteoinductive activity is a polypeptide prepared from BMP by limited proteolysis, and designated with the acronym BMP-p. Limited proteolysis is obtained with either pepsin or trypsin, under specified conditions of time, temperature, and ratio of substrate (BMP) and enzyme. The term "limited" is applicable when a product of proteolysis is stable, identifiable by a particular structure and specific function. The following procedures are applied to either the BMP protein aggregates obtained by step 10 (Tables II and III) or to purified BMP, and may be found to generate osteoinductive polypeptides with M_r values in the range of 7K to 4K. By means of sequential gel filtration, hydroxyapatite chromatography, and reversed-phase HPLC, BMP-p with M_r values as low as 4.1K have been isolated.

Pepsin-Limited Proteolysis. One gram of the BMP protein aggregate (step 10, Table II) or 10 mg of BMP purified by HA chromatography is digested for 2 hr at 37° in 2 liters of 0.01 N HCl containing pepsin (Sigma Co., St. Louis, MO), 10 μg/mg protein. The pepsin-cleaved HCl-insoluble and HCl-soluble proteins are separated into two parts as shown in Fig. 2. Part 1 consists of the soluble proteins found in the supernatant 0.01 N HCl solution of pepsin. Part 2 consists of HCl-insoluble protein aggregates.

[20] S. M. Seyedin, T. C. Thomas, A. Y. Thompson, D. M. Rosen, and K. A. Piez, *Proc. Natl. Acad. Sci. U.S.A.* **82,** 2267 (1985).

[21] T. A. Linkhart, S. Mohan, J. C. Jennings, J. R. Farley, and D. J. Baylink, "Hormonal Proteins and Peptides," p. 279. Academic Press, New York, 1984.

[22] M. A. Nathanson, *Clin. Orthop.* **200,** 142 (1985).

[23] N. Muthukumaran and A. H. Reddi, *Clin. Orthop.* **200,** 159 (1985).

TABLE V
PARTIAL LIST OF CHARACTERISTICS OF BMP

Acidic protein, pI 5.0 ± 0.2
M_r 18.5K (bBMP)
M_r 17.0K ± 0.5K(hBMP)
No carbohydrate detected
Soluble in aqueous media under dissociative conditions in 6 M urea or 4 M GuHCl
Expelled from solution by formation of complexes with hydrophobic molecules
Forms a supramolecular aggregate with bone matrix histone 2B (M_r 17.5K) calmodulin (M_r 17.0K) and matrix Gla protein (M_r 14K)
Molecule with a hydrophobic core and about 30% acidic amino acids
Insoluble in acetone, absolute alcohol, chloroform – methanol
Insoluble in Triton X-100
Forms insoluble complexes with matrix Gla protein (M_r 14K)
Disulfide bonded
Inactivated by:
  Heat, 70°
  Deamination in HNO$_3$
  β-ME reduction
  Lathyrism
  Pencillamination
  Ultrasonification
Binds to OH apatite
Resistant to:
  Collagenase
  Chondroitinases ABC
  Amylase
  Neuraminidase
  Hyaluronidase
  Alkaline phosphatase
  Acid phosphatase
  Chymopapain
  Tyrosinase
  Thermolysin
Limited pepsin or trypsin proteolysis yields BMP-p. One unit of BMP is 1 mg of a purified protein, which, after implantation in a mouse thigh muscle pouch, produces 1 g of wet weight of bone within 21 days; *in vitro*, 5 μg of BMP induces differentiation of neonatal rat muscle connective tissue into cartilage by day 14
Action is GH or SM dependent

Samples of each part are bioassayed for BMP activity at intervals of 1 to 30 hr to plot the rate of degradation of the osteoinductive products.

The HCl-insoluble proteins are collected by centrifugation and extensively washed in cold water. The HCl-soluble proteins are collected by terminating the hydrolysis by adding 0.01 N NaOH to reach neutral pH and by dialysis (Spectrapor tubing, pore size 2000 MW, Spectrum Co., Los Angeles, CA) against deionized water (3×) until formation of a water-insol-

TABLE VI
COMPOSITION AND OSTEOACTIVITY OF BY-PRODUCTS OF PURIFICATION OF BMP

By-Product	Derivation	SDS–PAGE, M_r	Induced Cartilage + Bone formation
Bone gelation peptides	GuHC CaC extract, citric acid soluble	90	0
High M_r proteins	4 M GuHCl soluble, water soluble	85 to 65	0
High and low M_r proteins	4 M GuHCl soluble, 0.5 M GuHCl insoluble	85 to 2K	±
High and low M_r proteins	1.5 M GuHCl soluble water insoluble	60 to 2	±
Low M_r proteins	1.5 M GuHCl insoluble Triton X soluble, osteonectin	34 to 38	0
Low M_r proteins	1.5 M GuHCl-insoluble residue	24	0
Not retained by 10K (Amicon)	6 M urea soluble, water insoluble	14 to 1	0
Proteins retained by 10K (Amicon)		60 to 2K	+
Unbound by HA columns	0.5 M GuHCl soluble	22	0
Eluted from HA columns, 0.02–0.10 phosphate ion	6 M urea soluble, water soluble	60,45,32,17, 16,15,14,12, 10 to 2	0
Eluted from HA column, 0.15 to 0.18 M phosphate ion	Water soluble	18.5, 17.5	+
Fractions I to V, S-200 gel filtration	4 M GuHCl	90 to 45	0
Fractions VI–VIII		45 to 14	+
Fractions IX–XI		15 to 1	0

uble precipitate is complete. The precipitate is separated from the supernatant by centrifugation at 50,000 g for 1 hr, and by washing 3× in cold water. The washed HCl-insoluble, water-insoluble, and the water-soluble proteins are lyophilized separately. The water-soluble supernatant of HCl-soluble products is collected by dialysis against water and lyophilized.

The lyophilized products are weighed and are either redissolved in 0.05 M phosphate buffer, pH 6.8, in 6 M urea (deionized in Dowex 50) for hydroxyapatite (HA) chromatography, or in 4 M GuHCl for S-200 gel filtration. The solutions are applied to a Sephacryl S-200 column (5 × 95 cm) with downward flow regulated by a peristaltic pump and collected in

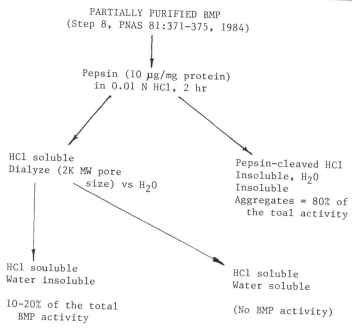

FIG. 2. Flow diagram of production of bone morphogenetic polypeptides (BMP-p) by pepsin-limited proteolysis. By 2 hr, there was an orderly conversion of nearly all of the proteins of high M_r to polypeptides of M_r less than 30,000 and peptides of M_r less than 1000. The conversion occurred with formation of HCl-insoluble, water-insoluble aggregates from a solution of 0.01 N HCl. The conversion also occurred with formation of an aggregate of low M_r/HCl-soluble, water-insoluble products. Provided that proteolysis did not extend beyond 2 hr of incubation, the total BMP activity was retained and partitioned between 0.01 N HCl-insoluble and HCl-soluble BMP-p. The water-soluble products had no osteoinductive activity.

five fractions. Two fractions with $M_r > 30,000$ are pooled, dialyzed against water, and relyophilized for bioassay for new bone formation induced per milligram of implanted protein. Three fractions with M_r proteins $< 30,000$ are pooled and similarly prepared. The proteins with $M_r < 30,000$ are further fractionated by hydroxyapatite (HA) chromatography as follows. One sample is applied to HA columns (2.5 × 40 cm) and eluted along a stepwise gradient of concentrations of 0.01 to 0.05, 0.05 to 0.2, 0.2 to 0.5 M phosphate buffer in 6 M deionized urea. The proteins are collected in nine fractions, examined by SDS–PAGE, desalted in sacs with a pore size of 2000 MW by dialysis against water, lyophilized, and implanted for BMP activity.

Polypeptides ($M_r < 10K$), isolated either by S-200 or HA chromatography, are subfractionated using Sephadex G-50 (Ultrafine) molecular sieve chromatography. Samples of freeze-dried polypeptides are dissolved in 0.01 M sodium phosphate (pH 7.0) containing 6 M urea and charged to a column (1.5 × 100 cm) equilibrated in the same buffer. Fractions of approximately 20 ml each are collected, combined within each peak, dialyzed extensively against cold, deionized water, and lyophilized. The resulting fractions are analyzed by SDS–PAGE.

Fractionation of the products of proteolysis isolates groups of bovine bone morphogenetic polypeptides (bBMP-p) with M_r values ranging from 7K to 4K. The molecular weight of a concentrate of bBMP-p, collected at the 2-hr interval of proteolysis, is 5.6K ± 0.9K. The BMP-p structure, isolated by hydrophobic reversed-phase HPLC, has an M_r of 4.7K. The M_r values ranged from 4.7K to 4.5K to 4.1K from one batch of bone to another. The pI is 7.6K ± 0.1K. Implants of bBMP-p in the mouse's thigh induce differentiation of cartilage and bone. In a tissue culture medium, 10 μg/ml of the M_r 4.7K BMP-p induces differentiation of muscle-derived connective cells into cartilage. The bBMP-p, M_r <7.0K and >3.0K, reacts with rabbit anti-bBMP antibodies. Bovine BMP, M_r 4.7K, cross reacts with mouse anti-human BMP. bBMP-p has an amino acid composition characteristic of a neutral or slightly basic polypeptide. Sequence determinations are in progress. Although these procedures on bovine BMP-p are reproducible on human BMP-p, differences in the amino acid composition and molecular weight are apparent but not yet analyzed in detail.

Ultrafiltration. The products of limited pepsin proteolysis are also dissolved in 4 M GuHCl in 0.01 M phosphate buffer (pH 7.0) and filtered through an HIP10-8 hollow fiber cartridge (Amicon, M_r 10,000, approximate cutoff). The retained concentrates are washed by passing an additional 6 liters of 4 M GuHCl through the hollow fibers and then dialyzed against deionized water at 4°. The insoluble concentrates are centrifuged and washed in cold water 3×. The washed insoluble proteins and the supernatants are separated and lyophilized.

Isolation of Slab Gel Slices. Polypeptides are also electrophoresed in duplicate disk and slab gel slices, one of which is stained to locate the individual bands. The gel slices corresponding to the major bands are eluted with 4 M GuHCl, dialyzed against water, and lyophilized.

SDS–PAGE on Peptide Gels. The above-described chromatographic fractions are solubilized in the buffered solution used for proteins, and analyzed by SDS–PAGE on gel compositions modified for peptides. Five microliters (2.5 μg/ml) of each sample is applied to a 37.5% acrylamide + 37.5% Bis and electrophoresed at 25 mA. The gels are stained either with silver or with both silver and 0.25% Coomassie Brilliant Blue R-250 in methanol/acetic acid/H_2O(5:1:5). The relative migration mass (M_r) is determined by using protein standards (Pharmacia) with a range of M_r

94,000 – 14,000, and by peptide standards (Pharmacia) with a range of M_r 17,200 – 1600. The M_r values are calculated by plotting the logarithm of molecular weight versus distance of migration of six standard peptides relative to the distance of the unknown.

Hydrophobic Column HPLC. Further purification of the polypeptides, isolated by G-50 gel filtration, is accomplished by HPLC using a hydrophobic column (Ultrapore, reversed-phase Spherogel, Beckman).

Isoelectric Focusing. The pI values of polypeptides isolated by HA chromatography and HPLC are determined by isoelectric focusing by the method of Righetti and Drysdale using a micro-pH probe (Microelectrodes, Inc.) for pH measurements on the cold gels (5% polyacrylamide and 2% ampholyte). Focusing occurs at 10 W at 2° for 3.5 hr.

Immunoassay of BMP-p. Rabbit polyclonal antibody and mouse monoclonal antibody to M_r 18.5K protein are applied by M_r 4.7K polypeptides and to BMP by the dot ELISA method to determine whether the biologically active segment and the immunoactive site are the same. The same principles are applied to the Western blot method detecting antigens and antibodies.

Amino Acid Analysis. Amino acid analysis is performed on an amino acid analyzer (modified Beckman 121). Cystine and half-cystine analysis is performed following performic acid oxidation.

Amino Acid Sequences. The M_r 4.7K polypeptide, purified by reversed-phase hydrophobic column chromatography, is analyzed with the aid of a gas/liquid phase protein sequenator.

Trypsin-Limited Proteolysis. The above-described procedures were performed with pepsin because the structure of BMP is remarkably stable in acidic solutions. The following procedures are performed to exploit the high specificity of trypsin for hydrolysis of carbonyl groups of arginyl and lysyl residues. Observations on limited trypsin proteolysis of crude GuHCl extracts and on purified BMP are interesting because the trypsin-cleaved BMP yields a BMP-p of slightly different molecular weight but equally high specific activity when calculated on a molar basis.

One gram of crude, partially purified BMP was digested for intervals of 15 min, 30 min, 1 hr, and 4 hr in solutions of trypsin (TPCK-treated, bovine pancreas, Sigma, St. Louis, MO), 5 mg/g of total protein in 0.1 M Tris, pH 7.20, at 37°. At the designated intervals, hydrolysis is terminated by the addition of 1 N HCl to lower the pH to 1.0. The solution is centrifuged at 20,000 rpm for 10 min to separate the acid-soluble from the acid-insoluble precipitates. The precipitate is washed in cold water 3×. The acid-soluble supernatant is transferred to Spectrapor tubing (pore size 2000 MW, Spectrum Co., Los Angeles, CA) and dialyzed against ionized water (3×). A water-insoluble precipitate that formed inside the sac is

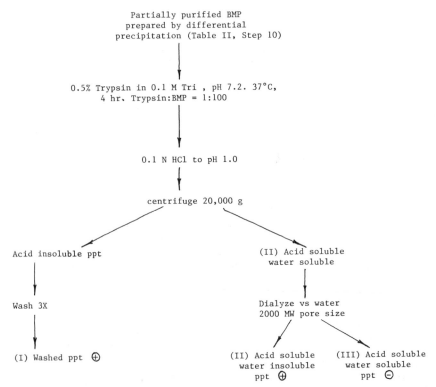

Partially purified BMP
prepared by differential
precipitation (Table II, Step 10)

0.5% Trypsin in 0.1 M Tri , pH 7.2. 37°C,
4 hr. Trypsin:BMP = 1:100

0.1 N HCl to pH 1.0

centrifuge 20,000 g

Acid insoluble ppt

(II) Acid soluble
water soluble

Wash 3X

Dialyze vs water
2000 MW pore size

(I) Washed ppt ⊕

(II) Acid soluble
water insoluble
ppt ⊕

(III) Acid soluble
water soluble
ppt ⊖

FIG. 3. Diagramatic representation of the distribution and solubility of the products of limited trypsin proteolysis. Precipitates (ppts) I and II had 10✕ higher osteoinductive activity (%, w/w) than the starting material. There was no osteoinductive activity in acid-soluble, water-soluble proteins (designated III).

separated from the water-soluble components by centrifugation at 50,000 g for 1 hr, and by washing the precipitate 3✕ in deionized cold water. The water-soluble supernatant and the washed precipitates are separately collected, lyophilized, and weighed. The procedure is diagrammatically shown in Fig. 3 along with the distribution of the osteoinductive activity. The purification procedures are comparable to the above-described methods for BMP-p generated by limited pepsin proteolysis.

Bioassay Methods in Vivo and in Vitro. Fractions with weights of 5 to 10 mg of BMP or BMP-p and associated proteins collected at each step in the purification process are lyophilized and implanted in the hindquarter muscles of Swiss-Webster strain mice. For controls, samples of the partially purified proteins with M_r values of 38,000, 24,000, 15,000, 12,000, 4000, 8500, 3000, and 2500 K are implanted in triplicate in the contralateral hindquarter muscle pouches (Tables III and VI).

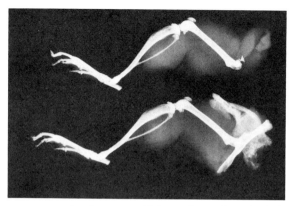

FIG. 4. Roentgenograms showing mouse thigh muscle pouch method of assay of implant of lyophilized protein fractions. Note control, 21 days after implant of 10 mg of bovine serum albumin (top), and the experimental thigh (bottom) at the 21-day interval postimplantation of 5 mg of an aggregate of 23K, 18.5K, and 14K proteins isolated by hydroxyapatite chromatography.

The quantity of new bone is measured by correlated observations on roentgenograms, excised wet ossicle weights, histological sections, histomorphometric data, and ^{45}Ca uptake by calcified new bone. The histomorphometric estimates are performed on roentgenograms (Fig. 4) of histologically valid deposits of bone (Fig. 5) using the random point analysis method with the following formula:

$$\frac{\text{Points on radiopaque bone } (\times 100)}{\text{Points on quadriceps muscle compartment}} = \frac{\% \text{ BMP-induced}}{\text{bone formation}}$$

Tissue Culture Assay. Samples are added to a CMRL (GIBCO) culture medium. The inductive activity per 0.1 to 1.0 mg of *partially* purified proteins or polypeptides is measured by counting the number of cartilage cells per low power field. Neonatal rat muscle connective tissue outgrowths onto a substratum of 4 M GuHCl-extracted BMP-free matrix are exposed to HA or HPLC chromatographic fractures in triplicate by the method of Sato and Urist.[16]

Methods of Isolation from Columns Which Exclude BMP and Retard Bone Growth Factors. Ion exchange chromatography can be useful for removal of non-BMP components of 4 M GuHCl extracts of demineralized bone matrix.[12,20] Under prescribed conditions, BMP is collected in the *unbound* fraction with DEAE and retarded only in trace amounts by CMC columns.[12] Bone TGF-β is isolated by elution from CMC columns.[24]

[24] S. M. Seyedin, A. Y. Thompson, H. Bentz, D. M. Rosen, J. M. McPherson, A. Conti, N. R. Siegel, G. R. Galluppi, and K. A. Piez, *J. Biol. Chem.* **261**, 5693 (1986).

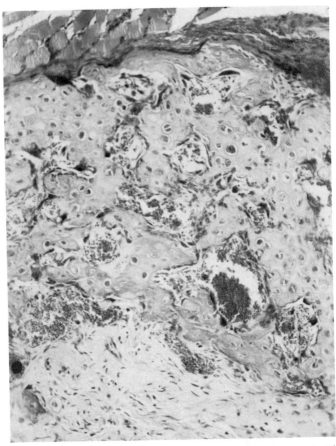

Fig. 5. Photomicrograph of deposit of woven bone (center) in the mouse's thigh muscle shown in Fig. 4. Note: muscle (top) and mesenchymal-type tissue (bottom). Hematoxylin, eosin, and azure II stain.

Calf and ox bone demineralized in EDTA and extracted with 4 M GuHCl contains at least six growth factors: platelet-derived growth factor (PDGF), anionic and cationic forms of fibroblast growth factor (FGF), cartilage growth factor (CGF), endothelial cell growth factor (CEGF), and bone growth factor (BGF).[25] All six are soluble in NaCl in 0.01 M Tris (pH 7.0) and most are heat stable. BMP is heat labile and forms water-insoluble

[25] P. V. Hauschka, A. E. Mavrakos, M. D. Iafrati, S. E. Doleman, and S. Klagsbrun, *J. Biol. Chem.* **261,** 12665 (1986).

aggregates with other bone matrix noncollagenous proteins. BMP dissociates from other insoluble proteins and becomes water soluble when separated from matrix Gla protein (MGP). The separation occurs under dissociative conditions in 6 M urea in 0.5 M EDTA, and the solution is passed through a 10K molecular weight pore size membrane by ultra filtration. When these water-soluble MGP-free proteins are fractionated on heparin–Sepharose columns by elution with NaCl by the method of Klagsbrun and Shing,[26] the BMP component with M_r values of 18.5K to 19.0K is collected in the unbound fraction with several other proteins. The heparin affinity is an empirical but effective method for fractionation of various growth factors, which excludes BMP along with other unbound acidic proteins. A less empirical and somewhat more discriminative exclusion of BMP can be observed with fractionation of bone matrix proteins in solutions of 2 M urea containing 0.002 M $CaCl_2$, applied to calmodulin–Sepharose columns by the method of Linden et al.[27] Under the conditions specified, calmodulin binds a protein with an M_r value of 14K to 15K and a protein with M_r of 23K but not BMP; the BMP is collected in the unbound effluent; the calmodulin-bound 14K and 24K proteins are eluted with 2 mM EDTA.

Sampath and Reddi[28] fractionated GuHCl crude extracts of HCl-demineralized bone powders by S-200 gel filtration. Protein fractions with M_r < 50K induced bone formation when redissolved in 4 M GuHCl, reprecipitated on surfaces of pulverized demineralized rat bone matrix, and the composites of matrix and proteins were implanted in rats. In the mouse thigh pouch assay, BMP and aggregated noncollagenous proteins induce bone formation without any bone matrix in the system. In fact, in the mouse thigh, composites of bone gelatin, collagen, or matrix reduce rather than increase the yield of new bone. Injections of BDGF prepared by the methods of Canalis and Centrella,[29] SGF by the procedure of Farley and Baylink,[30] or purified growth factors such as recombinant human EGF (manufactured by Chiron Corp., Emeryville, CA), TGF-β (gift of M. D. Sporn, U.S. Public Health Service, National Institutes of Health), FGF,[31] IGF (gift of Chiron Corp.), or IL-1 (gift of Cistron Corp., Morristown, NJ)

[26] M. Klagsbrun and Y. Shing, Proc. Natl. Acad. Sci. U.S.A. 82, 805 (1985).
[27] C. D. Linden, J. R. Dedman, J. G. Chafonleas, A. R. Means, and T. F. Roth, Proc. Natl. Acad. Sci. U.S.A. 78, 308 (1981).
[28] T. K. Sampath and A. H. Reddi, Proc. Natl. Acad. Sci. U.S.A. 78, 7599 (1981).
[29] E. Canalis and M. Centrella, Endocrinology 118, 2002 (1986).
[30] J. R. Farley and D. J. Baylink, Biochemistry 21, 3502 (1982).
[31] D. Gospodarowicz and J. S. Moran, Annu. Rev. Biochem. 45, 531 (1976).
[32] E. Canalis, M. Centrella, and M. R. Urist, Clin. Orthop. 198, 289 (1985).

do not induce bone formation *in vivo* in the mouse thigh muscle pouch (unpublished observations). Although these exogenous growth factors appear not to be' osteoinductive, some are ubiquitous components of serum and bone cell secretions,[25] and could act coefficiently or synergistically[5] to either stimulate or suppress bone cell differentiation. The BMP hypothesis assumes that BMP is a tissue-specific cell differentiation factor, or morphogen, rather than an unspecific growth factor or mitogen. Mitogens which are known to compete for receptor sites of each other may thereby modulate the response to BMP.

In contrast to growth factors,[25,32] which are active in nanogram concentrations, presently available preparations of BMP *in vitro* require microgram, and *in vivo* milligram, doses to produce a locally effective concentration gradient and grossly visible deposits of cartilage and bone. Differing from a typical bone-derived growth factor, BMP is growth hormone dependent *in vivo*. Accordingly, BMP is consistently effective only in media containing all systemic growth factors supplied by 10–15% fetal calf serum. Since methods are available for isolation of membrane receptors for local hormones including BMP, their functions under physiological conditions are bound to be elucidated in the near future. In this new field of autocrine physiology, knowledge of interaction of systemic and local hormonal factors including BMP may grow, and new methods of treatment are beginning to develop.

Acknowledgment

Supported in part by a grant-in-aid from USPHS, NIH DEO2103, and the Max Factor Family Foundation.

[29] Preparation of Cartilage-Derived Factor

By Fujio Suzuki, Yuji Hiraki, and Yukio Kato

Introduction

The process of replacing cartilage by bone is known as endochondral ossification. During the first stage of conversion into bone, epiphyseal cartilage cells show increased matrix synthesis, and proliferate rapidly, becoming enlarged and hypertrophic. During this series of events culminating in transformation into bone, the metabolism of growth cartilage is subject to hormonal control. However, the mechanism of endochondral ossification in growth cartilage remains unclear. To understand the function of growth cartilage, we tried to isolate and cultivate growth cartilage cells from the ribs of young rabbits and investigated their biochemical properties. We found that growth cartilage cells in culture have a marked osteogenic potential,[1] forming differentiated colonies producing abundant macromolecules[1,2] such as proteoglycans and type II collagen *in vitro*. We also found that these cells respond well to calcitonin,[2] parathyroid hormone,[2] multiplication-stimulating activity (MSA),[3] and thyroid hormone to stimulate proteoglycan synthesis, the differentiated cartilage phenotype. Besides these classic hormones, extracellular matrix components have long been suggested to play important roles in the control of proteoglycan synthesis.[4] Thus, we have searched for such regulatory factors in cartilage and found that fetal bovine cartilage contains growth factors which stimulate the incorporation of [^{35}S]sulfate into glycosaminoglycans (GAG) and proteoglycans in rabbit costal chondrocytes in culture.[5-7] We termed this peptide factor the cartilage-derived factor (CDF).[5]

Assay

The biological assay of CDF is based on the incorporation of [^{35}S]sulfate or [^{3}H]glucosamine into GAG synthesized by chondrocytes isolated from the ribs of young rabbits.

[1] Y. Shimomura, T. Yoneda, and F. Suzuki, *Calcif. Tissue Res.* **19**, 179 (1975).
[2] F. Suzuki, T. Yoneda, and Y. Shimomura, *FEBS Lett.* **70**, 155 (1976).
[3] Y. Kato, N. Nasu, T. Takase, and F. Suzuki, *J. Biochem.* **84**, 1001 (1978).
[4] Z. Nevo and A. Dorfman, *Proc. Natl. Acad. Sci. U.S.A.* **69**, 2069 (1972).
[5] Y. Kato, Y. Nomura, Y. Daikuhara, N. Nasu, M. Tsuji, A. Asada, and F. Suzuki, *Exp. Cell Res.* **130**, 73 (1980).

Separation and Culture of Growth Cartilage Cells

The costochondral junction is removed by aseptic technique from the ribs of young rabbits weighing 300–600 g. After removal of surrounding soft tissue, growth cartilage is cut off and chopped up with a scalpel. This material is incubated with 0.1% EDTA in Ca^{2+}- and Mg^{2+}-free, balanced salt solution (CMF) for 20 min at 37°. The material is digested first with 10 vol of CMF containing 0.2% trypsin for 1 hr. After washing with CMF to remove traces of trypsin, the material is transferred into a digestion chamber containing a small magnetic stirring bar and then digested with 10 vol of CMF containing 0.2% collagenase (Sigma Chemical Co., type I) under gentle stirring for 2–3 hr at 37°. The resulting cell suspension is filtered through a nylon sieve with a 120-μm pore size to remove debris and undigested cartilage fragments. The filtrate is centrifuged, and the pellet is washed three times with Eagle's minimum essential medium (MEM) supplemented with 10% fetal calf serum (FCS; previously heated for 30 min at 56°) and 60 μg/ml of kanamycin. After the third wash, cells are counted with a hemocytometer and serial dilutions made so that 0.1 ml contains 10^4 cells/ml for 6-mm plastic multiwell plates. The average yield of chondrocytes is 1×10^7 growth cartilage cells from five rabbits.

The cells are incubated for 6–8 days at 37° in a water-jacketed CO_2 incubator. Growth cartilage cells, when grown in Eagle's MEM supplemented with 10% FCS, are found to be polygonal and epithelial in nature. The culture exhibits typical properties of growth cartilage cells, including the formation of a refractile matrix and metachromasia staining with toluidine blue.

[^{35}S]Sulfate or [^{3}H]Glucosamine Incorporation

CDF activity is determined by measuring the ability of samples to stimulate the incorporation of [^{35}S]sulfate or [^{3}H]glucosamine into GAG of confluent growth cartilage cells in culture. When chondrocytes are grown to confluence, cultures are started in 0.1 ml of Dulbecco's modified Eagle's medium (DMEM) containing 0.3% FCS. After 24 hr, the medium is replaced by the same medium containing test samples. After 3 hr, the cells are exposed for 17 hr at 37° to 5 μCi of [^{35}S]sulfate (carrier free) or 5 μCi of [^{3}H]glucosamine in 0.1 ml of DMEM in the presence of test sam-

[6] Y. Kato, Y. Nomura, M. Tsuji, H. Ohmae, M. Kinoshita, S. Hamamoto, and F. Suzuki, *Exp. Cell Res.* **132**, 339 (1981).

[7] Y. Kato, Y. Nomura, M. Tsuji, M. Kinoshita, H. Ohmae, and F. Suzuki, *Proc. Natl. Acad. Sci. U.S.A.* **78**, 6831 (1981).

ples. The medium is then removed and the cell layers are solubilized with 0.15 ml of 0.5 M NaOH. The medium and cell fractions are combined and neutralized with 6 N HCl. To 0.5 ml of the mixture, 0.5 ml of Tris–HCl buffer, pH 7.8, containing 5 mM CaCl$_2$ and 0.5 mg of Pronase E (1000 tyrosine units/mg; Kaken Kagaku Co., Tokyo, Japan) is added, and the mixture is incubated for 12 hr at 55°. Then 50 μl of water containing 5 μg of chondroitin sulfate and 0.5 ml of 2 mM MgSO$_4$ are added. Polysaccharides are precipitated by addition of 0.5 ml of 1% cetylpyridinium chloride containing 20 mM NaCl and then the mixture is kept for 1 hr at 37°. The precipitate is collected on a Millipore filter disk (25 mm diameter, 0.45 μm pore size) and washed five times with 2.0 ml of 1% cetylpyridinium chloride–20 mM NaCl. The filter is air dried and solubilized in 10 ml of Insta-Gel emulsifier (Packard). The radioactivity is measured in a scintillation spectrometer (Rack-Beta, LKB-WALLAC, Sweden).

One unit of CDF activity is defined as the amount of CDF comparable to 5% FCS necessary to stimulate GAG synthesis in a 17-hr incubation period. For assays of column fractions, a constant volume of each fraction is chosen so that the peak fraction displays about 0.5–1.0 U of CDF activity.

Purification Procedure

Preparation of Cartilage Extract

Fetal calves weighing 10–13 kg are obtained from a public slaughter house. Cartilage is excised from their scapulae and limbs. Adherent soft tissue is removed and then the cartilage is sliced in a Robot Coupe R2 tissue cutter (Rosny, France) and homogenized in a Polytron (Kinematica with 10 vol of 1 M guanidine hydrochloride containing 0.1 M 6-amino-n-carproic acid, 0.02 M 2-(N-morpholino)ethanesulfonic acid, pH 6.0 (extraction buffer) at 4°. The homogenate is gently mixed with a stirrer for 48 hr and then centrifuged at 10,000 g for 20 min. Acetone is added to the extract to a final concentration of 45% (v/v). After 20 min at 4°, the insoluble material is removed by centrifugation, and further acetone is added to the supernatant to a final concentration of 65% (v/v). The insoluble material is collected by centrifugation and solubilized with extraction buffer. The solution is exhaustively dialyzed against distilled water at 4° and lyophilized. Approximately 6 g of acetone-fractionated preparation, obtained from about 800 g of cartilage, is dissolved in 2000 ml of 4 M guanidine hydrochloride/0.1 M 6-amino-n-caproic acid/0.01 M EDTA/0.02 M 2-(N-morpholino)ethanesulfonic acid (MES)/1.0 M NaCl.

The solution is filtered through an Amicon XM300 filter, which excludes molecules of more than 3×10^5 Da. The filtrate is then filtered through a UM20 filter (20,000 M_r cutoff). The filtrate is finally filtered through a UM10 filter (10,000 M_r cutoff). The fraction concentrated by UM10 filter (UM10 fraction) is dialyzed against distilled water and lyophilized. Approximately 40–60 mg of UM10 fraction was obtained from 6 g of acetone-fractionated preparation.

Heparin–Sepharose Affinity Chromatography

Columns of heparin–Sepharose CL-6B (Pharmacia; 6 ml bed volume) are equilibrated with 10 mM phosphate buffer containing 0.15 M NaCl, pH 7.3. Fifty milligrams of UM10 fraction (2 mg/ml) is applied to the column. After washing with 10 column volumes of equilibration buffer, CDF is eluted with a 0.15 M–2.0 M NaCl (50 ml/50 ml) gradient at a flow rate of 5–6 ml/hr. Fractions of 2 ml are collected and each fraction is monitored directly for absorbance, conductivity, and CDF activity. CDF activity is recovered in two peaks eluted at about 0.6 M NaCl (peak I) and 1.3 M NaCl (peak II). However, the recovery of peak II is variable from lot to lot and, hence, peak I is used for further purification. The recovery of peak I is highly reproducible. The active fractions containing CDF activity (peak I) are pooled, dialyzed against distilled water, and lyophilized. The recovery of CDF activity from columns of heparin–Sepharose is about 50–75%. About 4 mg of heparin–peak I was obtained from 50 mg of UM10 fraction.

Reversed-Phase High-Pressure Liquid Chromatography (RP-HPLC)

The final purification is achieved by RP-HPLC. A Gilson HPLC system is used which includes model 302 solvent delivery pumps, model 811 dynamic mixer, model 7125K sample injector, and a model 1001 UV-Master (M & S). All separations are performed on a Radial-pac μBondapak C_{18} reversed-phase column (Waters, 10-μm particle size, 8 mm $\times$ 10 cm).

The First RP-HPLC. Heparin–peak I (4.2 mg) is dissolved in 400 μl of 0.1% trifluoroacetic acid in water, filtered through Millipore filter (0.45 μm), and applied to the column. A linear 2-propanol-acetonitrile gradient composed of 0.1% trifluoroacetic acid in water as starting buffer and 0.1% trifluoracetic acid in 80% 2-propanol–acetonitrile (7:3 v/v) as limit buffer is used. All solutions are degassed before use. The column is operated at a flow rate of 1.0 ml/min at 25°; the effluent is monitored continuously at 280 nm. The column effluent is collected in 1-ml fractions. After lyophilization, the fractions are dissolved in 300 μl of distilled water and 5-μl aliquots are used for CDF assays. An active peak is eluted with 50% 2-propanol–acetonitrile and an elution time of 18–19 min, as shown in Fig. 1. Active fractions are pooled and concentrated by lyophilization.

The Second RP-HPLC. An active fraction (295 μl) eluted from the first RP-HPLC column is lyophilized and dissolved in 300 μl of 0.1% trifluoroacetic acid in 20% 2-propanol–acetontrile (7:3, v/v). The solution is then filtered through Millipore filter (0.45 μm) and applied to the same RP-HPLC column. A linear gradient composed of 0.1% trifluoroacetic acid in 20% 2-propanol-acetonitrile (7:3, v/v) as starting buffer and 0.1% trifluoroacetic acid in 50% 2-propanol-acetonitrile as limit buffer is used. Absorbance is measured at 214 nm. Other conditions are the same as described in

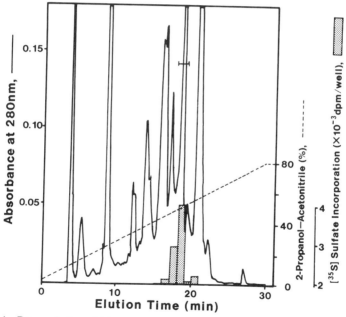

FIG. 1. Reversed-phase high-pressure liquid chromatography (RP-HPLC) of CDF (heparin–peak I). Chromatograms were performed with a flow rate of 1.0 ml/min. Absorbance was measured at 280 nm. Heparin–peak I (4.2 mg) was dissolved in 400 μl of 0.1% trifluoroacetic acid in water, filtered through a Millipore filter, and injected. The gradient was from 0 to 80% 2-propanol–acetonitrile (7:3, v/v) in 0.1% trifluoroacetic acid as indicated by the dashed line. The shaded bars indicated the CDF activity. The active peak indicted by the solid bar was further applied to the second RP-HPLC.

the first RP-HPLC. An active peak is eluted with 32% 2-propanol–acetonitrile, as shown in Fig. 2. The active peak fraction gives a single band of peptide on SDS–polyacrylamide gel electrophoresis and its molecular weight is approximately 16,000. This active peptide has the amino acid composition as shown in Table I.

This procedure yields 72 µg of electrophoretically pure CDF (16K peptide) from 800 g of fetal bovine cartilage.

Comments

As reported previously, various molecular species of CDF including a neutral 11K peptide, at least two basic 16K peptides, and others, were obtained by isoelectric focusing.[7] We have succeeded in purifying extensively a neutral 11K CDF by this procedure,[7] but its yield was very low. The improved scheme for this purification of CDF described here yields higher amounts of pure protein (16K CDF) in a short time.

As described above, the purified CDF enhances the synthesis of proteo-

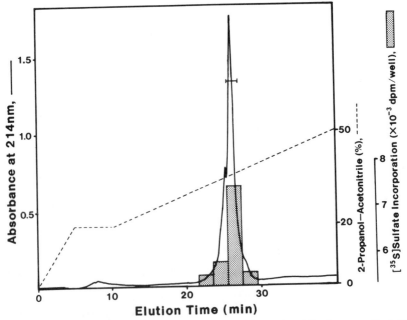

FIG. 2. The second RP-HPLC of the first RP-HPLC fraction. The linear gradient produced by 20 to 50% 2-propanol–acetonitrile in 0.1% trifluoroacetic acid was developed after the breakpoint in the dashed line. Absorbance was measured at 214 nm. The active peak indicated by the solid bar was used for the amino acid analysis.

TABLE I
AMINO ACID COMPOSITION OF CDF

Amino acid	Observed[a]	Integer
Asx	13.1	13
Thr[b]	7.4	7
Ser[b]	9.2	9
Glx	12.9	13
Pro	8.9	9
Gly	17.9	18
Ala	5.6	6
Cys[b]	4.3	4
Val[c]	7.8	8
Met	1.3	1
Ile[c]	12.2	12
Leu	9.6	10
Tyr	6.9	7
Phe	3.7	4
Lys	14.2	14
His	5.5	6
Trp[b]	0.9	1
Arg	4.9	5
Number of residues		147
Calculated M_r		16,128

[a] Values are averages of data obtained by two 30-μg samples hydrolyzed in 4 M methanesulfonic acid and 0.2% tryptamine for 24 and 72 hr.
[b] Based on extrapolation of data obtained by hydrolysis of 24 and 72 hr.
[c] Based on data obtained by hydrolysis of 72 hr.

glycans of rabbit costal chondrocytes in culture, but has relatively small effects on DNA synthesis and cell division of chondrocytes, although crude CDF stimulates not only proteoglycan synthesis but also DNA synthesis.[8-10] In contrast, epidermal growth factor (EGF) or fibroblast growth factor (FGF) stimulate DNA synthesis and cell division of chondrocytes, but do not stimulate nor inhibit proteoglycan synthesis.[9,10] The combination of CDF and EGF or CDF and FGF have additive or synergistic effects on the

[8] Y. Kato, Y. Nomura, M. Tsuji, H. Ohmae, T. Nakazawa, and F. Suzuki, J. Biochem. **90**, 1377 (1981).
[9] Y. Kato, R. Watanabe, Y. Hiraki, F. Suzuki, E. Canalis, L. G. Raisz, K. Nishikawa, and K. Adachi, Biochim. Biophys. Acta. **716**, 232 (1982).
[10] Y. Kato, Y. Hiraki, H. Inoue, M. Kinoshita, Y. Yutani, and F. Suzuki, Eur. J. Biochem. **129**, 685 (1983).

growth of chondrocytes.[10] Multiplication-stimulating activity (MSA) mimicks these effects of CDF.[8-10]

Properties of CDF

The biological activity of CDF is stable to heating at 95° for 15 min. The CDF activity is lost completely on 3 hr of incubation at 37° with trypsin or Pronase E, but is not affected by treatment with chondroitinase AC and collagenase at 37° for 3 hr. The purified CDF preparation can be stored indefinitely as a lyophilized powder or as a frozen solution.

Acknowledgment

We wish to thank Drs. S. Hara and T. Ikenaka for analyzing the amino acid composition of CDF.

[30] Purification of Cartilage-Derived Growth Factors

By JOACHIM SASSE, ROBERT SULLIVAN, and MICHAEL KLAGSBRUN

Introduction

In order to understand the local control of the hormonal regulation of cartilage metabolism research has been focused on the presence of growth and maturation factors found in normal or degenerated cartilage. Our laboratory has purified a cartilage-derived growth factor (CDGF) from normal cartilage[1-4] which was shown to be active at nanogram concentrations. By means of a variety of criteria CDGF is different from cartilage-derived factor (CDF), a peptide closely related to the somatomedins.[5-7]

[1] M. Klagsbrun, R. Langer, R. Levenson, S. Smith, and C. Lillihei, *Exp. Cell Res.* **105**, 99 (1977).

[2] M. Klagsbrun and S. Smith, *J. Biol. Chem.* **255**, 10859 (1980).

[3] M. C. Bekoff and M. Klagsbrun, *J. Cell Biochem.* **20**, 237 (1982).

[4] R. Sullivan and M. Klagsbrun, *J. Biol. Chem.* **260**, 2399 (1985).

[5] Y. Kato, Y. Nomura, M. Tsuji, M. Kinoshita, H. Ohmae, and F. Suzuki, *Proc. Natl. Acad. Sci. U.S.A.* **78**, 6831 (1981).

[6] E. Canalis, Y. Kato, Y. Hiraki, and F. Suzuki, *Calcif. Tissue Int.* **36**, 102 (1984).

[7] Y. Kato, R. Watanabe, Y. Hiraki, F. Suzuki, E. Canalis, L. G. Raisz, K. Nishikawa, and K. Adachi, *Biochem. Biophys. Acta* **716**, 232 (1982).

However, CDGF appears to be related to basic FGF. In this chapter we describe the purification of CDGF from a variety of cartilage sources by using affinity chromatography on heparin–Sepharose (Ref. 8; also see Klagsbrun et al. [8], Vol. 147, this series).

Extraction of CDGF from Cartilage

Sources of Cartilage

Extracts of CDGF are obtained from human costal cartilage, bovine articular and chick sternal cartilage. Three methods are used for the preparation of cartilage extracts: (1) Cartilage is extracted with 1 M guanidine–HCl, pH 6.5, for 24 hr at room temperature[2]; (2) cartilage is extracted with 1 M NaCl, 10 nM Tris–HCl, pH 7.0, for 48 hr at 4°; (3) cartilage is digested with 2 mg/ml of clostridial collagenase to produce an extracellar matrix fraction.[3]

In order to obtain bovine cartilage, the shoulders from newborn calves (1–14 days old) are purchased from a local slaughterhouse. There are two sources of cartilage in the shoulder. One is the scapula which has a piece of cartilage measuring 3 × 8 cm and the other are the articular surfaces. The shoulders are washed with the disinfectant betadine. Subsequently, the surrounding muscle and connective tissue are cut away using scalpels with #10 and #21 blades. The cartilage itself is cleaned of fibrous connective tissue and perichondrium by scraping with a #21 blade, and is then cut from the bone. The 0.5-cm region proximal to the bone is discarded because it is slightly vascular. The articular surfaces are free of perichondrium and therefore easily removed by cutting off slices with a #21 blade. One calf shoulder yields 25–35 g of pure cartilage.

Human costal cartilage (10–50 g per operation) is routinely obtained by our laboratory from children undergoing corrective surgery for pectus excavatum by the surgical service of Children's Hospital. Connective tissue, muscle, and perichondrium are removed in surgery.

Chick sterna are purchased by special order from Pel-freeze (Rogers, AK) free of connective tissue, muscle, and perichondrium. The cartilage is cut free of the sternum with a sharp blade, taking care not to include any bone. One pound of crude chick sterna yields approximately 150 g of cartilage.

Cleaned cartilage from all three sources is finely diced using a scalpel or homogenized using a food processor in preparation for the extraction.

[8] Y. Shing, J. Folkman, R. Sullivan, C. Butterfield, J. Murray, and M. Klagsbrun, *Science* **223**, 1296 (1984).

Extraction Procedures

Extraction with Guanidine Hydrochloride. Initially CDGF was extracted from cartilage with guanidine hydrochloride.[2] Briefly, 400 g of diced calf cartilage was extracted with 4 liters of 1.0 M guanidine hydrochloride buffered with sodium maleate, pH 6.0, for 24 hr at room temperature. The extract was filtered, concentrated by ultrafiltration, dialyzed against distilled water, and lyophilized. Subsequently, we have found that the harsh denaturing effect of guanidine destroys a considerable amount of the biological activity of CDGF. For this reason, we now favor extraction of cartilage with either NaCl or by collagenase digestion.

Extraction with Sodium Chloride. For extraction with NaCl, homogenized cartilage is suspended in 1.0 M NaCl, 0.01 M Tris–HCl, pH 7.5 (100–200 g cartilage/800 ml extraction buffer), and stirred at 4° for 48 hr. The extract is clarified by centrifugation at 10,000 g for 30 min. The resulting supernatant containing 100,000–200,000 U of growth factor is passed through a depth filter (AP 25 prefilter, Millipore) and stored frozen at −20°.

Digestion with Collagenase. For digestion with collagenase the cartilage is finely diced by using a scalpel and is resuspended in Dulbecco's phosphat-buffered saline (PBS), pH 7.5, supplemented with 0.2% clostridial collagenase (150 U/mg, Worthington, CLS II), penicillin (200 mg/ml), and streptomycin (2000 μg/ml), and incubated with constant agitation at 37°. The incubation times depend on the source of the growth factor. For example, bovine cartilage is incubated for 12–24 hr, human for 16–18 hr, and chick for 2 hr. The digests are passed through a 153-μm mesh Nitex filter (Tetko) to remove debris and undigested material, and clarified by centrifugation at 10,000 g for 30 min. The extracts are passed through a depth filter (AP 25 prefilter, Millipore) and are either used immediately for further purification or stored at −20°.

Extraction Strategy. Digestion of cartilage with collagenase produces the highest yield of biologically active CDGF and represents the method of choice for relatively small amounts of cartilage (about 100–500 g) obtained from the different scources. For the extraction of large quantities of cartilage (over 500 g) the cost of collagenase needed becomes prohibitive. In this situation, we prefer the extraction with NaCl—even though the total yield is usually 20–50% lower.

Growth Factor Assays. Growth factor activity is determined by measuring the ability of samples to stimulate the incorporation of [³H]thymidine into the DNA of confluent quiescent BALB/c 3T3 cells (Ref. 1; for details, see Shing *et al.* [4] this volume). One unit of activity is defined as the amount of growth factor necessary to stimulate half-maximal DNA synthesis in 96-well microliter plates.

Cation-Exchange Chromatography on Bio-Rex 70

CDGF is a very basic protein (pI 9.5–10) and therefore can be purified to a great extent using the cation exchange resin Bio-Rex 70. Bio-Rex 70 (1 lb, Bio-Rad Laboratories) is first converted to the sodium form by suspending the resin in 2000 ml of 0.5 M NaCl, 0.5 M Tris–HCl, pH 10. The slurry is stirred overnight using a motor-driven stirrer (Heller 4 blade propeller, model GT-21, Thomas Scientific), and then washed with batches of 0.1 M Tris–HCl, pH 7.5 (4000 ml), until the pH has stabilized.

The collagenase extracts of cartilage can be applied directly onto Bio-Rex 70 cation exchanger resin (90,000 U/100 ml resin) equilibrated in 0.1 M Tris–HCl, pH 7.5. Before applying the 0.01 M Tris–HCl, pH 7.5, 1 M NaCl extracts, their ionic strength has to be reduced to 0.1–0.2 M NaCl either by dilution with at least 4 vol of 0.01 M Tris–HCl, pH 7.5, or by dialyzing against a known volume of 0.01 M Tris–HCl, pH 7.5.

If relatively small amounts of cartilage extracts (100–300 ml, up to 10 g) are to be processed, the extracts are applied at 4° onto columns (2.5 × 20 cm) containing Bio-Rex 70 equilibrated with 0.1 M NaCl, 0.01 M Tris–HCl, pH 7.5. The Bio-Rex 70 columns are washed with five column volumes of equilibration buffer. Subsequently, the growth factor is eluted at a flow rate of 60 ml/hr with a linear gradient of 0.1 to 1.0 M NaCl, 0.01 M Tris–HCl, pH 7.5 (six column volumes). The concentration of NaCl required to elute CDGF (~0.4–0.5 M) is calculated by measuring the conductivity in milliohms and comparing the value to a calibrated curve of milliohms versus NaCl concentration.

For large-scale preparations (500–5000 ml of cartilage extract, over 10 g) a batch adsorption and elution technique is used. About 150–1500 ml of Bio-Rex 70 resin, equilibrated in 0.1 M NaCl, 0.01 M Tris–HCl, pH 7.5, is stirred overnight with the cartilage extract. The resin is poured into a column (6 × 70 cm), where it is washed with equilibration buffer until the absorbance at 280 nm reaches baseline. Subsequently, the growth factor is eluted from the resin with one column volume of 0.6 M NaCl, 0.01 M Tris–HCl, pH 7.5.

Affinity Chromatography on Heparin–Sepharose

CDGF has a marked affinity for heparin (Ref. 4; also see Klagsbrun et al. [8], Vol. 147, this series). As a result CDGF enriched by cation exchange chromatography on Bio-Rex 70 can be purified to homogeneity by affinity chromatography on immobilized heparin. Typically the active fractions from a Bio-Rex 70 column are pooled (50–100 ml; 100,000–200,000 U) and are directly applied to a column (0.9 × 8 cm, 5 ml) of heparin–Sepharose (Pharmacia) equilibrated with 0.6 M NaCl, 0.01 M Tris–HCl, pH 7.5. Chromatography is performed at 4° at a flow rate of 30 ml/hr. The resin is washed with the equilibration buffer until the absorbance at 280 nm reaches baseline, after which a gradient of 0.6 to 2.5 M NaCl, 0.01 M Tris–HCl, pH 7.5 (200 ml total volume), is applied. The CDGF elutes within a range of 1.6–1.8 M NaCl as a homogeneous preparation. The apparent molecular weight of CDGF isolated from human, bovine, and chick cartilage is 18,000–20,000 as determined by SDS–polyacrylamide gel electrophoresis (Fig. 1). Homogeneous CDGF has a specific activity of 5 U/ng. From 500 g of cartilage approximately 10 μg (= 50,000 U) of homogeneous CDGF is obtained.

Recycling Affinity Chromatography on Heparin–Sepharose

An alternative to the two-step purification scheme (Bio-Rex 70 chromatography, followed by heparin–Sepharose chromatography) is the method of recycling cartilage-derived growth factor on heparin–Sepharose two to three times. The strong affinity of CDGF to heparin makes it possible to produce a highly purified preparation of CDGF by using repeated chromatography on columns of heparin–Separose alone without previous cation exchange chromatography.[4]

Cartilage extracts (200–800 ml, 5–20 g) dialyzed against column equilibration buffer, 0.6 M NaCl, 0.01 M Tris–HCl, pH 7.5, are applied to a column of heparin–Sepharose (160 ml, 5 × 8 cm) at a flow rate of 60 ml/hr. Alternatively, the cartilage extract (15 g/300 ml of equilibration buffer) is adsorbed to heparin–Sepharose batchwise by shaking it with 160 ml of heparin–Sepharose overnight at 4°, after which the supernatant is discarded and the heparin–Sepharose is mixed with 1 vol of equilibration buffer and poured into a column (5 × 8 cm). In either case the column is washed with the equilibration buffer until the absorbance at 280 nm reaches baseline, after which 300 ml of 3.0 M NaCl, 0.01 M Tris–HCl, pH 7.5, is applied to elute the CDGF. Starting with 10 g of guanidine extract, approximately 10 mg of partially purified CDGF containing 100,000 U of growth factor activity is recovered.

The active fractions from the batch heparin–Sepharose column are pooled and the NaCl concentration is lowered to 0.6 M NaCl by diluting with 0.01 M Tris–HCl, pH 7.5, or by dialyzing against 0.01 M Tris–HCl, pH 7.5. The concentration of NaCl is ascertained by measuring the con-

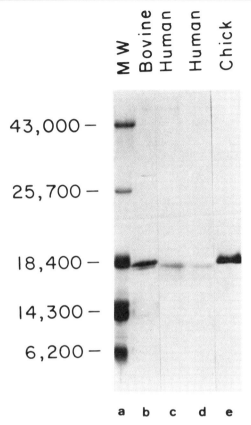

FIG. 1. Sodium dodecyl sulfate–polyacrylamide gel electrophoresis (SDS–PAGE) of purified cartilage-derived growth factors from different sources. (a) Molecular weight markers; (b) bovine CDGF; (c,d) human CDGF; (e) chick CDGF. Bovine CDGF (lane b) and human CDGF (lane c) were purified by cation exchange chromatography on Bio-Rex 70, followed by affinity chromatography on heparin–Sepharose. Human CDGF (lane d) and chick CDGF (lane e) were obtained by recycling affinity chromatography on heparin–Sepharose.

ductivity. The active pool is then applied to a second heparin–Sepharose column (0.9 × 8 cm, 5 ml) at a flow rate of 35–40 ml/hr. The column is washed with 250 ml of equilibration buffer and a 0.6 to 3.0 M NaCl gradient (400 ml total volume) is applied. The CDGF elutes with 1.6–1.8 M NaCl resulting in a 150,000-fold purification and a homogenous growth factor preparation (Fig. 1, lanes d and e) active at approximately 3–5 U/ng. The recovery is about 30%.

Section V

Techniques for the Study of Growth Factor Activity:
Assays, Phosphorylation, and Surface Membrane Effects

[31] Assay of Mitogen-Induced Effects on Cellular Incorporation of Precursors for Scavenger, *de Novo*, and Net DNA Synthesis

By ROBERT B. DICKSON, SUSAN AITKEN, and MARC E. LIPPMAN

Introduction

In recent years the study of the mechanisms of cellular growth regulation by steroid hormones and polypeptide growth factors has been a focal point of study in the areas of pharmacology, cell biology, endocrinology, and cancer research. Much of this research has centered on the regulation of mitogen production and on receptors for mitogen response. Another major area of importance is the determination of possible metabolic mechanisms mediating mitogen action. This consideration is virtually ignored in the rush to study the ligands and receptors themselves; most investigators rely on a very simple assay for mitogen-stimulated growth which involves radioactive thymidine incorporation into acid-insoluble cellular macromolecules.

Although such a thymidine incorporation assay probably remains the most common method employed to estimate DNA synthesis, interpretation of data obtained with this highly specific activity trace is complex.[1,2] First, the accurate determination of the true specific activity of labeled precursor in each experimental situation is critical. Second, thymidine nucleotides allosterically regulate a number of key enzymatic steps in pyrimidine synthesis and thus can affect its own utilization. Third, it is possible that metabolism of labeled precursor may result in incorporation into material which is not DNA. Fourth, deoxyribonucleotide pools may be compartmentalized intracellularly; consequently differential incorporation of exogenous salvage and *de novo* derived thymidine may obscure the determination of net DNA synthetic rates. Figure 1 illustrates some of the complexities of the pyrimidine synthetic and utilization pathways in mammalian cells.

A practical and accurate analysis of the role of the salvage pathway of pyrimidine synthesis and utilization is especially critical in the context of experiments directed toward assessing mitogenic responses. This chapter will be organized as a systematic analysis of the effects of a mitogenic hormone, 17β-estradiol (E$_2$), and its antagonist or antihormone, tamoxifen

---

[1] R. L. P. Adams, *Exp. Cell Res.* **56**, 55 (1959).
[2] D. Kuebbing and R. Werner, *Proc. Natl. Acad. Sci. U.S.A.* **72**, 3333 (1975).

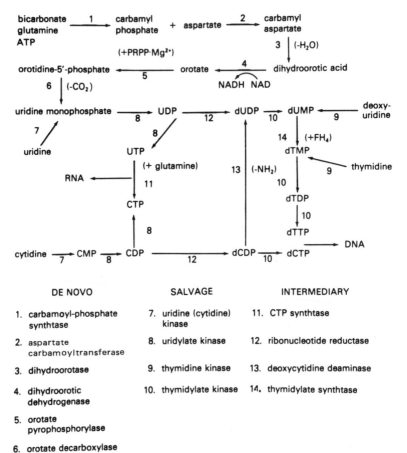

DE NOVO

1. carbamoyl-phosphate
   synthtase

2. aspartate
   carbamoyltransferase

3. dihydroorotase

4. dihydroorotic
   dehydrogenase

5. orotate
   pyrophosphorylase

6. orotate decarboxylase

SALVAGE

7. uridine (cytidine)
   kinase

8. uridylate kinase

9. thymidine kinase

10. thymidylate kinase

INTERMEDIARY

11. CTP synthtase

12. ribonucleotide reductase

13. deoxycytidine deaminase

14. thymidylate synthtase

FIG. 1. Pathways of pyrimidine synthesis in mammalian cells. PRPP, Phosphoribosyl pyrophosphate. From Ref. 19.

(tam),[3] on incorporation of various precursors into DNA in the MCF-7 human breast cancer cell line. These experiments logically followed initial demonstrations that E_2 and tam, respectively, stimulated and inhibited growth of these cells *in vivo* and *in vitro*.[4-7] We will first discuss conditions

[3] Tamoxifen, [1-(4-β-dimethylaminoethanoxyphenyl)-1,2-diphenylbut-1-ene.

[4] M. E. Lippman, G. Bolan, and K. Huff, *Cancer Res.* **36**, 4595 (1976).

[5] B. Katzenellenbogen, M. J. Norman, R. L. Eckert, S. W. Peltz, and W. F. Mangel, *Cancer Res.* **44**, 112 (1984).

[6] H. D. Soule and C. M. McGrath, *Cancer Lett.* **10**, 177 (1980).

[7] P. Darbre, J. Yates, J. Curtis, and R. J. B. King, *Cancer Res.* **43**, 349 (1983).

for valid use of [^{3}H]dThd incorporation in measuring the scavenger (salvage) pathway. Results will be compared to measurements of net DNA synthesis as monitored by ortho [^{32}P]phosphate incorporation into DNA. Finally we will develop an assay for *de novo* pyrimidine biosynthesis. These experiments should be useful to any investigator who wishes to establish the effects of virtually any mitogen on pyrimidine nucleotide pool size, salvage, or *de novo* synthesis.

Materials

A hormonally responsive subline of MCF-7 cells (originally obtained from the Michigan Cancer Foundation[6])

Richter's IMEM (Gibco, Grand Island, NY) containing phenol red and 10% fetal calf serum (Gibco)

Richter's IMEM containing phenol red, 2.5% charcoal-treated calf serum,[6] and 10^{-7} *M* insulin (Lilly Biologicals, Chicago, IL[8])

Richter's IMEM-X special formulation lacking asparagine and reduced in phosphate (to 10^{-5} *M*) and containing phenol red

Dulbecco's phosphate-buffered saline (Gibco)

Multiwell tissue culture dishes (Flow Laboratories, Inc., Rockville, MD)

17β-estradiol (Steraloids, Inc., Pawling, NJ) 10^{-5} *M* stock in ethanol

Tamoxifen (tam) (ICI Limited, Washington, DE) 10^{-3} *M* stock in ethanol

Insulin (Lilly Biologicals, Chicago, IL)

[methyl-^{3}H]Deoxythymidine (or [^{3}H]dThd) (NEN, Boston, MA)

[2-^{14}C]Acetate, 45–60 mCi/mmol (Amersham-Searle, Arlington Heights, IL)

Ortho [^{32}P]phosphate, carrier free (ICN, Irvine, CA)

Millipore filters (type HA) and filtration manifold (Millipore Corp, Bedford, MA)

Perchloric acid (Fisher Scientific Co., Fair Lawn, NJ)

RNase (types 1 and T$_1$) (Boehringer-Mannheim Biochemicals, Indianapolis, IN)

[^{14}C]Glucose, 40–60 mCi/mmol (New England Nuclear, Boston MA)

[^{14}C]NaHCO$_3$, 40–60 mCi/mmol (New England Nuclear, Boston MA)

[U-^{14}C]Aspartate, 200 mCi/mmol (Amersham-Searle, Arlington Heights, IL)

[6-^{14}C]Orotate, 40–60 mCi/mmol (Amersham-Searle, Arlington Heights, IL)

Pronase (Calbiochem-Behring Corp, Gaithersburg, MD)

Hydroxylapatite (Bio-Rad Laboratories, Rockville Centre, NY)

Aquasol (New England Nuclear, Boston, MA)

[8] A. Richter, K. K. Sanford, and V. J. Evans, *J. Natl. Cancer Inst. (U.S.)* **49**, 1705 (1972).

Hydrofluor (National Diagnostics, Inc., Somerville, NJ)
Isoton (Coulter Electronics, Inc., Hialeah, FL)

Equipment
Beckman liquid scintillation counter (Beckman Instruments, Inc., Irvine, CA)
Cell counter (Coulter model E; Coulter Electronics)
Sonicator (sonifier cell disrupter 350 with microtic probe; Branson Sonic Power Co., Danbury, CT)
Centrifuge (high-speed serofuge from Clay Adams, Parsippany, NJ)

Procedures

Tissue Culture Techniques

MCF-7 cells were maintained in monolayer culture in IMEM,[8] supplemented with twice the usual concentration of glutamine, penicillin, and streptomycin and with 10% fetal calf serum. These cells were repeatedly shown to be free of *Mycoplasma* contamination during the course of this study. Two passages prior to a given experiment, cells were placed in IMEM supplemented with 10^{-7} mol insulin/liter with 2.5% charcoal-treated calf serum shown to be essentially free of estradiol.[6] Upon reaching confluence, cells were suspended in trypsin solution (0.5% trypsin; 0.02% EDTA) and plated replicately in plastic multiwell tissue culture dishes in IMEM plus 2.5% charcoal-treated calf serum plus 10^{-7} mol insulin/liter. When plates became 60–70% confluent, medium was replaced with IMEM containing a reduced concentration of phosphate (10^{-5} mol/liter) and lacking asparagine. This nonessential amino acid was removed from medium to facilitate labeling of DNA with anticipated precursors of *de novo* pyrimidine synthesis used in some experiments. This medium will be referred to as IMEM-X. After 4–12 hr, medium is again replaced with IMEM-X plus or minus E_2 (5×10^{-9} M) or the antiestrogen tam (10^{-6} M). Cells are labeled *in situ* with 1–2 μCi/ml carrier-free ortho [^{32}P]phosphate (ICN, Irvine, CA) and 5–10 μCi [^{14}C]acetate (Amersham-Searle, Arlington Heights, IL) for 6 or 8 hr prior to harvest. [^{3}H]dThd (1–2 μCi) is administered 2 hr before harvest. In the experiments presented, cells were not deliberately synchronized in the cell cycle prior to labeling. However, the two serum-free medium changes induced a limited degree of synchrony.

The methods presented in this chapter should be generally applicable to cell culture systems other than MCF-7 provided that reduced phosphate, thymidine, and asparagine medium is not cytotoxic. However, it may be necessary to develop a serum-free test system for the study of some mitogens whose activities are also apparent in serum.

Analytical Methods

For cell counts, cells are suspended in Dulbecco's phosphate-buffered saline to 0.02% EDTA and vortexed. Aliquots (500 μl) are dispersed in 10 ml Isoton and counted in a cell counter. Alternatively, the cell pellet is suspended $(0-4°)$ in 0.6 to 1.0 ml of double glass-distilled water and disrupted with five 1-sec bursts of a sonicator at its lowest setting. Aliquots are taken from the sonicate for protein determination[9] and for collection of total acid-precipitable material on Millipore filters.[10] The acid-soluble fraction is obtained by treatment with 5% perchloric acid $(0-4°)$ and centrifugation at 1500 g for 10 min. Determination of [^{3}H]dThd (scavenger pyrimidine probe) incorporation could be based on total acid-precipitable radioactivity. However, purification of cellular DNA is required in studies of incorporation of [^{14}C]acetate (*de novo* pyrimidine pathway probe) or ortho[^{32}P]phosphate (net DNA synthesis probe). DNA purification is performed as follows: an aliquot of the cellular sonicate is incubated[11] sequentially with 75 μg pancreatic RNase (type I), 7.5 μg T_1 RNase, and 150 μg pronase per milliliter of sonicate $(0-4°,$ 1 hr each).[11] Following incubations, a portion of this material is analyzed fluorometrically for DNA content.[12] A second portion is chromatographed on hydroxylapatite columns,[11,13] precipitated on Millipore filters, and counted as for acid-precipitable material. Occasionally, the column eluate is acid precipitated, hydrolyzed, and specific nucleotides are isolated by thin-layer chromatography as described below. Protein and RNA account for less than 1% of radioactivity detected following acid precipitation; also, approximately 95% of label could be detected (following correction for recovery) in individual nucleotide components after thin-layer chromatography.[14] Cellular-soluble phosphate content must be measured in order to determine the specific activity of phosphorylated compounds incorporated into DNA. A 5% perchloric acid-soluble supernatant is prepared as described above. The supernatant is analyzed for total phosphate content by a modification of the Fiske–Subbarow procedure.[15] A portion of this fraction is also taken for determination of incorporation of ^{32}P. These data will provide a value for the specific activity of the intracellular phosphate pool.[16,17]

[9] O. H. Lowry, N. J. Rosebrough, A. L. Farr, and N. J. Randall, *J. Biol. Chem.* **193**, 265 (1951).

[10] C. Lee, L. Oliver, E. L. Coe, and R. Oyasu, *J. Natl. Cancr Inst. (U.S.)* **62**, 193 (1979).

[11] W. Meinke, D. A. Goldstein, and M. A. Hall, *Anal. Biochem.* **58**, 82 (1974).

[12] C. K. Brunk, K. C. Jones, and T. W. James, *Anal. Chem.* **92**, 497 (1979).

[13] G. G. Markov and I. G. Ivanov, *Anal. Biochem.* **59**, 555 (1971).

[14] S. C. Aitken and M. E. Lippman, *Cancer Res.* **42**, 1727 (1982).

[15] C. H. Fiske and Y. Subbarow, *J. Biol. Chem.* **66**, 375 (1925).

[16] Z. F. Chmielewicz and G. Rosen, *Arch. Biochem. Biophys.* **25**, 262 (1950).

[17] K. Randerth and E. Randerth, *J. Chromatogr.* **16**, 111 (1964).

Thin-layer chromatography is utilized to isolate dThd and cytidine derivatives from acid-soluble, DNA, and total acid-precipitable fractions.[1] Differential acid hydrolysis (5% perchloric acid, 90°, 1 hr) of acid-precipitable and DNA fractions is followed by neutralization with KOH, centrifugation, and lyophilization of the supernatant.[18] The acid-soluble fraction is directly neutralized and lyophilized. The resulting concentrates are suspended in 50 to 100 μl of 0.01 N HCl for thin-layer chromatography on polyethyleneimine cellulose plates. Appropriate standards are employed as markers. The procedures for elution, extraction, and quantification by UV absorption of dTMP, dUMP, and dCMP are described in detail elsewhere.[18,19] Determinations of specific activities of nucleotide fractins labeled with ortho[^{32}P]phosphate[14,20] or [^{14}C]acetate are then based on the ratios of radioactivity to total quantity of material.

Radioactivity associated with the above cellular fractions is determined in a Packard liquid scintillation counter with windows set to isolate ^{14}C, ^{3}H, and ^{32}P channels with maximum efficiency and minimal cross over. Aquasol or Hydrofluor are used as liquid scintillants.

Results

[^{3}H]dThd Incorporation and the Scavenger Pathway

The accurate use of radioactive dThd in monitoring rate of DNA synthesis is only possible under the following conditions: (1) exogenous dThd must be in equilibrium with intracellular precrusor pools; (2) the true specific activity of thymidine in the system must be known; and (3) the perturbation of the system due to addition of trace must be appropriately controlled. A series of experiments addressing these conditions will be presented below.

The equilibration of exogenous [^{3}H]dThd with precursor pools for DNA synthesis is examined by following the time course of incorporation of label into acid-soluble pools and DNA[19] (Fig. 2). At time 0 [^{3}H]dThd is added to all wells of multiwell culture dishes and cells are harvested at varying times thereafter. Constant radioactivity is seen in the acid-soluble intracellular pool within 20 min of exposure to radioactive trace. No significant differences in the amount of label accumulating at steady state due to E$_2$ or tam treatment are seen[20] (data not shown). Incorporation of

[18] K. K. Tsaboi and T. D. Price, *Arch. Biochem. Biophys.* **81**, 223 (1959).
[19] S. C. Aitken and M. E. Lippman, *Cancer Res.* **43**, 4681 (1983).
[20] M. E. Lippman and S. C. Aitken, in "Hormones and Cancer" (S. Iacobelli, R. J. King, H. R. Linder, and M. E. Lippman, eds.), Vol. 14, p. 3. Raven Press, New York, 1980.

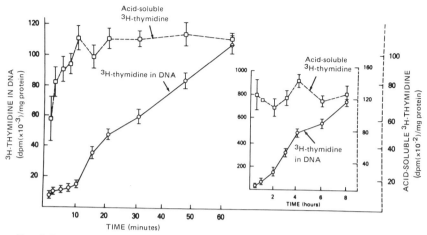

FIG. 2. Incorporation of [³H]dThd into DNA (acid-precipitable) and acid-soluble pool of MCF-7 cells. Cells are plated as described and placed in IMEM-X medium 24 hr prior to addition of fresh IMEM plus [³H]dThd (time 0). Two separate experiments representing a short period of study (1–60 min) and a longer time period (0.5–8 hr) are depicted. The latter is shown as an inset. In both experiments values are normalized per unit protein and are the average of three determinations ±SD. DNA (circles) and acid-soluble (squares) fractions were isolated as described. From Ref. 20.

[³H]dThd into acid-precipitable material is also linear within 20 min (normalizing to either DNA or protein); this observation indicates that equilibrium of exogenous dThd with endogenous dThd precursor pools for DNA synthesis is achieved within this interval. Equilibrium conditions are maintained thereafter for at least 8 hr. It should be noted that the medium employed in this study is otherwise devoid of thymidine.[8]

Under equilibrium conditions ortho[³²P]phosphate can also be used as a precursor for DNA synthesis and, indeed, as an index of net DNA synthesis. However, one might anticipate the equilibration at the α-phosphate position of deoxynucleotides will require a longer time period than is necessary with a [³H]dThd label. We could demonstrated that constant radioactivity in the acid-soluble fraction of MCF-7 cells is achieved after 6 hr under all experimental conditions (control, E_2, tam). Incorporation of ³²P into DNA becomes linear within approximately 4 hr; addition of E_2 or tam again failed to alter this time requirement.[14]

We next conducted an experiment to determine the effect of extended incubations of cells with E_2 and tam on incorporation of these precursors ([³H]dThd and ortho[³²P]phosphate) in DNA. Duration of labeling is based upon the steady state conditions established earlier. As anticipated, ortho[³²P]phosphate incorporation (net DNA synthesis) is cyclically stimulated by E_2 (at 24 and 48 hr; S phase) and inhibited by tam. Paradoxically, however, [³H]dThd incorporation is stimulated by both E_2 and tam. In addition, E_2 only stimulates [³H]dThd incorporation during the first wave of DNA synthesis. In order to confirm that ³²P incorporation accurately reflects DNA synthesis we also examined hormonal effects on total phosphate metabolism. Specific activity of [³²P]P_i is directly measured (see methods) in acid-soluble fractions at 24 and 48 hr in control, E_2, and tam groups. ³²P disintegrations per minute are then converted into P_i mass units. The results (not shown) indicate minimal effects of hormone treatment on intracellular ³²P specific activity. It appears that hormonal treatment more profoundly affects phosphate incorporation into DNA than total phosphate metabolism. E_2 stimulates net DNA synthesis 160% at 24 hr and 180% at 48 hr, while tam inhibits net DNA synthesis to 55% at 48 hr (no effect at 24 hr).[14] Thus after verification of the effects of hormones on net DNA synthesis by taking into account specific activity of ³²P within the cell, the [³H]dThd incorporation data remain paradoxical.

Two approaches can be taken to resolve the disparity between [³H]dThd incorporation and growth response. One can mathematically model additional kinetic data obtained with [³H]dThd as described immediately below. This method will, however, be presented only in a summary descriptive manner; a complete description may be found elsewhere.[20] The second approach will be discussed in detail: development of biochemical techniques to quantify *de novo* pyrimidine synthesis.[19] In the case of MCF-7 cells treated with E_2 or tam, paradoxical data are ultimately resolved by either technique as resulting from major effects on *de novo* pathways of pyrimidine utilization in addition to scavenger [³H]dThd utilization. The reader is thus alerted to the pitfalls of relying only on [³H]dThd incorporation assays as a marker for mitogen action.

Mathematical Modeling of [³H]dThd Utilization

Accurate analysis of mitogenic effects on [³H]dThd incorporation requires information on both the intracellular specific activity of [³H]dThd and the potentially variable equilibration period of exogenous label. Although linearity of DNA incorporation can be estimated in Fig. 2 by visual inspection of the data, more rigorous treatment of the data has been described.[20] Specifically, the data for acid-soluble pools in [³H]dThd experiments fit a two-compartment model developed by Cooper[21] in which the rate of entry and exit of label are approximatel equal. This computer-assisted model can be used to derive both the initial velocity of incorpora-

[21] H. L. Cooper, *Anal. Biochem.* **53,** 49 (1973).

tion and the time required for dThd pool equilibration.[20] In addition, a linear regression analysis can be used with raw data to determine the initial velocity of incoporation of label into DNA.[20] A modified isotope dilution technique can then be applied to determine the actual mass of dThd in intracellular (or extracellular) pools.[22] This method utilizes three to four concentrations of nonradioactive thymidine with constant [^{3}H]dThd to estimate intracellular pool size.[20]

In MCF-7 cells the addition of [^{3}H]dThd in the range of 1–2 μCi (2–5 × 10^{-8} M) contributes only 5–10% of total thymidine nucleotides in intracellular pools and DNA; furthermore, pool sizes vary considerably over time in serum-free medium.[20] It might be anticipated that treatment with mitogens will introduce further variability in this system. Consequently [^{3}H]dThd incorporation data expressed in disintegrations per minute are likely to provide inaccurate estimates of actual DNA synthesis unless pool sizes are considered.[20] Coupling of the above computer-assisted two-compartment model for pool size and linear regression analysis of incorporation into DNA with isotope dilution studies[23] may, however, resolve the question of whether extracellular salvage or *de novo* synthesized dThd are the primary targets of mitogen action. Results of such a study have been reported.[20] After tam treatment thymidine nucleotide pool size was reduced to as little as 20–25% control. In contrast, E$_2$ stimulated thymidine nucleotide pool size to 200–300% of control. Consideration of these findings helps to explain paradoxical effects of tam (a growth inhibitor) on stimulation of [^{3}H]dThd incorporation. In addition, these findings suggest that E$_2$ and tam may chiefly regulate *de novo* pyrimidine synthesis. This analysis also led to the conclusion that substantial metabolism of the acid-soluble pool (to non-DNA products) occurred.[20] Calculation of the rate of exit of total thymidine from the intracellular pool into DNA shows that E$_2$ stimulates incorporation 2 to 3-fold and tam inhibits incorporation 60%. A direct labeling technique to verify this prediction is presented in the next section.[19]

[^{14}C]Acetate as a Precursor for de Novo Pyrimidine Biosynthesis

A number of conventional precursors such as orotate, aspartate, and bicarbonate have been used to measure *de novo* pyrimidine synthesis.[24-26]

---

[22] D. S. Scudiero, E. Henderson, A. Novin, and B. Strauss, *Mutat. Res.* **29**, 473 (1975).
[23] S. C. Aitken and M. E. Lippman, unpublished.
[24] S. Guller, P. C. Smith, and G. C. Trembley, *Biochem. Biophys. Res. Commun.* **56**, 934 (1974).
[25] M. G. Ord and L. A. Stocker, *Biochem. J.* **129**, 175 (1972).
[26] R. A. Yates and A. B. Pardee, *J. Biol. Chem.* **221**, 743 (1956).

TABLE I
INCORPORATION OF *de Novo* PRECURSORS INTO MCF-7 CELLS[a]

		Incorporation		
			DNA	
Precursors	Acid soluble (dpm)	Acid precipitable (dpm)	dpm	pmol
NaH[14]CO₃	1,430	10,080	315	4,970
[6-^{14}C]Orotate	13,700	15,990	1,920	16
[U-^{14}C]Aspartate	58,830	100,240	2,160	63
[2-^{14}C]Acetate	274,420	658,340	26,540	214

[a] MCF-7 cells are labeled for 12 hr with each of the indicated precursors at 10 μCi/ml. Acid-soluble, acid-precipitable, and DNA fractions are isolated as described (Materials and Methods). Incorporation into DNA is converted to picomoles, correcting for the specific activity of each label in IMEM-X. Values are the mean of three determinations, normalized per unit of protein.[19]

Unfortunately, the MCF-7 cell line fails to incorporate enough of these tracers to detect potential experimental differences (Table I). Only acetate label appears in the DNA fraction to any significant degree. On conversion of dpm units to mass units (based on the calculated specific activity of these tracers in IMEM-X), it is clear that large quantities of bicarbonate do, in fact, accumulate in DNA. However, the experimental detection of these precursors would be extremely difficult in small amounts of cellular material. In addition, bicarbonate cannot be removed from the medium without affecting cellular viability. [^{14}C]Acetate might, however, be a potential marker for *de novo* pyrimidine synthesis and utilization. Acetate enters the cell, is rapidly converted to CO_2 (bicarbonate), enters the urea cycle, and is utilized in nucleotide biosynthesis as aspartate (Fig. 1).

We first established the time requirements for equilibration of acetate with DNA precursor nucleotide pools.[19] Incorporation into the acid-soluble pool in all cases reaches a constant level after approximately 6 hr of exposure to trace. However, incorporation into the specific precursor pool for DNA synthesis requires only 2 to 3 hr to reach steady state (indicated by a linear rate of incorporation into DNA). Examination of incorporation into MCF-7 cells between 6 and 8 hr of exposure to trace (as has been described previously for ortho[^{32}P]phosphate) will allow measurable incorporation following equilibrium of label in precursor pools.

Soluble intracellular acetate pools must be saturated with [^{14}C]acetate before the specific activity of tracer can be considered constant.[1,27] We first

[27] D. A. Sjostrand and D. R. Forsdyke, *J. Biochem.* **138**, 253 (1974).

examine the relationship between specific activity of exogenous acetate and the small pool of pyrimidine end product. If the effective specific activities of cytidine and dThd labeled with [^{14}C]acetate are identical with specific activity of added trace, a one-to-one relationship exists (1 nmol of acetate utilized = 1 nmol of pyrimidine based synthesized = 0.5 nmol of DNA base pairs synthesized). If not, appropriate corrections are necessary in experimental measurements of DNA synthesis. Duplicate samples (2×10^8 cells/sample) are labeled with [^{14}C]acetate (1 μCi/ml, 56 mCi/ mmol, 8 hr). The acid-soluble fraction is processed to isolate dTMP and dCMP components as described previously. The specific activites of the cytidine [41.6 ± 13.1 cpm/pmol (SD)] and dThd (47.5 ± 13.4 cpm/pmol) fractions are approximately 86% of those predicted on the basis of the specific activity of [^{14}C]acetate [52.1 cpm/pmol]. This direct comparison of specific activity suggests that (1) on the average, one acetate molecule is incorporated into each pyrimidine molecule and (2) one can extrapolate directly from the given specific activity of [^{14}C]acetate administered to the mass of pyrimidine synthesized. In addition, it can be demonstrated that under these conditions nearly 100% of [^{14}C]acetate label recovered from purified DNA is localized in pyrimidine nucleotides.[19]

Additional independent experiments verify precursor pool saturation with label. Variable concentrations of nonradioactive acetate (0–200 μM) may be employed in an isotope dilution technique to evaluate if [^{14}C]acetate derived from exogenous [^{14}C]glucose decreases proportionally to the mass of exogenous, competing nonradioactive acetate and is very low at concentrations of acetate used in exogenous [^{14}C]acetate labeling experiments. In these experiments the lipid fraction is isolated as a suitable metabolic product of acetate and glucose. We observe[19] (data not shown) that both of these conditions are met; dilution of specific activity of [^{14}C]acetate derived from [^{14}C]glucose is linearly related to appearance of label in end product. In addition, 50 μM acetate (the amount used in [^{14}C]acetate labeling experiments) could effectively compete (80–90%) for incorporation of [^{14}C]glucose into the lipid fraction.[19,25] Finally, a two-compartment, kinetic computer-assisted analysis may be conducted (as with [^{3}H]dThd incorporation, in conjunction with further acetate isotope dilution experiments), showing the specific activity of [^{14}C]acetate to be constant over time and independent of E_2 or tam treatment. The effective specific activities of soluble acetate label and pyrimidine label are approximately 80% of initial specific activity of trace by this analysis.

We finally demonstrate directly the correlation between net DNA synthesis (^{32}P incorporation into DNA) and de novo pyrimidine synthesis ([^{14}C]acetate incorporation into DNA over time). At various intervals after hormone treatment, we label MCF-7 cells with both precursors for either 6 or 8 hr. The data indicate that inhibition of net DNA synthesis is closely associated with inhibition of *de novo* pyrimidine synthesis. Conversely, E_2 stimulation of net DNA synthesis corrrelates with increases in rate of *de novo* biosynthesis-derived pyrimidines.[19] These data should be contrasted with the previously paradoxical data obtained comparing [³H]dThd and ³²P labels.

Subsequent studies[25,28] have shown that E_2 induces at least five enzymes in the *de novo* pathway (carbamoyl-phosphate synthase, aspartate carbamoyltranferase, orotate pyrophosphorylase, thymidylate synthase, and orotidylate decarboxylase). Tam inhibits all of these enzymes except aspartate carbamoyltransferase. It should also be noted that antiestrogen effects in cultured breast cancer cells depend on competition at the estrogen receptor for the weakly estrogenic phenol red present in the culture medium as a pH indicator.[30]

Summary and Conclusion

In summary we have presented data on [³H]dThd incorporation into DNA for a positive and negative growth modulator. These data clearly do not correspond to those based on net DNA synthesis (ortho[³²P]phosphate incorporation into purified DNA) and previous knowledge of the effects of these hormones on cell number and cellular DNA accumulation. The paradox was resolved by directly analyzing effects of hormones on *de novo* pyrimidine biosynthesis. E_2 stimulated both *de novo* and scavenger pathways in the first wave of DNA synthesis and only *de novo* in the second (and subsequent) waves. In contrast, tam progressively inhibited *de novo* pyrimidine biosynthesis. These hormonal effects on intracellular pyrimidine pool sizes rendered [³H]dThd incorporation data by itself uninterpretable. [³H]dThd is a useful measure of DNA synthesis only if verified by independent measures of net DNA synthesis for the same time course and treatment conditions. In addition it may prove beneficial in various experimental systems to develop direct assays of *de novo* pyrimidine biosynthesis to assess mitogen effects. The experiments presented may also prove to be useful in evaluating the effects of mitogens and antimitogens on cells synchronized in the cell cycle by any of a variety of means.[29]

Acknowledgments

The authors would like to thank Phyllis Rand for typing the manuscript and Dr. James D. Moyer (National Cancer Instiute, Bethesda, MD) for helpful comments.

[28] S. C. Aitken and M. E. Lippman, *Cancer Res.* **45**, 1611 (1985).
[29] T. Ashihava and R. Baserga, this series, Vol. 58, p. 248.
[30] Y. Berthois, J. A. Katzenellenbogen, and B. S. Katzenellenbogen, *Proc. Natl. Acad. Sci. U.S.A.* **83**, 2496 (1986).

[32] Soft Agar Growth Assays for Transforming Growth Factors and Mitogenic Peptides

By ANGIE RIZZINO

Many different tumor cells have been shown to produce and release growth regulatory factors that are currently referred to as transforming growth factors (TGF). TGFs are widely believed to influence the growth and behavior of the cells that produce them (autocrine growth control).[1] Consequently, the production of TGFs may have important effects on the growth of tumors *in vivo*. TGFs are so named because of their capacity to reversibly induce nontransformed cells to behave as tumor cells *in vitro*. Most notably, TGFs can induce the soft agar growth (anchorage-independent growth) of nontransformed cells. This is an important property of TGFs, since the ability to grow in soft agar is one of the best indicators of a cell's tumorigenic capacity. TGFs were first identified in medium conditioned by virally transformed 3T3 cells and were originally referred to as sarcoma growth factor (SGF).[2] The soft agar growth-promoting activity of SGF is now known to be due to two TGFs, TGFα and TGFβ, which are heat and acid stable.[3,4] Together, TGFα and TGFβ induce normal rat kidney fibroblasts to grow in soft agar, whereas each alone has little or no ability to induce anchorage-independent growth. More recently, it has been determined that (1) other growth factors can modulate the effects of TGFα and TGFβ; (2) the effects of TGFs on growth are cell line dependent; (3) TGFβ can inhibit, rather than promote, the growth of some cells; and (4) other TGFs appear to exist.

TGFs are usually assayed by their ability to induce the soft agar growth of nontransformed cells. This chapter will describe the methods employed to examine TGFs and related growth factors for their ability to influence anchorage-independent growth. Specifically, the following topics are discussed: the indicator cell lines most commonly used, the general methodology employed, the selection of appropriate culture media, and a rapid assay, which is suitable for testing small numbers of cells for the release of factors that influence anchorage-independent growth.

[1] M. B. Sporn and A. B. Roberts, *Nature (London)* **313**, 745 (1985).
[2] J. E. De Larco and G. J. Todaro, *Proc. Natl. Acad. Sci. U.S.A.* **75**, 4001 (1978).
[3] A. B. Roberts, M. A. Anzano, L. C. Lamb, J. M. Smith, and M. B. Sporn, *Proc. Natl. Acad. Sci. U.S.A.* **78**, 5339 (1981).
[4] M. A. Anzano, A. B. Roberts, J. M. Smith, M. B. Sporn, and J. E. De Larco, *Proc. Natl. Acad. Sci. U.S.A.* **80**, 6264 (1983).

Indicator Cell Lines Used

Selection of the cell lines used to assay for growth factors that influence anchorage-independent growth should be considered carefully, since the cell line used is likely to determine whether or not a TGF-related growth factor will be detected. The indicator cell line most commonly used to assay TGFs is the rat kidney fibroblast cell line (clone 49-F), which is usually referred to as normal rat kidney (NRK) cells. These cells were cloned from a mixed population by De Larco and Todaro[5] and most of the procedures described in this chapter have been developed for this cell line. However, these procedures, in particular the methods used to set up the assays, are directly applicable to other indicator cell lines. In addition, the general principles discussed in this chapter, e.g., the choice of the culture medium, are relevant when other cell lines are employed.

NRK cells were first used to monitor the purification and characterization of TGFα and TGFβ. TGFα exhibits approximately 35–40% amino acid sequence homology with EGF[6] and competes with EGF for binding to membrane receptors. In this regard, it is important to note that *EGF and TGFα are functionally equivalent in the TGF assay,* since they bind to the same receptor. Individually, TGFα and EGF are unable to induce NRK cells to form large colonies in soft agar but each is able to induce the formation of small colonies when the assay is performed in serum-supplemented medium (SSM). Large colonies are only obtained when TGFβ is added with EGF or TGFα. TGFβ binds to a receptor distinct from that of EGF and is unable to promote the anchorage-independent growth of NRK cells in the absence of TGFα or EGF.[4,7] In SSM, NRK cells form numerous colonies when EGF and TGFβ are added to the assay, whereas in plasma-supplemented medium (PSM) significantly fewer colonies form in response to EGF and TGFβ. However, if EGF, TGFβ, and PDGF are added to PSM, the number of colonies that form is increased.[8] More recently, this laboratory has determined that FGF will also potentiate the effect of EGF and TGFβ in PSM.[9] In this regard, it is noteworthy that PDGF added alone to PSM, but not FGF, induces NRK cells to form small colonies, and the same effect has been observed in a serum-free medium developed to assay TGFs[9,10] (see below).

[5] J. E. De Larco and G. J. Todaro, *J. Cell. Physiol.* **94,** 335 (1978).
[6] H. Marquardt, M. W. Hunkapiller, L. E. Hood, and G. J. Todaro, *Science* **223,** 1079 (1984).
[7] A. Rizzino, *In Vitro* **20,** 815 (1984).
[8] R. K. Assoian, G. R. Grotendorst, D. M. Miller, and M. B. Sporn, *Nature (London)* **309,** 804 (1984).
[9] A. Rizzino, E. Ruff, and H. Rizzino, *Cancer Res.* **46,** 2816 (1986).
[10] A. Rizzino, *In Vitro* **20,** 274 (1984).

Although it is sufficient under most circumstances to monitor the production of TGFs by employing only one indicator cell line, there are several good reasons for employing more than one indicator cell line. First, tumor cells often produce more than one growth factor and, therefore, the use of more than one indicator cell line will help detect the presence of the different growth factors. Similarly, the use of several indicator cell lines can help evaluate the purity of growth factor preparations. This is particularly important when working with growth factors such as TGFβ and PDGF, since both are purified from human platelets. Second, the effects of TGFs are cell dependent. Consequently, the use of different indicator cells will not only help characterize the types of TGFs that are present, but may also prove helpful when examining the molecular mechanisms of TGFs.

Several cell lines, besides NRK cells, have been used to assay TGFs, including AKR-2B cells[11] and 3T3 cells.[12] Moses and associates use AKR-2B (clone 84A) cells to monitor TGF activity. In SSM, these cells are induced by TGFβ to form colonies in the absence of added EGF,[11] which contrasts with the results obtained with NRK cells. More recently, AKR-2B cells have been shown to form colonies in response to TGFβ in a serum-free medium.[13] This serum-free medium is very similar to one previously developed for NRK cells,[7] and apparently EGF must be added with TGFβ to induce soft agar growth in serum-free medium. Thus, the growth factor requirements for soft agar growth of both cell lines in serum-free medium appear to be very similar.

At present, several other cell lines are used in this laboratory to monitor the purification and/or activity of TGFs. One is a variant of the human epidermoid carcinoma cell line A431. We have isolated and cloned a variant, A431R-1, whose growth in soft agar is *stimulated* by EGF.[14] A431R-1 cells also respond to TGFβ, but do not respond to PDGF since they lack PDGF receptors. Therefore, A431R-1 cells provide a useful indicator system for detecting TGFβ and EGF-related growth factors in preparations of PDGF. A second useful indicator cell line is a clone of NR-6 cells that was selected in this laboratory for its ability to respond to PDGF and FGF by growth in soft agar.[15] This clone, which is referred to as NR-6-R, does not respond to EGF, since it lacks EGF receptors, and exhibits little or no response to TGFβ. Our data indicate that TGFβ is able to potentiate the soft agar growth response of NR-6-R cells to suboptimal

[11] H. L. Moses, E. L. Branum, J. A. Proper, and R. A. Robinson, *Cancer Res.* **41**, 2842 (1981).

[12] K. Yamaoka, R. Hirai, A. Tsugita, and H. Mitsui, *J. Cell. Physiol.* **119**, 307 (1984).

[13] C. B. Childs, G. D. Shipley, and H. L. Moses, *J. Cell Biol.* **99**, 272a (1984).

[14] A. Rizzino, E. Ruff, and H. Rizzino, *In Vitro* **22**, 31A (1985).

[15] A. Rizzino and E. Ruff, *In Vitro* **22**, 749 (1986).

concentrations of PDGF and FGF.[15] Thus, the presence of PDGF or FGF in preparations of TGFβ (or other growth factors) may be detected by employing NR-6-R cells as an indicator cell line. (PDGF can be distinguished from FGF, since FGF is heat and acid labile but PDGF is not.) A third useful indicator cell line is A549, a human lung tumor cell line, which forms colonies in soft agar in either SSM[16] or PSM (A. Rizzino and E. Ruff, unpublished results, 1985). Under both conditions, EGF stimulates the soft agar growth of these cells, whereas TGFβ, at concentrations as low as 150 pg/ml, completely inhibits their growth, even in the presence of EGF.[16] Thus, A549 cells provide a sensitive indicator cell line for the presence of TGFβ. It is also noteworthy that these cells grow more slowly in PSM than in SSM. Unlike the results with NRK cells, we have not observed a stimulatory effect of PDGF in PSM.

NRK 49-F cells and A549 cells can be obtained from the American Type Culture Collection. A431R-1 cells and NR-6-R cells were isolated and cloned in this laboratory from A-431 cells and NR-6 cells, respectively. A431 cells were obtained from Joseph De Larco (National Cancer Institute, Bethesda, MD), as were NRK 49-F cells. NR-6 cells were obtained from Harvey Herschman (University of California at Los Angeles, Los Angeles, CA).

Methodology for Testing TGFs and Other Mitogenic Factors for the Capacity to Induce Anchorage-Independent Growth

In most cases, the TGF assay is used to test whether concentrated serum-free conditioned medium prepared from cells suspected of producing TGFs has the ability to induce soft agar growth of nontransformed cells. This section outlines the methodology involved, which generally parallels the methodology used by De Larco and Todaro.[2] This section also discusses some of the problems often encountered and offers possible solutions to these problems. In the next section, we discuss the composition of the culture medium itself.

The soft agar assay is set up using two layers of agar, which are added sequentially to a culture dish or other suitable container. The base layer is generally 0.5% agar (agar noble—Difco) in culture medium. It is prepared from a 5% sterile solution of agar (i.e., agar dissolved in water of the highest purity available and sterilized by autoclaving for 20–30 min). As needed, this solution is remelted using a boiling water bath. From this point on, the assay is set up in a tissue culture hood. After the melted agar has cooled to

[16] A. B. Roberts, M. A. Anzano, L. M. Wakefield, N. S. Roche, D. F. Stern, and M. B. Sporn, *Proc. Natl. Acad. Sci. U.S.A.* **82,** 119 (1985).

42–45°, culture medium is added to achieve an agar concentration of 0.5%. (This results in a 10% dilution of the culture medium, but does not appear to have any significant effect on the assay. However, dilution of the culture medium by more than 10% does appear to affect the growth of the cells.) After the agar and culture medium have been mixed, the base layer is added to the culture container (base layers of 1, 2–3, and 0.5 ml are used for 35-mm dishes, 60-mm dishes, and 16-mm wells, respectively). Once the base layer has solidified (usually by allowing the base layer to cool for 30 min at 4° and 10–15 min at room temperature), the second layer is added. The second layer is prepared by adding to the 0.5% agar solution, which is dispensed in polypropylene tubes, a small volume of cells, the conditioned medium/factors to be tested and additional culture medium to achieve the desired agar concentration (usually 0.3%, but see below). Since the agar will begin to solidify as the temperature drops, it is convenient to keep the tubes in a test tube rack that is partially immersed in water maintained at 42–44°. The cells are added at a final density of 1000 to 5000 cells/ml of agar. In this laboratory, the cells, which are maintained for up to 20 min at room temperature prior to use, are added last. Once the components are mixed (by rocking the tubes), the top layer is added by pipetting it on top of the solidified base layer. (Top layers of 1, 2–3, and 0.5 ml are used for 35-mm dishes, 60-mm dishes, and 16-mm wells, respectively.) It is recommended that the top layer be added to the center of the culture dish. If the top layer is added along the edges of the culture dish, some of the indicator cells may appear on the bottom of the culture dish, which will compromise the accuracy of any quantitative measurements. (This situation probably results from a slight contraction of the base layer after it solidifies.) Once the top layer has been added, it is allowed to solidify by incubating the assay container at room temperature for 20–30 min (when 35- or 60-mm dishes are used) or at 4° for 5 min and 10–20 min at room temperature (when 16-mm wells are used). The assay container is then placed at 37° in a humidified atmosphere of 95% air and 5% CO_2. It is noteworthy that in most laboratories the assay is performed in 35- or 60-mm dishes. In this laboratory, the assay is usually performed in 16-mm wells (Linbro dishes or other trays that contain 24 wells). The use of 16-mm wells has several advantages: (1) It uses a total of 1 ml for the top and base layers and this reduces the amounts of the factors used, which is important when working with expensive growth factors; (2) we use only the eight interior wells and fill the remaining wells with sterile water or PBS, which reduces the possibility of the cultures drying out during prolonged assay periods.

Under most circumstances, small colonies begin to appear after several days and continue to grow for 2–3 weeks. In the case of NRK cells, TGFβ and EGF together induce colonies of more than 60–100 cells over a period of 7–14 days. Once the colonies have formed, their number can be determined by counting enough microscopic fields to obtain statistically significant numbers. Although the number of colonies is often the main indicator of anchorage-independent growth, in some cases it is also important to determine the sizes of the colonies. This can be achieved with the aid of a grid or micrometer in the eyepiece of the microscope. The actual sizes of the colonies can be determined by comparison to the size of a known object, such as the grid on a hemocytometer, which can be easily visualized by inverting the hemocytometer on the stage of a phase-contrast microscope. (The counting chamber, which is 10^{-4} ml, is 1 mm in length.) One can also quickly count and size the colonies using an image analyzer, such as an Omnicon image analysis system or Quantimet 800.

It is clear from work with various growth factors that some growth factor combinations affect the number and size of the colonies differently than others. Specifically, some growth factor combinations induce the formation of relatively few colonies, which reach a very large size, whereas other growth factor combinations induce the formation of many colonies, whose size remains relatively small. In some cases this is due, at least in part, to limitations imposed by the culture medium, either the availability of growth factors, or of the nutrients, or both. For this reason, it is important to use caution when interpreting the results from soft agar assays. Since these effects generally become more apparent with time, it is advisable to score the assay more than once for both size and number.

Agar Selection. In most assays for TGFs, agar (agar noble) is used at 0.5% in the base layer and at 0.3% in the top layer. We have also obtained satisfactory results with agarose (SeaPlaque — FMC), but the use of agarose does have one drawback. In our experience, the colonies that form tend to accumulate in the center of the culture dish and this complicates quantitation. This problem can be minimized by incubating the assay for 5–10 min at 4° and 30 min at room temperature before the assay is placed in a CO_2 incubator. This appears to reduce the fluidity of the agarose, which fixes the cells in their original positions. Although we have not observed any significant advantages with agarose when working with NRK cells, agarose may improve the cloning efficiency of other cells. Therefore agarose should be considered as a possible alternative to agar for some cells.

Another important consideration in performing soft agar assays is the percentage of agar employed. In most laboratories, the indicator cells are suspended in 0.3% agar. However, we have determined that TGFβ plus EGF induces NRK cells to form colonies in agar concentrations at least as high as 0.6%.[17] This is of interest since clones from a fibrosarcoma cell line

[17] A. Rizzino and E. Ruff, J. *Tissue Cult. Meth.,* in press (1987).

that exhibit different metastatic potentials also exhibit different abilities to grow in soft agar: clones of high metastatic potential will grow in 0.6% agar, whereas clones of low metastatic potential form colonies in 0.3% agar but form relatively few colonies in 0.6% agar.[18] It is also noteworthy that a wider colony size distribution is observed with NRK cells in 0.3% agar than in 0.6% agar. At the lower concentration both large and small colonies are observed, whereas only large colonies are observed at the higher concentration. This may be of use when attempting to reduce the background observed in the TGF assay. The problem of background activity is discussed in the next section.

Selection of an Appropriate Culture Medium

Selection of the culture medium is an important step when first setting up an assay for TGFs. It is important since some culture media support significant soft agar growth of indicator cells in the absence of TGFs. This *background* activity is observed with most serum lots when NRK cells are used. In addition, this background becomes more pronounced when EGF (in the absence of other growth factors, such as TGFβ) is added to the assay. In this section, several ways to avoid or minimize the problem of high backgrounds are discussed.

Synthetic Portion of Culture Medium. In many laboratories, Dulbecco's modified Eagle's medium (DME—Gibco, #430-2100) is used as the synthetic portion of the culture medium. In this laboratory, DME is replaced with a 1:1 mixture of DME and Ham's F12 (nutrient mixture—Gibco, #430-1700) to which we add 2.4 g/liter of sodium bicarbonate, HEPES buffer (15 mM, pH 7.4), sodium selenite (25 nM), and antibiotics (penicillin, 50 U/ml, and streptomycin, 50 μg/ml). This 1:1 DME/F12 mixture was initially employed during development of the serum-free medium used to assay TGFs.[7] For consistency, this synthetic medium is used in this laboratory for all assays performed in serum-supplemented, plasma-supplemented, and serum-free media. It is prepared from powdered medium, stored at 4°, and used within 2 weeks.

Serum-Supplemented Medium (SSM). The TGF assay is usually performed in medium supplemented with bovine calf serum. Since most serum lots (>80%) support relatively high backgrounds, one way to minimize the background problem is to screen several serum lots and select a lot that supports little background activity. Fetal bovine serum is rarely used, since it almost always has a higher background activity than calf serum when NRK cells are used as the indicator cells. Although it is far from clear, it is likely that the background activity is due, at least in part, to

[18] M. A. Cifone and I. J. Fidler, *Proc. Natl. Acad. Sci. U.S.A.* 77, 1039 (1980).

the presence of low, but significant, levels of TGFβ. (Serum is expected to contain someTGFβ, since it is present in platelets and would be released during the preparation of serum.)

A related issue is the finding that some clones of NRK cells exhibit less background growth than others. Thus, it is useful to carefully select the cells, as well as the serum, to be used in the assay. One can quickly screen either the cells or the serum for high backgrounds by growing the cells under three conditions: (1) culture medium not supplemented with growth factors, (2) culture medium plus EGF (10 ng/ml), and (3) culture medium plus EGF and TGFβ (250 pg/ml). Those clones and serum lots that support the fewest colonies under the first two conditions, while responding strongly to EGF and TGFβ, are selected for further use. Once a clone has been selected, it should be stored under liquid nitrogen in a large number of vials. (NRK cells can be frozen at $1-2 \times 10^6$ cells/ml in medium supplemented with 10% DMSO.) It is also noteworthy that as the passage number of certain clones increases so does their background soft agar growth. Our preliminary evidence indicates that this is due, at least in part, to the release of a growth factor(s), possibly TGFβ.[17] Thus, it is recommended that the indicator cells be periodically checked (every 5-10 passages) to be certain that their properties have not changed. In our experience, some change in the response of the cells to EGF and/or TGFs occurs after 10-15 passages.

Once a low background serum lot has been identified, it is best to purchase enough of the serum for the entire project. For prolonged periods (>6 months) the serum should be stored at $-80°$. Before the serum is used in the assay, it is heat inactivated at 56° for 30 min. It is also advisable to determine the optimum serum concentration to be used in the assay. This can help reduce the background growth of the indicator cells even further. Our recent results with one serum lot illustrate this point. At high serum concentrations (e.g., >20% calf serum), EGF induced the same number and size of colonies as did EGF plus TGFβ. At 10% serum, EGF alone induced far fewer and smaller colonies. At 6-8% serum, there was relatively little response to EGF alone but the response to EGF and TGFβ was essentially the same as that observed at higher serum concentrations. For the purpose of the TGF assay, the serum concentration giving the greatest differential between EGF alone and EGF plus TGFβ probably provides the best conditions.

Plasma-Supplemented Medium (PSM). Another method of reducing the background activity in soft agar growth assays is to replace SSM with medium supplemented with bovine calf plasma (Irving Scientific). (Since plasma is likely to contain far less TGFβ than serum, assays in PSM should exhibit less background activity than assays in SSM.) NRK cells in PSM do not form any colonies unless an appropriate combination of TGFs is added.[9] NRK cells in PSM also do not form colonies in soft agar when EGF and TGFβ are added individually. When EGF and TGFβ are both added to PSM, some colonies appear, and this response is significantly potentiated if PDGF[8] and/or FGF[9] is added. We have also observed that PDGF added alone to PSM induces the formation of small colonies, and the size of these colonies is increased when EGF is added; whereas FGF, alone or in combination with EGF, does not induce the formation of NRK cell colonies in PSM.[9] Thus, PSM plus FGF provides a useful alternative to SSM for assaying TGFs such as TGFβ and TGFα.

Plasma should be heated (56°, 30 min) and dialyzed prior to its use in TGF assays. Dialysis not only removes the sodium citrate added during the preparation of the plasma, but also eliminates the low background response of the cells to EGF that is observed in undialyzed plasma. We dialyze the plasma at 4° against five changes of DME containing antibiotics and sodium bicarbonate. Dialysis against PBS should be avoided, since it appears to yield plasma that is toxic in the TGF assay. After dialysis, the plasma is spun to remove a noticeable precipitate, filtered with a 0.22-μm filter, and stored at −20°. Just prior to use, an appropriate amount of plasma is refiltered with a 0.22 μm filter to remove any remaining debris. Any refiltered plasma not used in the assay is discarded. (It is also of interest that one can reduce the background observed with some batches of calf serum by dialyzing the serum against culture mdium.) As is the case for serum, it is best to test several plasma lots, since some are inhibitory.

Serum-Free Medium. Another means of eliminating the background activity in the TGF assay is to use a serum-free assay[7] that was developed to aid in the characterization of TGFs and to help identify all the growth factors required for the response to TGFs. In place of serum, the medium is supplemented with bovine insulin (10 μg/ml—Sigma), human transferrin (5 μg/ml—Sigma), human high-density lipoprotein (HDL, 200–400 μg/ml—Meloy Laboratories, Springfield, VA), and FGF (10 ng/ml—R & D Systems Inc.). The methodology for performing the serum-free assay is the same as that employed for assays in SSM or PSM with the exception of how stock cultures of the indicator cells are trypsinized. For cells to be used in serum-free medium, the action of trypsin is stopped by the addition of soybean trypsin inhibitor instead of serum. As usual, the cells are collected by centrifugation to remove the inactivated trypsin and the cells are resuspended in culture medium. In serum-free assays, the factors used in place of serum are added just before the cells are added, in order to reduce the time that the factors are exposed to the temperature of the melted agar. Although it has not been determined whether prolonged periods at 42–44°

will affect the activity of these factors, short periods are unlikely to have any significant effects.

The preparation of insulin, which is prepared weekly and stored at 4°, and the preparation of transferrin have been described previously.[19] Insulin (Sigma, bovine) is prepared as a 100× stock solution. Insulin is first dissolved at 10 mg/ml in 0.1 M HCl. After the insulin is dissolved, it is diluted to 1 mg/ml with water and filter sterilized with a 0.2 μm filter. Transferrin (Sigma, human) is prepared as a 100× stock solution by dissolving it at 500 μg/ml in water and then filter sterilizing the solution with a 0.2 μm filter. FGF is used as suggested by the commercial supplier. Once it is redissolved, it is stored at −20° and subjected to no more than five cycles of freezing and thawing. HDL is prepared from fresh unfrozen human plasma by sequential flotation using potassium bromide for adjustments of density.[20,21] The HDL fraction (1.087−1.21) is washed once, dialyzed (0.15 M NaCl and 0.01% EDTA at pH 7), and concentrated by ultrafiltration (20−40 mg/ml). The HDL density of 1.087−1.21 is used rather than the broader density range of 1.063−1.21, to avoid the toxicity associated with HDL of the latter density. It is also very important to use water of the highest quality when preparing the HDL, as well as during the preparation of the serum-free medium itself. Three times distilled water has not always given consistent results. Water purified by systems that use combinations of ion exchange, carbon adsorption, and ultrafiltration may provide better results. Once prepared, HDL should be stored at 4° and used within 4−6 weeks. HDL stored for longer periods begins to lose activity. HDL should not be frozen, since this significantly reduces its activity.

The serum-free assay has several important advantages over the assay performed in SSM. First, colonies do not form in the serum-free medium unless other growth factors are added, nor do EGF or TGFβ alone induce the formation of even small colonies. Second, clones of NRK cells that exhibit high backgrounds in our best lots of bovine calf serum fail to form colonies in the serum-free medium unless TGFs are added to the medium. Thus, the serum-free assay eliminates the problem of background activity observed with SSM and NRK cells. Last, and most important, the serum-free assay provides the best culture conditions currently available for determining the hormone and growth factor requirements for anchorage-independent growth. However, use of the serum-free assay does have some disadvantages. The major disadvantage is the need for HDL, which is

[19] A. Rizzino and C. Crowley, *Proc. Natl. Acad. Sci. U.S.A.* **77**, 451 (1980).
[20] R. J. Havel, H. A. Eder, and J. H. Bragdon, *J. Clin. Invest.* **34**, 1345 (1955).
[21] M. S. Brown, S. E. Dana, and J. L. Goldstein, *J. Biol. Chem.* **249**, 789 (1974).

expensive to purchase and time-consuming to prepare. This is unfortunate, since HDL is essential for growth of the NRK cells in the serum-free assay and a substitute for it has not been found. Furthermore, we have recently determined that some lots of commercially prepared HDL are inactive or even toxic. For this reason, it is necessary to perform a dose–response curve for each lot of HDL (in the presence of factors known to have TGF activity in the serum-free assay, e.g., TGFβ, EGF, and FGF). We have recently determined that the results obtained with some lots of HDL are improved by adding bovine serum albumin (1 mg/ml BSA; Pentax — Miles) to the serum-free medium.[9] The reason why BSA is needed is not clear. No requirement for BSA was observed during development and testing of the serum-free assay, which involved the use of over 10 different lots of HDL. Although the problems encountered with different HDL lots is not understood at this time, it is possible that some donors are better than others. Alternatively, some aspect of shipping the HDL over long distances, which was not involved during the earlier work, may affect its activity in this assay. This problem is currently under investigation. For these reasons, and due to the expense of purchasing HDL, it is recommended that HDL be prepared in-house when it is to be used on a large scale (1 U of blood will provide approximately 100 mg of HDL).

Rapid Assay for Detecting the Production of TGFs

The assays described thus far are used to test conditioned media and growth factor preparations for TGF activity. Since the preparation of conditioned medium usually involves several steps, it is sometimes useful to employ a modified assay in which the cells are directly tested. Such an approach was used to determine that early mouse embryos[22] and a wide range of embryonal carcinoma cells[23] release factors that exhibit TGF activity. The assay is performed by plating the test cells as a monolayer. After the cells have attached, they are overlaid with a basal layer of agar (0.5%). Once this layer has solidified, a second layer of agar (0.3%) is added, followed by a third layer (0.3%) containing the indicator cells (e.g., NRK cells). Under these conditions, if the test cells release TGFs, these factors will diffuse to the third agar layer and induce the soft agar growth of the indicator cells. As a control, the indicator cells are omitted from the top layer to verify that the colonies formed have not arisen from the cells that were plated as a monolayer. When performing this assay, it is best to test several different densities of the cells under investigation (500–5000 cells/

---

[22] A. Rizzino, *In Vitro* **21**, 531 (1985).
[23] A. Rizzino, L. S. Orme, and J. E. De Larco, *Exp. Cell Res.* **143**, 143 (1983).

35-mm dish). Generally, the indicator cells will begin to form colonies after 2–3 days. However, these colonies usually do not become very large. This is especially true when the test cells proliferate rapidly and quickly deplete the culture medium. For this reason, it is best to monitor the colonies daily and score them as soon as the colonies appear.

This assay provides a rapid means of determining whether the cells under study are releasing TGFs. However, this assay is not well suited for quantitative studies. The cells plated as a monolayer tend to grow more rapidly than the indicator cell colonies and they exhaust the nutrients present in the culture medium and significantly reduce the pH of the culture. Consequently, it is difficult to compare the relative amounts of TGFs produced by two different cell lines using this assay, especially if the two cell lines exhibit significantly different growth rates. Despite this drawback, this assay is very useful for rapidly determining whether the cells in question are producing TGFs and whether further study is warranted. Last, this assay is useful for examining cells that are available in limited amounts, such as primary tissue explants, early embryos, or their early embryonic tissues.

Conclusions

This chapter has dealt with the methods needed to assay TGFs and other growth factors for their ability to influence anchorage-independent growth. The two major considerations in setting up these assays are (1) the choice of indicator cell line and (2) the choice of culture medium. If NRK cells are selected as the indicator cells, there are three major choices for the culture medium. Serum-supplemented medium is suitable for most work but requires careful screening of the serum lots and clones of NRK cells employed in order to minimize the problem of background. Alternatively, serum-supplemented medium can be replaced with either plasma-supplemented or serum-free medium. Which option is most appropriate depends on the goals of the project, and more than one approach may be needed. Given the time and expense of preparing HDL, the serum-free medium is probably best used to identify the hormones and growth factors that influence anchorage-independent growth, rather than for performing routine TGF assays.

Acknowledgments

The author thanks Heather Rizzino for excellent editorial assistance and for her many useful comments during the preparation of this chapter. Eric Ruff is also thanked for exceptional technical assistance and for helping in the assessment of this chapter. This work was supported by grants from the Nebraska Department of Health (85-58) and the National Cancer Institute (CA 36727).

[33] Assay of Growth Factor-Stimulated Tyrosine Kinases Using Synthetic Peptide Substrates

By Linda J. Pike

Introduction

In 1980, Cohen and co-workers first reported that EGF stimulated the phosphorylation of proteins on tyrosine residues.[1] Subsequently, work from a number of different laboratories indicated that in addition to EGF, PDGF,[2] insulin,[3] and IGF-I[4] all stimulated the phosphorylation of proteins on tyrosine residues. The data suggested that each of these growth factor receptors possessed intrinsic tyrosine kinase activity that was regulated by the growth factor. These tyrosine kinase activities were first identified by the ability of these kinases to carry out an autophosphorylation reaction.[1-4] While the use of the growth factor receptors themselves as specific substrates for these kinases was helpful in determining the relationship between the receptor and the kinase, the lack of a suitable exogenous substrate limited the types of experimental questions that could be approached. For example, determination of the kinetic mechanism and estimation of the specific activity of the growth factor-stimulated kinases was impossible in the absence of an exogenous substrate.

In developing a synthetic acceptor substrate for the tyrosine kinases several criteria for suitability were set forth:

1. The substrate should be specific enough to allow the tyrosine protein kinases to be assayed in crude extracts containing serine and threonine protein kinases.

2. Quantitation of the phosphorylated substrate should be straightforward and easy to perform on large numbers of samples.

The synthetic peptide shown in Table I was found to meet these criteria.[5] The sequence of this synthetic peptide was based on the site of

[1] H. Ushiro and S. Cohen, *J. Biol. Chem.* **255**, 8363 (1980).

[2] B. Ek, B. Westermark, A. Wasteson, and C.-H. Heldin, *Nature (London)* **295**, 419 (1982).

[3] M. Kasuga, Y. Zick, D. L. Blithe, M. Crettaz, and C. R. Kahn, *Nature (London)* **298**, 667 (1982).

[4] S. Jacobs, F. C. Kull, H. S. Earp, M. E. Svoboda, J. J. Van Wyk, and P. Cuatrecasas, *J. Biol. Chem.* **258**, 9581 (1983).

[5] J. E. Casnellie, M. L. Harrison, L. J. Pike, K. E. Hellstrom, and E. G. Krebs, *Proc. Natl. Acad. Sci. U.S.A.* **79**, 282 (1982).

TABLE I

SEQUENCE OF A SYNTHETIC PEPTIDE SUBSTRATE FOR GROWTH FACTOR-STIMULATED
TYROSINE KINASES[a]

Sample	Sequence
Synthetic peptide	Arg-Arg-Leu-Ile-Glu-Asp-Ala-Glu-Tyr-Ala-Ala-Arg-Gly
pp60[src]	Ala-Arg-Leu-Ile-Glu-Asp-Asn-Glu-Tyr-Thr-Ala-Arg-Gln-Gly

[a] The sequence of the synthetic peptide used in the author's laboratory is compared with the sequence present at the site of autophosphorylation in pp60[src].

autophosphorylation in the viral tyrosine protein kinase, pp60[src].[6] To make the peptide substrate useful in a general assay procedure, the natural sequence was modified in several ways. The threonine carboxy-terminal to the tyrosine was replaced with an alanine. This left tyrosine as the only phosphorylatable amino acid in the sequence, making the peptide a specific substrate for tyrosine kinases. In addition, the asparagine at position 7 and the glutamine at position 13 were replaced with alanine or deleted, respectively, to simplify the synthesis of the peptide and preclude a change in sequence of the peptide upon storage due to deamidation at these residues. Last, an additional arginine not found in the natural sequence was added to the amino-terminus to enable the phosphorylated peptide to be isolated and quantified using a simple phosphocellulose paper assay. This approach of adding arginines to facilitate substrate isolation has been used previously in generating synthetic peptide substrates for assay of the cAMP-dependent protein kinase.[7]

Following the initial description of this synthetic peptide substrate[5,8] a number of other low-molecular-weight natural and synthetic peptides were also reported to serve as substrates for the tyrosine-specific protein kinases. These include other peptides based on the sequence of the autophosphorylation site of pp60[src],[9] peptides related to gastrin,[10] and peptides related to angiotensin[11] (Table II). The common feature of all these peptides is the presence of at least one acidic residue amino-terminal to the phosphorylat-

[6] A. P. Czernilofsky, A. D. Levinson, H. E. Varmus, and J. M. Bishop, Nature (London) 287, 198 (1981).

[7] D. B. Glass, R. A. Masaracchia, J. R. Feramisco, and B. E. Kemp, Anal. Biochem. 87, 566 (1978).

[8] L. J. Pike, B. Gallis, J. E. Casnellie, P. Bornstein, and E. G. Krebs, Proc. Natl. Acad. Sci. U.S.A. 79, 1443 (1982).

[9] C. Erneux, S. Cohen, and D. L. Garbers, J. Biol. Chem. 258, 4137 (1982).

[10] G. S. Baldwin, A. W. Burgess, and B. E. Kemp, Biochem. Biophys. Res. Commun. 109, 656 (1982).

[11] T. W. Wong and A. R. Goldberg, J. Biol. Chem. 258, 1022 (1983).

TABLE II

SEQUENCES OF SOME SYNTHETIC PEPTIDES KNOWN TO BE PHOSPHORYLATED BY
GROWTH FACTOR-STIMULATED KINASES

Peptide	Sequence	Reference
$L_{12}G_1$	Leu-Glu-Asp-Ala-Glu-Tyr-Ala-Ala-Arg-Arg-Arg-Gly	a
Gastrin	Arg-Arg-Leu-Glu-Glu-Glu-Glu-Glu-Ala-Tyr-Gly	b
[Val⁵]Angiotensin II	Asp-Arg-Val-Tyr-Val-His-Pro-Phe	c

[a] C. Erneux, S. Cohen, and D. L. Garbers, *J. Biol. Chem.* **258,** 4137 (1983).
[b] G. S. Baldwin, A. W. Burgess, and B. E. Kemp, *Biochem. Biophys. Res. Commun.* **109,** 656 (1982).
[c] T. W. Wong and A. R. Goldberg, *J. Biol. Chem.* **258,** 1022 (1983).

able tyrosine residue. While the acidic nature of the peptide substrates seems to be important for the utilization by the growth factor-stimulated kinases, studies comparing the kinetic parameters of protein versus peptide substrates have indicated that primary sequence is only part of the information used by the tyrosine kinases in selecting these substrates.[12]

Assay Method

Because the author's principal experience in assay of the growth factor-stimulated tyrosine kinases has been with the synthetic peptide shown in Table I, all assay procedures described here will deal specifically with the use of this peptide. In general, the same conditions may be applied for assay of the growth factor-stimulated tyrosine kinases regardless of the peptide substrate used. However, the phosphocellulose paper method for quantitation of phosphorylated peptide may not be applicable in all cases since this requires the presence of at least two positively charged residues in the peptide sequence.[7] In these instances, high-voltage paper electrophoresis or SDS gel electrophoresis may be required to isolate the phosphopeptide for quantitation.

Principle

Assay of the growth factor-stimulated tyrosine kinases is based on their ability to catalyze the transfer of the γ-phosphate of ATP to a suitable acceptor substrate, in the present case a synthetic, tyrosine-containing peptide. Use of [γ-³²P]ATP permits the radioactive labeling of the phosphorylated peptide. Contaminating endogenous protein acceptors are se-

[12] L. J. Pike, E. A. Kuenzel, J. E. Casnellie, and E. G. Krebs, *J. Biol. Chem.* **259,** 9913 (1984).

lectively precipitated with trichloroacetic acid. The peptide is separated from unreacted [γ-^{32}P]ATP by adsorption onto phosphocellulose paper, then quantitated by liquid scintillation counting.

Because crude preparations of cells or tissues usually contain a number of different tyrosine kinases, the activity of a specific growth factor-stimulated tyrosine kinase can only be accurately measured by addition of the appropriate growth factor and calculation of the amount of activity stimulated by the growth factor. Even in crude membrane preparations, a 50–100% stimulation of most growth factor receptor/kinase activities can be obtained if the appropriate conditions are used.

Reagents

Stock solutions:
  10 mM ATP
  1 M MgCl$_2$
  1 M MnCl$_2$
  6 mM sodium orthovanadate
  1.2 M p-nitrophenyl phosphate
  12 mM solution of synthetic peptide dissolved in water with the pH adjusted to 7 with NaOH
  Growth factor dissolved in buffer A at a concentration 6-fold higher than the desired final concentration in the assay
  5% Trichloroacetic acid
  75 mM H$_3$PO$_4$ (washing solution)
Buffer A:
  40 mM imidazole, pH 7.2
  250 mM NaCl
  10% Glycerol
  Detergent as needed (see below)

Using the stock solutions, a radioactive ATP mix (referred to hereafter as hot ATP mix) is prepared which contains in final concentrations: 1.2 mM ATP, 72 mM MgCl$_2$, 12 mM MnCl$_2$, 600 μM sodium orthovanadate, 120 mM p-nitrophenyl phosphate, and 0.1 to 0.5 mCi/ml [γ-^{32}P]ATP. Due to dilution of this solution into the reaction mixture, the final concentration of these components is 6-fold lower in the final assay than given above.

Procedure

The following procedure has been found to be the most generally applicable for the assay of growth factor-stimulated tyrosine kinases using a synthetic peptide substrate. Once the presence of a growth factor receptor/kinase has been demonstrated in a particular preparation, the assay conditions can be optimized for that particular system as described in the next section. The final assay volume is 30 μl but this may be scaled up or down according to the amount of material available. Volumes of additions are as follows: 10 μl growth factor receptor preparation in buffer A, 5 μl buffer A or growth factor in buffer A (for basal and stimulated activities, respectively), 5 μl peptide substrate solution, 5 μl water or other addition if desired, and 5 μl hot ATP mix. All reagents should be kept on ice while the assay is being set up.

The reagents are pipetted into the tubes and the assay is begun by the addition of an aliquot of the hot ATP mix followed by vortexing and placement of the tube in a 30° water bath. The reaction is allowed to proceed for 5 min at which time the assays are terminated by the addition of a 30-μl aliquot of 5% trichloroacetic acid.

The tubes are spun in a microfuge for 5 min at 10,000 rpm to pellet the precipitated proteins. An aliquot of the supernatant (30 to 50 μl) is then spotted on a 1- to 2-cm square of Whatman P81 phosphocellulose paper that has been numbered in pencil and the paper placed into a beaker containing 75 mM H_3PO_4. At least 10 ml of phosphoric acid per square should be used. The papers are then washed three times for 2 min in phosphoric acid and dried before counting for [32]P.

This process is facilitated by the use of a wire basket which fits into the wash beaker and leaves room underneath the basket for a stir bar. If the papers are placed in the basket, the acid solution can be stirred to improve washing without risking damage to the papers. The basket and filters can also be easily removed to change the washing solution. The papers can be dried by placing the wire basket beneath a hot air gun, though care should be taken not to overheat the filters as they burn easily, obscuring the pencilled numbers. The dried papers are placed in a vial with scintillation cocktail and counted for [32]P.

The counts on the papers represent peptide phosphorylation plus a variable amount of background. Background in the assay arises from two sources: (1) the phosphorylation of endogenous proteins that are not precipitated by the trichloroacetic acid step, and (2) contaminants in the [32P]ATP preparation that are retained by the P81 paper. The former can be corrected for by performing assays in the absence of peptide and subtracting these counts from the counts observed in the presence of peptide. The latter is automatically corrected for if the endogenous protein phosphorylation is subtracted since part of this phosphorylation represents counts due to the ATP. However, an estimate of the background due to contaminants of the [32P]ATP can be obtained by spotting a reaction mix which contains only hot ATP mix on P81 paper, washing, and counting as above. The background due to this source should not exceed 0.1% of the input counts. If a substantial increase in this parameter is evident, the purity of the [³²P]ATP should be checked.

Figure 1 shows a sample time course of the phosphorylation of the synthetic peptide by crude A431 cell membranes using the above procedure. The reaction was linear for at least 10 min under these conditions but showed evidence of becoming nonlinear thereafter. The background of endogenous phosphorylation is shown in the dotted line. The number of input counts was 2,473,490 cpm and the background due to ATP alone was 402 cpm.

Optimizing the Assay

General Considerations

Interfering Reactions. The principal reactions which inhibit accurate measurement of growth factor-stimulated tyrosine kinase activity are ATP degradation (mainly by ATPases) and substrate dephosphorylation by

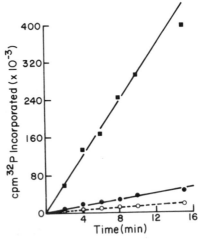

FIG. 1. Time course of peptide phosphorylation by A431 cell membranes. A431 cell membranes were prepared by homogenizing cells in 10 mM Tris, pH 7.4, 2 mM EGTA, 1 μg/ml leupeptin. The homogenate was spun 30 min at 35,000 rpm in a Ti 75 rotor and the pellet resuspended in buffer A containing no detergent. Triton X-100 was added to the assay at a final concentration of 0.05%. Assay conditions were as described in the text. Closed circles denote activity in the absence of EGF. Closed squares, activity in the presence of 300 nM EGF. Open circles show background counts due to phosphorylation of endogenous proteins obtained from assays carried out in the absence of added peptide.

phosphoprotein phosphatases. These reactions are particularly troublesome in crude enzyme preparations where kinase activity is low and competing reactions are high.

The use of relatively high concentrations of ATP coupled with short incubation times generally helps preserve ATP levels during the course of the assay. ATPases appear to be more sensitive than kinases to inhibition by App(NH)p. Thus this compound can be added to assays at concentrations at or below the concentration of ATP and will substantially reduce ATPase activity while having little effect on kinase activity.

Phosphatase inhibitors, such as p-nitrophenyl phosphate, should generally be included in the kinase assay to limit dephosphorylation of the product. However, the relatively high concentration needed to produce effective inhibition of the phosphatase has occasionally been found to inhibit the kinase. Thus, it is possible to observe an increase in tyrosine kinase activity in crude preparations upon addition of p-nitrophenyl phosphate due to inhibition of phosphatases and to note a reduction in kinase activity by addition of high concentrations of this compound to more highly purified preparations due to inhibition of the kinase.

Phosphatases may also be inhibited preferentially by lowering the assay temperature to 4°. The tyrosine kinases are slow, but active, at this temperature while the phosphatases are essentially inactive. While this is sometimes the best solution to the phosphatase problem, it is not always practical, especially if one wishes to do experiments under physiological or near-physiological conditions.

Sodium orthovanadate is thought to preferentially inhibit phosphotyrosine phosphatases and has no effect on kinase activity at levels of 100 μM or less. In addition, this compound is known to inhibit ATPase activity. Thus, inclusion of orthovanadate in all assays is recommended.

Concentration of Reactants. The final concentration of ATP in the above protocol is 200 μM, which is essentially saturating for most growth factor-stimulated tyrosine kinases. Use of lower concentrations of ATP permits an increase in the specific activity of the ATP and may yield more counts per assay, depending on the concentration of ATP used and the K_m of the kinase for ATP. This may facilitate the assay of a growth factor-stimulated kinase present in very low levels. However, as mentioned above, complications due to ATP degradation are likely to arise if too low a concentration of ATP is utilized with crude enzyme preparations.

All growth factor-stimulated tyrosine kinases require divalent metal ions to carry out the phosphorylation reaction. In general, low concentrations of Mn^{2+} (<2 mM) have been found to support good activity whereas much higher concentrations of Mg^{2+}, on the order of 10 to 50 mM, are required to support kinase activity. In some instances a combination of

Mg^{2+} and Mn^{2+} has been found to produce the highest levels of growth factor-stimulated kinase activity in a given preparation.[12] Hence, this combination is likely to be the best condition for initial assays of a kinase with unknown metal ion preferences.

The K_m of the growth factor-stimulated kinases for the synthetic peptide has been found to be approximately 2 mM regardless of the kinase assayed or the degree of purity. This value is certainly higher than what has been observed using protein substrates for these enzymes, which exhibit K_m values in the micromolar range.[12] What this means in practice is that (1) relatively large amounts of peptide must be added to each assay, and (2) it is essentially impossible to attain saturating levels of this substrate. A concentration of 2 mM peptide has generally been found to produce optimal results. Lower concentrations are well below K_m and yield slower reaction rates. Higher concentrations of peptide do not significantly increase the rate of phosphorylation apparently due to substrate inhibition. Because the assays are run at essentially K_m levels of peptide substrate, the actual activity measured does not reflect the maximum velocity of the reaction. If these data are required, a Lineweaver–Burk plot can be constructed from velocities obtained at varying concentrations of peptide and the V_{max} can be extrapolated from these data.

The growth factor-stimulated tyrosine kinases have all been found to be intrinsic membrane proteins. Thus, it is not surprising that some detergent seems to be required for solubility and stability of these enzymes. For kinases that have been solubilized and purified, a concentration of 0.1% of a nonionic detergent is normally sufficient. Adding much higher concentrations of detergent ($>0.5\%$), particularly in the case of the insulin receptor, reduces hormone binding and inhibits kinase activity.

Assaying Enzyme Preparations of Varying Degrees of Purity

Crude Particulate Fractions. This material is very difficult to assay accurately as the amount of tyrosine kinase activity is generally low while the amount of competing ATPase and phosphatase activity is high. Assay of dilute membrane preparations (100 μg/ml) in the presence of the previously mentioned ATPase and phosphatase inhibitors is usually the only successful approach. The addition to the assay of low concentrations of Triton X-100 ($<0.2\%$) generally improves tyrosine kinase activity, possibly by permeabilizing membrane vesicles to the entrance of substrate molecules.

Detergent-Solubilized Membrane Fractions. This material has the same problems with competing ATPase and phosphatase activities as crude particulate fractions; however, the solubilized nature of the preparation generally reveals more growth factor-stimulated tyrosine kinase activity. The material must usually be diluted for assay not only to help reduce interfering reactions but also to bring down the concentration of detergent to less than 0.5% for nonionic detergents.

Purified Preparations. The growth factor-stimulated tyrosine kinases have all been successfully purified, most usually by chromatography on lectin–agarose followed by affinity chromatography on a growth factor–agarose column. Once partially purified, these kinases are relatively simple to detect using the peptide phosphorylation assay because many of the competing activities have been removed. With long reaction times, phosphatases and ATPases continue to present problems even in partially purified preparations so that each preparation should be tested for linearity with assay time.

In the procedure described above, the phosphorylation assays are stopped by the addition of a dilute solution of trichloroacetic acid. This step induces the precipitation of most proteins in the assay but leaves the peptide in the supernatant. When working with crude preparations of enzyme, this step is crucial since, if not precipitated, the endogenous proteins phosphorylated during the reaction produce a high background on the phosphocellulose paper, making it difficult to observe low levels of peptide phosphorylation. As the enzyme is purified, this step becomes increasingly less important and can be replaced with an addition of 30% acetic acid with no subsequent centrifugation when highly purified kinase is used.

Differences among the Growth Factor-Stimulated Kinases

The insulin-, PDGF-, and EGF-stimulated tyrosine kinases have been successfully detected using the synthetic peptide assay.[8,13,14] In all cases, the K_m for the peptide was in the millimolar range. Utilizing a similar procedure, the IGF-I receptor/kinase has been assayed using a synthetic polymer as the acceptor substrate.[15] Thus the general conditions for assaying these different receptor/kinases are quite similar.

Some differences have been noted among these kinases in terms of nucleotide requirements. The insulin, EGF, and IGF-I receptors have all been found to phosphorylate synthetic substrates in the presence of concentrations of ATP ranging from 1 to 200 μM.[8,13,15] By contrast, PDGF-stimulated kinase activity in 3T3 cell membranes could not be demon-

[13] L. A. Stadtmauer and O. M. Rosen, *J. Biol. Chem.* **258**, 6682 (1983).
[14] L. J. Pike, D. F. Bowen-Pope, R. Ross, and E. G. Krebs, *J. Biol. Chem.* **258**, 9383 (1983).
[15] Y. Zick, N. Sasaki, R. W. Rees-Jones, G. Grunberger, S. P. Nissley, and M. M. Rechler, *Biochem. Biophys. Res. Commun.* **119**, 6 (1984).

strated when 10 μM ATP was used but was apparent when 100 μM ATP was used as substrate.[14] Thus the use of several concentrations of ATP when screening for one of these kinases in a new tissue is recommended.

Several studies have shown that certain acceptor substrates are utilized preferentially by some of the growth factor receptor/kinases.[12,16] These differences have generally been noted with protein substrates, not peptides, but if a specific synthetic peptide appears to be functioning poorly in a given assay, use of a slightly different peptide substrate might improve detection of the kinase activity.

One obvious difference among these kinases is the requirement for different concentrations of the appropriate growth factor to elicit maximal enzyme activity. For the insulin receptor/kinase, 30 nM insulin appears to be a saturating dose while 10 times this much EGF was required to observe maximal activity of the enzyme.[12] Optimal activity of the PDGF-stimulated kinase required 540 nM PDGF.[14] These differences obviously reflect the varying affinities of the ligands for their receptors. In addition, slight differences are likely to arise due to differences in tissue source, degree of purity of the enzyme, and assay conditions. Thus, once conditions for detecting kinase activity have been ascertained, a dose–response to growth factor should be performed to determine the optimal concentration of hormone to use in that particular system.

[16] S. Braun, W. E. Raymond, and E. Racker, J. Biol. Chem. **259**, 2051 (1984).

[34] Separation of Multiple Phosphorylated Forms of 40 S Ribosomal Protein S6 by Two-Dimensional Polyacrylamide Gel Electrophoresis

By Michel Siegmann and George Thomas

Introduction

Two methods have been employed to measure the extent of 40 S ribosomal protein S6 phosphorylation in cells grown in culture. In the first, cells are labeled to equilibrium with $^{32}PO_4$ and the amount of ^{32}P incorporated into S6 is determined after the addition of specific growth factors, hormones, or drugs to the culture. Measurements made using this method must assume that the total intracellular specific activity of P_i or ATP is equivalent to that of the S6 phosphate donor. This assumption has not always proved to be correct. In the second method the extent of S6 phosphorylation is determined by the electrophoretic shift of the protein on two-dimensional polyacrylamide gels. The more phosphorylated the protein becomes the slower its mobility in both dimensions of electrophoresis. This method not only allows for the unambiguous measurement of the extent of S6 phosphorylation but also makes it possible to separate the differentially phosphorylated forms of the protein. The separation of proteins by this method was initially developed by Kaltschmidt and Wittmann[1] to identify individual ribosomal proteins of *Escherichia coli;* later it was applied to the ribosomal proteins of both plants and animals. Its value in measuring the extent of S6 phosphorylation was first reported by Gressner and Wool.[2] Since this time the method has been altered and refined in a number of ways to optimize its application toward the analysis of S6 and its phosphorylated derivatives. Here we describe how this method is presently carried out in our laboratory.

Solutions

Hypotonic buffer contains 1.5 mM KCl, 2.5 mM MgCl$_2$, and 5 mM Tris–HCl, pH 7.4; Buffer A contains 100 mM KCl, 5 mM MgCl$_2$, 1 mM dithiothreitol, 0.7 M sucrose, 20 mM Tris–HCl, pH 7.4. Buffer B is identical to Buffer A except that the KCl concentration is 500 mM and the sucrose concentration is 1.6 M. First-dimensional polyacrylamide separating gel solution contains 6 M urea, 8% acrylamide, 0.3% N,N'-methylenebisacrylamide, 0.3% N,N,N',N'-tetramethylethylenediamine, 21 mM EDTA, 0.52 M boric acid, and 0.4 M Tris, pH 8.6 (3 μl of 10% ammonium persulfate/ml of gel solution is used for polymerization). First-dimensional polyacrylamide stacking gel solution contains 6 M urea, 4% acrylamide, 0.2% N,N'-methylenebisacrylamide, 0.06% N,N,N',N'-tetramethylethylenediamine, 2.3 mM EDTA, 52 mM boric acid, pH 8.6 (5 μl of 10% ammonium persulfate/ml of gel solution is used for polymerization). Gel sample solution is identical to first-dimensional polyacrylamide stacking gel solution except it has no acrylamide or N,N'-methylenebisacrylamide and contains 5% 2-mercaptoethanol. First-dimensional electrophoresis buffer contains 6.5 mM EDTA, 156 mM boric acid, and 120 mM Tris adjusted with 10 M NaOH to pH 8.6. Second-dimensional polyacrylamide separating gel solution contains 6 M urea, 18% acrylamide, 0.375% N,N'-methylenebisacrylamide, 0.058% N,N,N',N'-tetramethylenediamine, 0.92 M acetic acid, 48 mM KOH, pH 4.0 (30 μl of 10% ammonium persulfate/ml of gel solution is used for polymerization). Sec-

[1] E. Kaltschmidt and H. Wittmann, *Anal. Biochem.* **36,** 401 (1970).
[2] A. Gressner and I. Wool, *J. Biol. Chem.* **249,** 6917 (1974).

ond-dimensional electrophoresis buffer contains 0.186 M glycine adjusted with acetic acid to pH 4.0.

Preparation of Ribosomes

Swiss mouse 3T3 cells are seeded at 3×10^5 cells/15-cm tissue culture plate in 10 ml of Dulbecco's modified Eagle's medium containing 10% (v/v) fetal calf serum. After 72 hr, cultures are refed with the same medium containing 15 μCi/ml of [^{35}S]methionine. Cultures become confluent after an additional 24 to 48 hr and are judged quiescent at 7 days following seeding when no mitotic cells are observed. At this point there are approximately 6×10^6 cells per plate. In the case of cells where large quantities of ribosomes can be obtained radioactive labeling of ribosomal proteins as well as the later steps of adding 80 S ribosomes as carrier and gel fluorography can be omitted. At this point, or following the addition of growth factors or hormones to the culture, the plate is transferred to a sheet of crushed ice, the medium removed, the plate rinsed twice with 10 ml of ice-cold hypotonic buffer, and the cells lysed with 2 ml of hypotonic buffer containing 1% Triton X-100 (Serva) and 1% sodium deoxycholate (Merck). The lysate is scraped to one edge of the plate with a rubber policeman, and the material from three plates (the number of cells required for a single gel) is pooled into a 15-ml Corex tube and centrifuged at 5000 rpm for 5 min in an SS 34 rotor (Sorvall) to remove nuclei. The pellets are discarded, the pooled supernatant is transferred to a 40-ml Beckman Quick Seal tube, and brought to 28 ml final volume with hypotonic buffer, 500 μg of rat liver 80 S ribosomes[3] are added as carrier, and the mixture is underlayed with 6 ml of buffer A followed by 6 ml of buffer B. The tubes are then sealed and centrifuged at 55,000 rpm for 16 hr at $0-2°$ in a Ti 60 rotor (Beckman). The high-speed supernatant is discarded, the ribosomal pellet resuspended in 1 ml of buffer A containing no sucrose, and either used immediately as described below, or frozen in a dry ice–acetone bath and stored at $-80°$.

Isolation of Ribosomal Proteins

Prior to extraction of ribosomal proteins, the 1-ml sample is made 10 mM in MgCl$_2$ and 0.7 vol of $-20°$ ethanol is added. The ribosomes immediately precipitate, and are centrifuged at 7000 cpm for 10 min at $0°$ in an HB-4 rotor (Sorvall). The supernatant is discarded, the excess ethanol carefully removed with a cotton swab, and the ribosomes resuspended in 25 μl of buffer A containing no sucrose. The ethanol precipitation step is

---

[3] G. Thomas, J. Gordon, and H. Rogg, *J. Biol. Chem.* **253**, 1101 (1978).

not only necessary to concentrate the ribosomes, which must be between $200-400$ $OD_{260}U/ml$ before beginning the extraction (see below), but in our hands it also enhances the resolution of the proteins during electrophoresis. The method of Hardy et al.[4] is followed for the extraction of ribosomal proteins. One-tenth volume of 1 M $MgCl_2$ is added simultaneously with 2 vol of glacial acetic acid to the ribosomes in buffer A while stirring on an ice-water bath. After stirring for 45 min, the precipitated RNA is removed by centrifugation at 10,000 rpm for 10 min in an HB-4 rotor. The supernatant (which contains the protein) is then transferred to a 7-ml conical tube, and, following the procedure of Barritault et al.,[5] 4 vol of $-20°$ acetone is added. At the initial concentration of $200-400$ OD_{260} U/ml of ribosomes the protein does not always precipitate immediately. If this is the case the acetone–protein mix is placed at $-20°$ for approximately 10 min, and then gently mixed by inverting the tube upside down several times. At this point the precipitate will begin to form and the mix is placed again at $-20°$ for 2 hr to allow the precipitation process to come to completion. After this time the protein precipitate is removed by centrifugation at 10,000 rpm for 10 min in an HB-4 rotor, washed two times with 1 ml of acetone, once with 1 ml of ether, and then lyophilized to ensure the pellet is completely dry. Then just prior to application to the first-dimensional polyacrylamide gel the dried pellet is resuspended in 60 μl of gel sample buffer and heated at $60°$ for 10 min. At this point 1 μl of each sample is removed in duplicate and the amount of [35S]methionine incorporated into protein determined. Normally 1×10^6 cpm are recovered per experimental point.

First-Dimensional Polyacrylamide Gel Electrophoresis

Employing 1.5 ml of first-dimensional polyacrylamide separating gel solution per gel, the first-dimensional gel is cast with a 200 mm $\times$ 16 gauge needle into a 2 $\times$ 150 mm Plexiglas tube to a height of 125 mm and gently overlayed with water. It is important that Plexiglas be used for the first-dimensional gel as during electrophoresis the gel swells and is extremely difficult to remove from glass tubes, even when siliconized. Following polymerization of the separating gel, a 5-mm stacking gel (200 μl of first-dimensional polyacrylamide stacking gel solution per gel) is cast over the first-dimensional gel and gently overlayed with water. The sample, resuspended in 60 μl of gel sample buffer as described above, is applied to the surface of the stacking gel and gently overlayed with first-dimensional

[4] S. Hardy, C. Kurland, P. Voynow, and C. Mora, *Biochemistry* **8**, 2897 (1969).
[5] C. Barritault, P. Expert-Bezancon, M. F. Guerin, and D. Hayes, *Eur. J. Biochem.* **63**, 131 (1976).

electrophoresis buffer, and both reservoirs of the electrophoretic chamber are filled with the same buffer. Electrophoresis is initially carried out for 15 min at 100 V or until the pyronine G tracking dye focuses into a sharp band at the interphase between the stacking and separating gel, and then continued for 14.5 hr at 225 V. The first dimension of electrophoresis gives the largest separation of the S6 derivatives but is also extremely sensitive to the way in which the protein sample is prepared. If all the ribosomal RNA is not removed during the acetic acid extraction step a number of proteins will migrate toward the anode along with the RNA rather than into the gel toward the cathode. At this step it is essential that the ribosome concentration be between 200–400 OD_{260} U/ml (as pointed out above), otherwise there can be selective losses of S6. Also if the protein pellets are not completely dry following lyophilization they will be difficult to resuspend in gel sample buffer and will lead to elongated or streaked protein spots rather than sharp round protein spots. An obvious sign of an incorrectly prepared protein sample is that the pyronine G running dye (present in the gel sample buffer) does not concentrate into a sharp band at the interface of the stacking gel and separating gel during the first 15 min of electrophoresis, but instead moves as a broad band through the gel. Following completion of first-dimensional gel electrophoresis the gels are gently extruded with a 31-gauge needle injecting water as a lubricant between the gel and the tube wall. At this point one can continue with the second dimension of electrophoresis or store the first-dimensional gel for up to 48 hr at 0–2° or up to a week at −70° without loss of resolution.

Second-Dimensional Polyacrylamide Gel Electrophoresis

Prior to casting the second-dimensional polyacrylamide gel the first-dimensional gel is soaked in a solution containing 6 *M* urea and 0.3 *M* HCl for 5 min at 0° to lower the pH of the gel, then soaked in second-dimensional polyacrylamide separating gel solution for an additional 5 min at 0°. The gel is then transferred to a glass plate and 2.5 cm is trimmed off the cathode end of the gel so that it can be cast into the second-dimensional polyacrylamide gel as depicted in Fig. 1 and as first described by Howard and Traut.[6] The second-dimensional Plexiglas spacers are thinly coated with Vasoline and laid out on the slotted glass plates as shown. The spacer along the top is 16 cm × 1.5 cm × 1.5 mm and the two spacers along the side are 18 cm × 1 cm × 1.5 mm. The first-dimensional gel is laid against the upper spacer and a small amount of second-dimensional polyacrylamide gel solution is applied to the gel such that it is slightly wetted. This wetting of the first-dimensional gel is necessary to prevent the gel from

[6] G. Howard and R. Traut, *FEBS Lett.* **29,** 172 (1973).

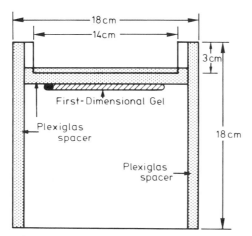

FIG. 1. Two-dimensional gel sandwich. Plexiglas spacers and first-dimensional polyacrylamide gel are laid out on a slotted glass plate as described in the text.

sticking to the second gel plate when it is now placed over the first plate. If the gel is dry at this point it can crack and break. This is due to the fact that the first-dimensional gel is thicker (2 mm) than the spacers (1.5 mm). The plates are now clamped together at the sides and top with binder clips (Tucker Products, England), inverted 180°, and the second-dimensional polyacrylamide gel solution (approximately 30 ml per gel) is applied through the opening at the top. After polymerization the single spacer at the bottom of the second-dimensional polyacrylamide gel is removed, the gel sandwich is again turned 180° so that the first-dimensional gel is again at the top, and the second-dimensional gel sandwich is mounted into a conventional slab gel apparatus. During the polymerization of the second-dimensional polyacrylamide gel it is possible that some proteins may become oxidized in the first-dimensional gel. Oxidation provents proteins from migrating out of the first-dimensional polyacrylamide gel into the second-dimensional polyacrylamide gel and leads to streaking. To reduce proteins a mixture of second-dimensional polyacrylamide gel solution containing 7% 2-mercaptoethanol is layered over the first-dimensional gel for 10 min at room temperature. Following removal of the reducing solution both reservoirs are filled with second-dimensional electrophoresis buffer. The first-dimensional gel is then overlayed with second-dimensional electrophoresis buffer containing 5% glycerol and 0.04% pyronine G. Electrophoresis is initially for 30 min at 50 V or until the pyronine G tracking dye passes through the first-dimensional gel; the voltage is then raised to 150 V for 19 hr. On completion of electrophoresis the gel is stained in a solution of 50% methanol and 10% acetic acid containing 0.1%

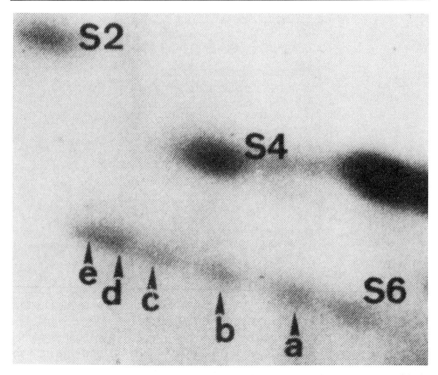

FIG. 2. Fluorogram of two-dimensional polyacrylamide gel. Quiescent cultures of Swiss mouse 3T3 cells labeled to equilibrium with 15 μCi/ml [^{35}S]methionine have been stimulated for 2 hr with 2.5% serum and processed for two-dimensional polyacrylamide gel analysis as described in the text. That area showing S6 and its increasingly phosphorylated derivates S6a through S6e are depicted. [Reprinted from G. Thomas, J. Martin-Pérez, M. Siegmann, and A. M. Otto, *Cell* **30**, 235 (1982).]

Coomassie Blue (R-250). After 45 min the gel is destained electrophoretically in a solution of 20% methanol and 10% acetic acid, and the area containing S6 is cut out of the gel and analyzed by fluorography (Fig. 2) as described by Bonner and Laskey.[7]

Other Modifications

There are a number of modifications which have been applied to this system which we have yet to examine. These include "mini" polyacrylamide gels, which greatly speed up the times of electrophoresis and require lower amounts of total protein[8]; longer first-dimensional polyacrylamide

[7] W. M. Bonner and R. A. Laskey, *Eur. J. Biochem.* **46**, 83 (1974).
[8] A. Lin, E. Collatz, and I. Wool, *Mol. Gen. Genet.* **144**, 1 (1976).

gels, which allow a better resolution of the most highly phosphorylated derivatives of S6, S6d, and S6e[9]; the use of agarose to attach the second-dimensional gel to the first, enabling the precasting of the first-dimensional gel.[9] In addition we have developed a monoclonal antibody against S6 which can distinguish the more highly phosphorylated derivatives of the protein from the less phosphorylated ones on a single-dimensional gel.[10] If this method could distinguish individual derivates it would be a useful technique since it would not involve radioactive labeling of cells, isolation of ribosomes, or two-dimensional gel electrophoresis.

Acknowledgments

The authors are indebted to Drs. I. Novak-Hofer and B. Rudkin as well as Joachim Krieg for their critical reading of the manuscript.

[9] K.-D. Sharf and L. Nover, *Cell* **30**, 427 (1982).
[10] P. Nielsen, K. Manchester, H. Towbin, J. Gordon, and G. Thomas, *J. Biol. Chem.* **257**, 12316 (1982).

[35] Phosphopeptide Analysis of 40 S Ribosomal Protein S6

By JORGE MARTIN-PÉREZ and GEORGE THOMAS

Introduction

As pointed out in the preceding chapter,[1] 40 S ribosomal protein S6 is multiply phosphorylated. Based on experiments *in vivo*[2] and *in vitro*[3] it has been argued that on two-dimensional polyacrylamide gels each increasingly phosphorylated derivative of S6 represents an additional mole of phosphate incorporated per mole of the protein. On the basis of this measurement as much as 5 mol of phosphate has been reported to be incorporated into S6 during regeneration of rat liver,[2] meiotic maturation of *Xenopus* oocytes,[4] and growth factor-induced mitogenesis of mouse 3T3 cells in culture.[5] To identify the kinases and phosphatases which regulate

[1] M. Siegmann and G. Thomas, this volume [34].
[2] A. M. Gressner and I. G. Wool, *J. Biol. Chem.* **249**, 6917 (1974).
[3] R. W. Del Grande and J. A. Traugh, *Eur. J. Biochem.* **123**, 421 (1982).
[4] P. J. Nielsen, G. Thomas, and J. L. Maller, *Proc. Natl. Acad. Sci. U.S.A.* **79**, 2937 (1982).
[5] G. Thomas, J. Martin-Pérez, M. Siegmann, and A. M. Otto, *Cell* **30**, 235 (1982).

the extent of S6 phosphorylation it will be necessary to sequence the sites which become phosphorylated *in vivo*. In addition, because the protein is multiply phosphorylated this raises the possibility that it may be phosphorylated in an ordered or random fashion. If the process is ordered or sequential it will be important to establish the amino acid sites of phosphorylation, since it may be that phosphorylation or dephosphorylation of one site by a kinase or phosphatase requires the phosphorylation or dephosphorylation of another site. As a first step toward identifying these kinases and phosphatases at least four approaches[6-9] have been used to separate the tryptic peptides which become phosphorylated. In our laboratory we have developed a two-dimensional thin-layer electrophoretic method to separate the digested protein into 10 to 11 major phosphopeptides.[9] This method has also been combined with the two-dimensional polyacrylamide gel system described in the preceding chapter[1] to show that *in vivo* these phosphopeptides appear in a specific order.[9] These two-dimensional thin-layer electrophoretograms have already proved valuable in the search for an S6 kinase[10-12] and should prove useful as a preparative step in the purification of the phosphopeptides for sequencing. The method as it is presently applied in our laboratory is described below.

Solutions

Gel elution buffer contains 100 mM (NH$_4$)HCO$_3$, 0.1% SDS, pH 8.3. First-dimensional electrophoresis buffer contains formic acid:acetic acid:water, pH 1.5, in a ratio of 1:3:16. Second-dimensional electrophoresis buffer contains pyridine:acetic acid:water, pH 3.5, in a ratio of 1:10:89. The alternative second-dimensional electrophoresis buffer at higher pH contains 1% (NH$_4$)HCO$_3$, pH 8.9.

Radioactive Labeling of Cells

Swiss mouse 3T3 cells are grown, maintained, and brought to quiescence as described in the preceding chapter.[1] To label cells to high specific activity with ^{32}P, quiescent cultures of 3T3 cells in 15-cm tissue culture plates are rinsed two times with 10 ml of phosphate-free Dulbecco's modi-

[6] S. M. Lastick and E. H. McConkey, *J. Biol. Chem.* **256**, 583 (1981).

[7] R. E. H. Wettenhall and P. Cohen, *FEBS Lett.* **140**, 263 (1982).

[8] O. Perisic and J. A. Traugh, *J. Biol. Chem.* **258**, 13998 (1983).

[9] J. Martin-Pérez and G. Thomas, *Proc. Natl. Acad. Sci. U.S.A.* **80**, 926 (1984).

[10] I. Novak-Hofer and G. Thomas, *J. Biol. Chem.* **259**, 5995 (1984).

[11] J. Blenis, J. G. Spivack, and R. L. Erikson, *Proc. Natl. Acad. Sci. U.S.A.* **81**, 6408 (1984).

[12] J. L. Maller, J. G. Faoulkes, E. Erikson, and D. Baltimore, *Proc. Natl. Acad. Sci. U.S.A.* **82**, 272 (1985).

fied Eagle's medium, and then incubated with 15 ml of the same medium containing 0.3 to 0.5 mCi/ml of $^{32}PO_4$ and the appropriate mitogenic or stimulatory agent. When serum is employed as the mitogen, it is first dialyzed against 150 mM NaCl, 10 mM Tris–HCl, pH 7.4, to remove phosphate. The absence of phosphate from the medium has no effect on either serum-induced S6 phosphorylation or the activation of protein synthesis.[9]

Isolation of S6

Cell lysis, preparation of ribosomes, isolation of total ribosomal protein, and the separation of S6 on two-dimensional polyacrylamide gels are performed as described in the preceding chapter.[1] Following staining and destaining of the two-dimensional polyacrylamide gel it is wrapped in cellophane and autoradiographed. The autoradiogram is then superimposed over the gel and S6 is excised from the gel with a scalpel by cutting directly through the autoradiogram. In the case where individual derivatives are taken only the central 50% of the protein spot is removed, leaving 25% at both the leading and trailing edge of the protein in the horizontal or first dimension of separation (see Fig. 1).[1] Elution of the protein is carried out in an Isco electrophoretic sample concentrator (model 1750). Prior to use, the sample well, the concentrator well, and the connecting bridge are loaded with gel elution buffer containing 0.05% Coomassie Blue dye (R-250) and 3 W is applied to the system for 15 min. Any leaks are

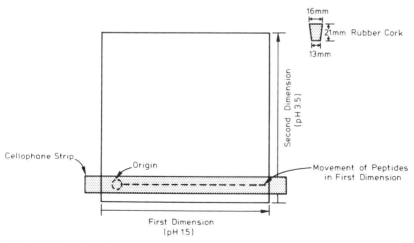

FIG. 1. Schematic diagram of the thin-layer Plate for two-dimensional separation of S6 tryptic peptides. Application of rubber cork and cellophane strip are explained in the text.

immediately evident by the streaming of the dye from the concentration well. Following this check the sample cup is extensively washed with elution buffer, and 100 μl of elution buffer containing 1 mg/ml phosvitin (Sigma) is applied to the concentration well. A perforated plastic disk is set on the stage above the concentration well, the sample cup is carefully filled with elution buffer, and the gel piece to be eluted is applied with forceps to the perforated disk. Elution is carried out at 3 W, and the extent of release of S6 from the gel piece is monitored by Cerenkov counts. Under these conditions 90 to 100% of the protein is released after 15 hr. To recover S6 from the concentration well, the gel elution buffer is first aspirated from all the other chambers and the gel piece together with the perforated plastic disk are removed from the stage lying over the concentration well. Next the 200-μl sample is collected with a Pasteur pipet to which a small piece of siliconized tubing has been attached to the end to avoid damaging the dialysis membrane. The sample well is then washed two times with 200 μl of water, and the washes combined together with the concentrated sample in a 1.5-ml Eppendorf tube. The sample plus the washes are frozen in a liquid nitrogen bath, lyophilized to dryness in a Speed Vac Concentrator (Savant), and washed three times with 1 ml each of acetone–water (9:1) at $-20°$. This washing procedure removes SDS and Coomassie Blue without loss of protein. However, it is essential that the acetone:water mixture remain below $0°$, otherwise S6 becomes partly soluble.

Oxidation of S6

Prior to trypsin digestion of the protein, to ensure that it is present in a homogeneous state it is oxidized with performic acid. Performic acid is prepared by mixing 98% formic acid (Merck) plus 30% H_2O_2 (Merck) at a ratio of 9:1 and allowing the mixture to stand for 1 hr at room temperature. The dry acetone pellet of protein is resuspended in 200 μl of performic acid and incubated for 2 hr at $0°$.[13] Following oxidation the sample is diluted with water up to 1.2 ml, quick frozen in a dry ice–acetone bath, and lyophilized to dryness in the Speed Vac concentrator. This process is repeated twice to ensure all formic acid is removed from the sample.

Trypsin Digestion

Trypsin has been used to cleave the protein into smaller peptides because of the abundance of arginine and lysine residues present in S6.[14] To guard against possible cross-contamination of trypsin with chymotryp-

[13] K. Beemon and T. Hunter, *J. Virol.* **28**, 551 (1978).
[14] E. Collatz, I. G. Wool, A. Lin, and G. Stoffler, *J. Biol. Chem.* **251**, 4666 (1976).

sin, trypsin treated with L-1-tosylamido-2-phenylchloromethyl ketone (Worthington) has been employed. Usually a stock of 1 mg/ml in 0.1 mM HCl is kept at $-20°$, it is stable under these conditions[15] for long periods of time. The performic acid-treated pellet is resuspended in 500 μl of 0.1 M (NH$_4$)HCO$_3$, pH 8.3, and 25 μl (25 μg) of trypsin is added to the sample, giving a final molar ratio of trypsin to protein of 1:4. (The amount of S6 is considered insignificant.) Incubation is carried out at room temperature for 18 hr while gently agitating the sample which is fixed to a Vortex Genie (Scientific Industries). After addition of a second 25-μg aliquot of trypsin the incubation is continued for another 7 hr under the same conditions. The sample is then brought to 1 ml with water, frozen, and lyophilized as described above. This is followed by resuspension of the sample in 1 ml of water, and then again freezing it and lyophilizing it to dryness as described above. This process is repeated three times to remove all traces of (NH$_4$)HCO$_3$. Following the washing steps, the pellet is resuspended in 500 μl of water and centrifuged at 12,000 rpm for 10 min in an Eppendorf centrifuge. The supernatant is set aside, and the pellet is washed with an additional 1 ml of water and then recentrifuged. The pellet, which usually contains less than 5% of the total Cerenkov counts, is discarded, the two supernatants are combined, frozen, and lyophilized as described above.

First-Dimensional Electrophoresis

The dried sample is resuspended at about 500 Cerenkov cpm/μl of water. A 10-μl aliquot (5000 Cerenkov counts) of the digested S6 is applied in two 5-μl portions to the origin of a 20 × 20 cm precoated cellulose thin-layer plate (Merck) 2 cm from each border of the plate (Fig. 1). To ensure that the peptides migrate as tight spots during electrophoresis the 5-μl aliquots are applied slowly, constantly bringing them to dryness by passing a stream of nitrogen over the application spot. The circle or origin formed by the sample is covered with a rubber stopper to one end of which a small circular piece of cellophane has been attached. It is attached to the end of the stopper that comes in contact with the sample. The circumference of the piece of cellophane should be as nearly as possible equivalent to that of the rubber stopper. First-dimensional electrophoresis buffer is then applied to the plate with a nitrogen bomb, connected to a 100 ml-Pyrex sprayer, being careful to apply the buffer equally on the plate. During this operation the buffer is sprayed around the cork, covering the origin as evenly and rapidly as possible. This is because as the buffer migrates beneath the cork it concentrates the sample to a tight circular spot or point.

[15] G. Allen, "Sequencing of Proteins and Peptides: Laboratory Techniques in Biochemistry and Molecular Biology." North-Holland, Amsterdam, 1981.

Uneven distribution of buffer leads to other geometric shapes at the origin which affects later resolution of peptides. After spraying the plate, excess buffer is removed by blotting the plate with 3M paper (Whatman). The thin-layer plate is then placed on the cooling plate of a Camag thin-layer electrophoretic chamber, covered with a plastic sheet having a rubber border 1 cm wide and 0.4 cm deep (to prevent the plastic sheet from coming in contact with the plate), and both reservoirs are filled with first-dimensional electrophoresis buffer. Electrophoresis is carried out at 1.5 kV for 1 hr with the sample moving toward the cathode (Fig. 1). The plastic sheet applied over the thin-layer plate prevents loss of buffer by condensation.

Second-Dimensional Electrophoresis

Following completion of electrophoresis in the first dimension, the thin-layer plate is dried with a fan for approximately 2 hr or until no solvent can be detected by smell. A 2-cm-wide cellophane strip is placed over the thin-layer plate such that it should lie 1 cm on each side of the predicted line the sample would have progressed in the first dimension (Fig. 1). The entire plate is then sprayed with second-dimensional running buffer, being careful to move rapidly along both edges of the plastic strip. When the application of the second-dimensional electrophoresis buffer is carried out correctly the buffer moves under the cellophane strip with approximately the same speed from both edges, such that the peptides are concentrated into a thin line. The plates are then blotted with 3M paper, the reservoirs filled with second-dimensional electrophoresis buffer, and the system is assembled as described above. Electrophoresis is at 1.0 kV for 2 hr, and the thin-layer plate is dried and autoradiographed. A typical autoradiogram is depicted in Fig. 2. Recently we have applied an alternative separation system at pH 8.9 in the second dimension. In this case the sample in the first dimension is applied in the center of the plate 2 cm from the border. Under these conditions the peptides are spread throughout the plate no longer along a diagonal. In addition we have reduced and carboxymethylated the protein prior to digestion with trypsin and have seen no difference in the peptide maps. We have also added 0.1 mM CaCl$_2$ to retard autodigestion of trypsin and have seen no noticeable effect.

Other Methods

At least two other ascending chromatographic methods have been applied to the separation of the peptides in the second dimension[6,8]; in our hands electrophoresis has proved a better method for moving the peptides.

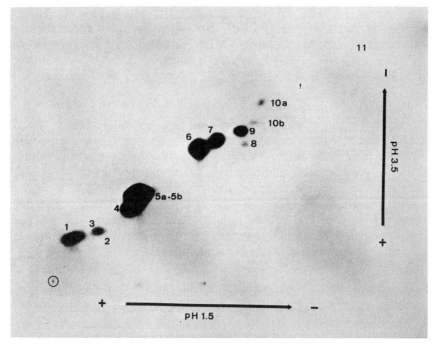

FIG. 2. Two-dimensional autoradiogram of S6 tryptic phosphopeptides. Quiescent cultures of Swiss mouse 3T3 cells have been stimulated with 10% serum for 2 hr in the presence of $^{32}PO_4$. Following lysis S6 phosphopeptides are analyzed as explained in the text. (Reprinted from J. Martin-Pérez, M. Siegmann, and G. Thomas, *Cell* **36**, 287 (1984).

Recently both isoelectric focusing[7] and HPLC[7] have also been used. These methods appear to be much more rapid and should save a great deal of time.

Acknowledgments

The authors are indepted to Drs. P. Jenö, C. Kruse, and I. Novak-Hofer for their critical reading of the manuscript.

[36] Growth Factor Effects on Membrane Transport:
Uptake Studies Using Cell Cultures and Isolated
Membrane Vesicles

By JULIA E. LEVER

The interaction of hormones or growth factors with their target cells is usually associated with effects on membrane transport processes. These responses fall either in the category of early events, occurring within 60 min of growth factor addition to cells, or transport stimulation may occur as a later response, requiring *de novo* protein synthesis.

This chapter describes methods for measurement of membrane transport in cell cultures and in their isolated plasma membrane vesicles. The latter approach offers a cell-free system for measurement of the membrane translocation step dissociated from intracellular effects.[1]

Procedure for Mixed Vesicle Preparation from Fibroblast Cultures[2]:
3T3 or SV3T3 Mouse Fibroblast and A431 Human Epidermoid
Carcinoma

Cell monolayers grown in roller bottles or 15-cm plastic dishes are rinsed twice with PBS and harvested by scraping with a rubber blade. Typically an 850-cm^2 roller bottle yields 10^8 cells at confluence. The cell suspension in PBS is collected by centrifugation in a Sorvall GS-3 rotor for 10 min at 5000 rpm.

The cell pellet is resuspended in 0.25 M sucrose, 1 mM MgCl$_2$, and 10 mM Tris–HCl, pH 7.4. An aliquot of cells is removed for cell count. The suspension is centrifuged at 6000 rpm for 30 min in a Sorvall GS-3 rotor. The clear supernatant is carefully removed and the cell pellet is resuspended in 100 ml of 0.25 M sucrose, 0.2 mM CaCl$_2$, and 10 mM Tris–HCl, pH 7.4. This suspension is transferred to a round-bottom 250-ml plastic centrifuge tube which has been shortened to fit in an Artisan nitrogen cavitation apparatus. The suspension is stirred during the cavitation procedure using a small stirring bar. The apparatus is sealed and pressurized to 600 psi using nitrogen gas, as per manufacturer's instructions. Optimal pressures for lysis of different cell types can be determined empirically but a 15-min equilibration time is sufficient. This step and all subsequent steps are at 4°. After 15 min, the homogenate is collected

[1] J. E. Lever, *CRC Crit. Rev. Biochem.* **7**, 187 (1980).
[2] J. E. Lever, *J. Biol. Chem.* **252**, 1990 (1977).

dropwise, with care to avoid foaming of the sample. Aliquots (1 ml) of homogenate and each subsequent fraction are retained at 4° for subsequent assay of recovery of marker enzyme activities. Samples of homogenate diluted in PBS containing a drop of Trypan Blue can be examined by phase microscopy to ensure that at least 90% cell lysis has occurred. Intact cells do not take up the dye while isolated nuclei will stain blue. If lysis is incomplete, the suspension can be immediately treated a second time by the cavitation procedure. The homogenate is immediately centrifuged in a Sorvall HB-4 swinging bucket rotor for 10 min at 5000 rpm. This step removes intact nuclei and any remaining intact cells. Although inclusion of Ca^{2+} in the homogenization buffer stabilizes nuclei, lysis of nuclei occurs over a period of several hours. The resulting increase of viscosity after nuclear lysis would prevent optimal membrane fractionation if nuclei were not removed before nuclear lysis. The nuclear pellet can be washed by repeating the centrifugation step. The supernatant is layered over a 35% sucrose cushion in a 50-ml centrifuge tube and centrifuged for 30 min at 10,000 rpm in a Sorvall HB-4 swinging bucket rotor. The supernatant above the sucrose cushion is centrifuged at 35,000 rpm for 60 min in a Beckman 50.2 Ti rotor. The "mixed vesicle" pellet is suspended at 5–10 mg protein/ml in 0.25 M sucrose, 10 mM Tris–HCl, pH 7.4, by repeated passage through a fine needle in a 1-ml disposable syringe. Vesicle suspensions are stored in 1-ml aliquots in Nunc ampoules in liquid nitrogen.

Aliquots of each subcellular fraction retained at 4° are assayed for marker enzyme activities in order to determine membrane yield and purity (Table I). Results of a typical fractionation are shown in Table II. The mixed vesicle fraction is enriched 3.5-fold in 5'-nucleotidase, a plasma membrane marker, and 3.6 fold in 2-aminoisobutyric acid (AIB) transport activity. For comparison, enrichment of these activities in a more purified plasma membrane fraction, obtained by the method of Quinlan and Hochstadt,[3] is shown. It is useful to compare transport activities of mixed vesicle fractions with more purified plasma membrane fractions, in order to detect possible contributions to the transport estimation from contaminating membrane fractions. However, due to the low membrane yields associated with highly purified plasma membrane preparation, mixed vesicle preparations are more practical for most studies providing lack of interference from contaminating membranes has been established.

Marker Enzyme Assay Procedures

The following are procedures for assay of the most widely used marker enzymes.

---

[3] D. C. Quinlan and J. Hochstadt, *J. Biol. Chem.* **251**, 344 (1976).

TABLE I

SOME MARKER ENZYMES USED TO EVALUATE MEMBRANE FRACTIONATION

Membrane marker	Membrane	Assay reference
5'-Nucleotidase	Plasma membrane	a
Na+,K+-ATPase	Plasma membrane or	
	basolateral membrane	b
Trehalase	Apical membrane	c
Succinate cytochrome c	Mitochondrial inner membrane	
reductase		a
NADH oxidase	Endoplasmic reticulum	
	and mitochondrial outer	
	membrane	a
β-Hexosaminidase	Lysosome	d
DNA	Nuclei	e

[a] J. Avruch and D. F. H. Wallach, *Biochem Biophys. Acta* **233**, 334 (1971).
[b] B. Forbush III, *Anal. Biochem.* **128**, 159 (1983).
[c] A. Dahlqvist, *Anal. Biochem.* **22**, 99 (1968).
[d] A. J. Barrett, in *"Lysosomes: A Laboratory Handbook"* (J. T. Dingle, ed.), pp. 46–126. Elsevier, New York, 1972.
[e] T. R. Downs and W. W. Wilfinger, *Anal. Biochem.* **131**, 538 (1983).

TABLE II

SODIUM GRADIENT-STIMULATED AIB UPTAKE ACTIVITY AND MARKER ENZYME DISTRIBUTIONS OF MEMBRANE FRACTIONS FROM SV40-TRANSFORMED MOUSE FIBROBLASTS

	Specific activity (nmol/min/mg)[a]		Distribution (%)[a]	
Membrane	AIB uptake	5'Nucleotidase	NADH oxidase	Succinate cytochrome c reductase
Homogenate	0.13 (100)	0.24 (100)	100	100
Crude mixed vesicles	0.47 (72)	0.84 (66)	38	7
Plasma Membranes	1.16 (20)	5.10 (20)	2.6	0.10
Crude endoplasmic reticulum	0.023 (1.4)	0.82 (12)	16	1.2

[a] Data taken from Ref. 2. Numbers in parentheses refer to percentage recovery of activity compared to that observed in the homogenate.

Na+,K+-ATPase. Na+,K+-ATPase activity is assayed in detergent-activated samples by the method of Forbush.[4] Typically, 10 μl of a membrane fraction containing approximately 5 mg protein/ml is incubated 10 min at room temperature with 50 μl of 0.6 mg/ml SDS (Bio-Rad), 1% bovine serum albumin, pH 7.0. Then, 300 μl of 0.3% bovine serum albumin, 25 mM imidazole buffer, pH 7.0, is added. Reaction mixtures containing 50 μl of detergent-activated samples and 250 μl of a mix containing 144 mM NaCl, 30 mM KCl, 4.8 mM disodium ATP, 4.8 mM MgCl$_2$, 72 mM Tris–HCl, and 1.2 mM disodium EDTA, pH 7.5, are incubated for 20 min in the presence or absence of 0.1 mM ouabain at 37°. Triplicate samples are assayed. Blank values are determined from samples prepared as above but lacking membranes. The reaction in terminated by addition of 500 μl of ice-cold stop solution. Stop solution is prepared freshly by adding 1 g ascorbic acid to 17 ml of 1 N HCl, 1.7 ml of 10% ammonium molybdate, 5 ml of 20% SDS, and 12 ml water. After a 10-min incubation on ice, inorganic phosphate liberated by ATP hydrolysis is determined by addition of 750 μl of 2% sodium arsenite, 2% sodium citrate, and 2% acetic acid. After a 10-min incubation at 37°, absorbance at 705 nm is determined. A standard curve is included with each set of assays, using a range of 0.05–0.2 mM phosphate in 0.05% bovine serum albumin. Since bovine serum albumin influences color development it is added to standards at the same concentration as present in unknowns. The phosphate standard curve is linear up to 2.5 OD and the slope is usually 12.4 OD units/μmol phosphate.

Succinate Cytochrome c Reductase.[5] Succinate cytochrome c reductase activity is assayed within 24 hr after membrane fractionation and frozen samples cannot be used. To a cuvette containing 0.9 ml of phosphate-buffered saline at 37°, 75 μl of 660 mM sodium succinate, 20 μl of 100 mM KCN, and 50 μl of a freshly prepared solution of 12.5 mg/ml cytochrome c (Sigma, type VI) are added in the stated order. The solution is mixed, then a sample of 5 to 50 μl of a membrane fraction is added, immediately mixed, and the optical density change at 550 nm is recorded on a strip chart recorder. Blank values determined in mixtures lacking membrane fractions are subtracted. Enzyme units are expressed as OD$_{550}$ change/hr at 37°.

NADH Oxidase.[5] This activity is stable in samples stored up to 4 days at 4°. To a cuvette is added 0.7 ml of 0.01 M Tris–HCl, pH 7.5, and 0.2 ml of 6.6 mM K$_3$Fe(CN)$_6$ at 37°. After mixing, 0.1 ml of a freshly prepared solution of 1 mg/ml NADH in 0.01 M Tris–HCl, pH 7.5, is

[4] B. Forbush III, *Anal. Biochem.* **128,** 159 (1983).
[5] J. Avruch and D. F. H. Wallach, *Biochim. Biophys. Acta* **233,** 334 (1971).

added and the solution is mixed again. Then $10-20~\mu l$ of a membrane fraction is added and the absorbance change at 340 nm is recorded as a function of time. Blank values determined in mixtures without membranes are subtracted. Enzyme units are expressed as the change in OD_{340}/hr at 37°.

5'-Nucleotidase.[5] 5'-Nucleotidase activity (EC 3.1.3.5) is used as a marker of plasma membrane in fibroblast cultures. An assay mix containing 1 ml of 0.5 M Tris–HCl, pH 8, 0.5 ml of 0.1 M $MgCl_2$, 0.1 ml of 0.1 M AMP (Sigma, type II), 40 μCi of [³H]AMP, and 3.36 ml of water can be stored frozen in aliquots. Membrane fractions in volumes made to 50 μl are incubated at 37° with 50 μl of assay mix for 20 min. Blank values are determined in samples lacking protein. The reaction is terminated by addition of 0.5 ml of $ZnSO_4$ and tubes are placed in an ice bath. Then 0.2 ml of a freshly prepared suspension of 0.25 M $Ba(OH)_2$ is added and the tube is mixed using a vortex mixer. The barium hydroxide does not dissolve but is used as a suspension. After 10 min on ice, tubes are mixed and centrifuged for 10 min in a microfuge to remove the precipitate. Unreacted AMP will remain bound to the precipitate while adenosine liberated from AMP hydrolysis will be in the supernatant. Radioactivity of the supernatant is counted by scintillation counting. Enzyme units are nmol/hr at 37°.

Isolation of Apical Membranes from Renal Epithelial Cell Cultures (LLC-PK₁)[6]

Cells from confluent cultures are harvested by scraping, washed, disrupted by nitrogen cavitation, and a postnuclear supernatant obtained, as above. In this case, nitrogen cavitation conditions are 900 psi for 15 min and $CaCl_2$ is omitted. The membrane suspension is then adjusted to 10 mM $MgCl_2$ and stirred for 15 min at 4°. The suspension is then diluted 1:1 in the same solution and centrifuged at 2000 rpm for 10 min in an SS-34 rotor. The supernatant is centrifuged at 35,000 rpm for 1 hr in a Beckman Ti 45 rotor. The final pellet (apical membrane vesicles) is resuspended by syringe in the sucrose–Tris solution at 5–10 mg protein/ml and stored in 1-ml aliquots in liquid nitrogen.

Transport Assay Using Membrane Vesicles.[2,6]

Incubations are carried out in 0.1-ml volumes at 21°. Typical incubation mixtures contain 50 to 200 μg protein membrane vesicles, 0.125 M sucrose, 5 mM $MgCl_2$, 10 mM Tris–HEPES, pH 7.4, 100 mM NaCl, and

[6] J. E. Lever, *J. Biol. Chem.* **257**, 8680 (1982).

labeled substrate. For Na^+-free incubation, NaCl is replaced by choline chloride. Uptake is initiated by adding vesicles to this solution and is terminated after the indicated time by dilution with 5 ml of ice-cold 0.8 M NaCl, 10 mM Tris–HCl, pH 7.4 (wash buffer), filtration through a 0.45-μm Schleicher and Schuell BA-85 nitrocellulose filter, followed by an additional 5 ml of wash buffer. Filters are dried under a heat lamp and counted in a liquid scintillation counter.

Blank (t_0) values are determined using aliquots of vesicles to which labeled substrate is added after dilution in wash buffer followed by immediate filtration and a second 5-ml wash. Typically, time points in the range of 15 sec to 15 min are taken. Time points shorter than 15 sec require the use of an automated device.[7]

Comments on Vesicle Isolation Methods. Several considerations regarding the choice of membrane isolation methods suitable to obtain transport-competent vesicles are as follows: The use of isolation methods which employ fixatives, ligands, or inactivating agents, e.g., Zn^{2+} and concanavalin A, should be avoided as these may affect transport activity. Trypsin should not be used to harvest cells. Na^+ is omitted from the homogenization step since it would interfere with subsequent determination of Na^+-dependent uptake. The internal contents of vesicles will be the same as the composition of the suspension medium during nitrogen cavitation. Alternately, vesicles can be preloaded with ions or substrates by preincubation, then collected by centrifugation in a Beckman airfuge and resuspended in the desired uptake medium. Vesicles are stored in aliquots to avoid repeated freezing and thawing.

Comments on Transport Assay in Vesicles. Leakage of vesicle contents is negligible during the 20 sec required for filtration steps. If wash buffer is at room temperature, leakage is greatly increased. Excessive amounts of protein (>500 μg) clog the filter, resulting in slower filtration and leakage of vesicle contents. Intravesicular volumes are small, typically 0.5–3 μl/mg vesicle protein, limiting the sensitivity of detection of uptake. As another consequence, the time for half-filling of vesicles is rapid (~30 sec or less) necessitating the use of short time points (5–15 sec) for the determination of initial rates of uptake. At longer time points (10–15 min) a plateau value of uptake is observed, which reflects a quasi-steady of accumulation. For Na^+-coupled uptake systems, this represents accumulation above the external concentration in response to existing Na^+ and electrical gradients across the membrane (Fig. 1). This value is collapsed to the plateau level observed in Na^+-free media after dissipation of these gradients or addition of the Na^+ ionophore monensin.

[7] M. Kessler and G. Semenza, *J. Membr. Biol.* **76,** 27 (1983).

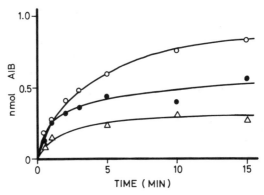

FIG. 1. Uptake of 0.2 mM 2-amino-[³H]isobutyric acid (AIB) by membrane vesicles from SV40-transformed mouse fibroblasts. Standard incubation mixtures contained either 50 mM NaCl (O) or 50 mM choline chloride (△). (●), Uptake of vesicles preloaded with 50 mM NaCl was assayed in Na⁺-containing medium.

Assay of Transport in Intact Cell Monolayers[6,8]

Solution A (10X): 0.14 g NaH$_2$PO$_4$ · H$_2$O; 6.8 g NaCl; 2.2 g NaHCO$_3$. Make up to 100 ml.

Solution B (10X): 0.13 g KH$_2$PO$_4$ · H$_2$O; 16.4 g choline chloride; 3.45 ml choline bicarbonate (80%). Make up to 100 ml.

Solution C (10X): 0.2 g CaCl$_2$; 0.4 g KCl, 0.2 g MgSO$_4$ · 7H$_2$O; 0.01 g phenol red. Make up to 100 ml.

Solution D: 4.5 g D-glucose/100 ml.

To make Earle's salts: Mix 10 ml each of solutions A, C, and D and make up to 100 ml. To make sodium-free formulation of Earle's salts, mix 10 ml each of solutions B, C, and D and make up to 100 ml. For measurement of glucose uptake, Earle's salts are formulated without glucose and 2-deoxy[³H]glucose or 3-0-methyl[³H]glucose is used as tracer. For ⁸⁶Rb⁺ uptake measurements, solution C is modified to contain 0.04 g KCl. Ouabain-sensitive Rb⁺ uptake is used as an *in vivo* measure of Na⁺,K⁺-ATPase activity. For measurement of amino acid transport via the *A system* the nonmetabolizable analog 2-aminoisobutyric acid, ³H- or ¹⁴C-labeled, is used. A more specific analog for this system, α-[1-¹⁴C]methylaminoisobutyric acid, is now available.

Dishes are equilibrated with 0.5 ml of the indicated Earle's salts about 15 min in a CO$_2$ incubator. Labeled substrate is added in 50 µl at zero time

[8] J. E. Lever, D. Clingan, and L. Jimenez de Asua, *Biochem. Biophys. Res. Commun.* **71**, 136 (1976).

to initiate uptake and dishes are immediately returned to the incubator. Uptake is terminated by aspirating labeled medium from the dish, followed by three rapid washes with ice-cold PBS. Then monolayers are treated for 15 min at 4° with 10% trichloroacetic acid. One milliliter of this extract is mixed with 10 ml of Triton–toluene and counted by liquid scintillation. Blank values are determined from dishes to which isotope is added then immediately removed followed by wash steps.

Transport rates can be normalized to protein content per dish. For determination of protein content per dish, trichloroacetic acid is removed by aspiration followed by three washes with ethanol. Then the cell monolayer is incubated 2 hr at 37° with 2% Na_2CO_3 in 0.1 N NaOH and protein is determined by the Lowry method. Alternately, results can be normalized to DNA content per dish. Cells are harvested from a 35-mm dish by trypsinization, centrifuged in an Eppendorf microfuge, and the cell pellet is washed with 1% bovine serum albumin in phosphate-buffered saline. The pellet is resuspended in 50 μl of 1 N NH_4OH, 0.2% Triton X-100, the microcentrifuge tube is then capped and incubated 10 min at 37°. Then 2 ml of 100 mM NaCl, 10 mM Na^+ EDTA, and 10 mM Tris–HCl, pH 7.0 (sample diluent), is added. DNA content of the extract is estimated fluorometrically by the method of Downs and Wilfinger.[9] To 50 μl of DNA standard or unknown samples in sample diluent is added 1.5 ml of bisbenzimide (prepared fresh by dilution of 50 μl of a 200 μg/ml stock solution to 100 ml of sample diluent buffer). After equilibration at 30°, fluorescence is measured, with excitation at 350 nm and emission at 455 nm.

Comments on Uptake in Cell Monolayers. Nonmetabolizable analogs should be used as uptake substrates where possible, to avoid measurement of contributions from intracellular metabolism. Wash steps should be carried out using a wide-orifice pipet to avoid detachment of the cell monolayer. Preequilibration of monolayers before addition of labeled substrate is necessary to achieve the pH and temperature uniformity required for reproducible uptake measurements. Alternately, uptake media employing a buffer other than bicarbonate, e.g., HEPES, can be used, in a non-CO_2-type incubator.

[9] T. R. Downs and W. W. Wilfinger, *Anal. Biochem.* **131,** 538 (1983).

[37] Measurement of Ion Flux and Concentration in Fibroblastic Cells

By STANLEY A. MENDOZA and ENRIQUE ROZENGURT

Introduction

Normal untransformed cells in culture become reversibly arrested at the G_0/G_1 phase of the cell cycle when they deplete the serum of essential growth factor(s). The addition of fresh serum or purified growth factors to these quiescent cells stimulates a complex array of biochemical events and subsequently DNA synthesis and cell division.[1,2] Since the demonstration that serum rapidly stimulates the activity of the Na–K pump in quiescent Swiss 3T3 cells[3] there has been considerable interest in the possible role of changes in ion transport and/or intracellular ionic concentration as regulators of the proliferative response (see Rozengurt and Mendoza[4] for review). Subsequent studies have demonstrated that stimulation of the Na–K pump is caused by enhanced entry of Na into the cells increasing intracellular Na.[5]. The Na entry is in part mediated by an amiloride-sensitive Na/H antiport which also causes intracellular alkalinization.[6] The resulting stimulation of the Na–K pump increases intracellular K and restores the electrochemical gradient for Na. This Na cycle is shown schematically in Fig. 1. In addition to the changes in monovalent ion fluxes and intracellular concentrations, certain mitogens also stimulate Ca efflux[7] and transiently increase the cytosolic concentration of Ca. The techniques which are required to measure cell volume, Na–K pump activity, intracellular Na and K concentrations, intracellular pH, Ca influx and efflux, and cytosolic Ca concentration are the subject of this chapter.

Cell Culture

Swiss 3T3 cells[8] are maintained in 90-mm Nunc Petri dishes in Dulbecco's modified Eagle's medium (DME) containing 10% fetal bovine

---

[1] A. B. Pardee, R. Dubrow, J. L. Hamlin, and R. F. Kletzein, *Annu. Rev. Biochem.* **47**, 715 (1978).

[2] E. Rozengurt, *Horm. Cell Regul.* **8**, 17 (1984).

[3] E. Rozengurt and L. A. Heppel, *Proc. Natl. Acad. Sci. U.S.A.* **72**, 4492 (1975).

[4] E. Rozengurt and S. A. Mendoza, *Curr. Top. Membr. Transp.,* **27**, 163 (1985).

[5] J. B. Smith and E. Rozengurt, *Proc. Natl. Acad. Sci. U.S.A.* **75**, 5560 (1978).

[6] S. Schuldiner and E. Rozengurt, *Proc. Natl. Acad. Sci. U.S.A.* **79**, 7778 (1982).

[7] A. Lopez-Rivas and E. Rozengurt, *Biochem. Biophys. Res. Commun.* **114**, 240 (1983).

[8] G. J. Todaro and H. Green, *J. Cell Biol.* **17**, 299 (1963).

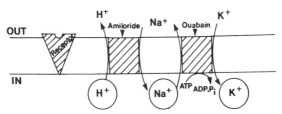

FIG. 1. Schematic representation of the Na cycle in Swiss 3T3 cells. The interaction of a mitogen with its receptors stimulates the activity of the Na^+/H^+ antiport. This increases intracellular pH and Na concentration. The increased intracellular Na in turn stimulates the Na–K pump, increasing intracellular K and reestablishing the electrochemical gradient for Na.

serum (FBS), 100 U/ml penicillin, and 100 μg/ml streptomycin in a humidified atmosphere of 10% CO_2–90% air at 37°. For experiments, 10^5 cells are subcultured into 33-mm Nunc Petri dishes with the same medium. Studies are performed 6–8 days later when the cells are confluent and quiescent.[7] The technique required to produce reversible growth arrest is different in other cell lines.[9,10]

Cell Volume

Intracellular volume is measured with [^{14}C]urea. Urea distributes in both extracellular and intracellular space. It can be used to measure intracellular volume if the urea is removed selectively from extracellular fluid. Four or five rapid washes (<10 sec) with cold 0.1 M $MgCl_2$ removes extracellular [^{14}C]urea with a negligible loss of [^{14}C]urea from the cells. In fact, there is little efflux of urea after 2 min if the $MgCl_2$ is cold (0°). If the $MgCl_2$ is at 37°, efflux of [^{14}C]urea from the monolayer is more rapid · (Fig. 2).

Method

Cultures are incubated for 1 hr with 0.925 ml fresh DME (or electrolyte solution whose composition is described under Intracellular pH) at 37°. Then, 75 μl of [^{14}C]urea is added to produce a final concentration of 0.1 mM with approximately 10^6 cpm/dish. After 15 min, the medium is aspirated and the dish washed rapidly four times with $MgCl_2$. This is done by dipping the dish sequentially into four beakers containing 150 ml of 0.1 M $MgCl_2$ placed in ice. Washing is generally complete 5–6 sec after

[9] S. A. Mendoza, N. M. Wigglesworth, P. Pohjanpelto, and E. Rozengurt, *J. Cell. Physiol.* **103,** 17 (1980).
[10] V. M. Reznik, J. Villela, and S. A. Mendoza, *J. Cell. Physiol.* **117,** 211 (1983).

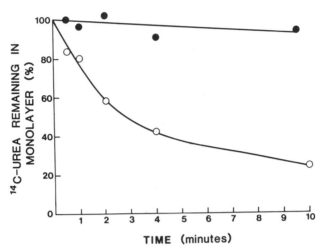

FIG. 2. Effect of time and temperature on efflux of [^{14}C]urea from Swiss 3T3 cells. In this experiment, quiescent cultures of Swiss 3T3 cells were preincubated for 20 min in 925 μl of medium containing insulin (1 μg/ml), epidermal growth factor (EGF) (5 ng/ml), and vasopressin (20 ng/ml). At that point, 75 μl of [^{14}C]urea was added containing 0.1 mM urea and 1 × 10^6 cpm of [^{14}C]urea. After 20 min, cultures were washed rapidly three times with 0.1 M MgCl$_2$ at 0° (O) or 37° (●). The washing was interrupted after the addition of the second wash for the period of time indicated.

aspirating the medium. After washing, the dishes are inverted. Three to 5 min later, the MgCl$_2$ which remains on the side of the dish is carefully removed with a tissue. A small amount of extracellular [^{14}C]urea is "trapped," i.e., not removed by washing. The magnitude of this trapped [^{14}C]urea is estimated by precooling dishes to 0°, adding [^{14}C]urea, and immediately washing with MgCl$_2$ as described above.

When the cultures are dry, the cells are dissolved by adding 1 ml of 0.1 M NaOH/2% Na$_2$CO$_3$/1.0% sodium dodecyl sulfate (SDS). After 30 min, a 200-μl aliquot is taken from each dish to estimate cellular protein.[11] The remaining extract is transferred to a liquid scintillation counter. The values obtained for trapped [^{14}C]urea are subtracted from the total counts obtained from each culture to obtain intracellular cpm. Cell H$_2$O is estimated by dividing intracellular cpm by the extracellular cpm/μl. Intracellular H$_2$O can be expressed as a function of cell protein or simply as μl/dish. In quiescent Swiss 3T3 cells, intracellular H$_2$O is 5.2 ± 0.2 μl/mg protein ($n = 36$).

[11] O. H. Lowry, N. J. Roseborough, A. L. Farr, and R. J. Randall, *J. Biol. Chem.* **193**, 265 (1951).

Comments

1. In general, four or five replicate dishes are used for each experimental condition.
2. It is difficult to study more than seven experimental conditions in a given set of studies.
3. The isotope is added to dishes at 20-sec intervals to allow the washing to begin exactly 15 min later.
4. After washing five to seven dishes, the first beaker of $MgCl_2$ is discarded, the remaining beakers are rotated, and a fresh fourth beaker of $MgCl_2$ is added.

Intracellular Na and K

Intracellular K content and/or concentration can be measured in a 33-mm dish but that of Na is too low to be measured accurately unless the cells are grown in 55- or 90-mm dishes. In general, intracellular Na and K are measured after a relatively short (< 1 hr) exposure to the experimental condition although the measurements can be made accurately at any time. After the desired preincubation, for example in the absence or presence of growth factors, the cultures are washed six times with 0.1 M $MgCl_2$ buffered with 10 mM Tris–HCl, pH 7.0, at 0°. After the final wash, the dishes are tipped to facilitate the removal of the final drop of the wash solution. The cultures are then allowed to dry. When they are dry, 1 ml of 5% trichloroacetic acid (TCA) containing 15 mM LiCl is added. This solution is read directly in the flame photometer with the LiCl serving as the internal standard. Representative values in quiescent Swiss 3T3 cells were as follows: intracellular K — 1.01 ± 0.02 μmol/mg protein; intracellular Na — 0.20 ± 0.03 μmol/mg protein.[9]

Comments

1. Isotonic $MgCl_2$ at 0° is the best wash solution tested. In 1 hr, only 10–15% of intracellular K is lost from dishes incubated with cold $MgCl_2$ (Fig. 3). In contrast, when cold isotonic NaCl is used as a wash solution 15% of intracellular K is lost within 5 min and 85% is lost in 1 hr (Fig. 3). Also, washing with NaCl precludes the measurement of intracellular Na. Isotonic sucrose and choline chloride were also tested as possible solutions for washing the cultures. Each gave a result intermediate between $MgCl_2$ and NaCl (data not shown).
2. The cells can also be lysed with a hypotonic solution of 15 mM LiCl containing 1% toluene. Alternatively, the cultures can be dissolved with 0.1

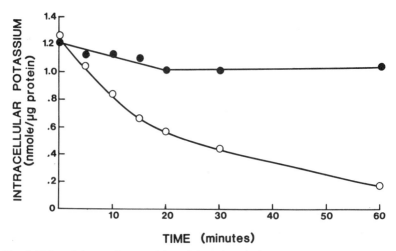

FIG. 3. Effect of time and composition of the wash solution on the measured intracellular K content. Quiescent cultures of Swiss 3T3 cells were incubated in Dulbecco's modified Eagle's (DME) medium for 1 hr. They were then washed with either 0.1 M MgCl$_2$ (●) or 0.14 M NaCl (○) at 0°. Both wash solutions were buffered with 10 mM Tris at ph 7.0. The washing was interrupted after the addition of the third wash for the period of time indicated.

N NaOH/2% Na$_2$CO$_3$/1.0% SDS if only the intracellular K is to be measured.

3. After washing, cultures should be handled only while wearing gloves. If this precaution is not taken, an occasional culture will give a reading for intracellular Na which is 10–100 times the expected value. Presumably this artifact is a result of contamination with Na from the skin. In general, K readings are unaffected.

4. Most flame photometers provide a reading for Na and K concentration in mEq/liter of the unknown solution before a 1:200 dilution. The solution in this technique is not diluted; therefore the actual concentration is 200 times less than the reading of the flame photometer.

5. Intracellular Na and K are usually reported as μmol/mg protein. This can be converted to a concentration by measuring cell volume under the same conditions. Protein can be measured on the same dish by washing twice with 5% TCA and twice with 95% ethanol to remove the TCA. When the dishes are dry, the precipitated material is dissolved in 0.1 N NaOH/2% Na$_2$CO$_3$/1.0% SDS and the protein measured colorimetrically.[11]

6. Since electrolyte loss is very slow when the cultures are washed as described, it is possible to wash a number of dishes (as many as 20) simultaneously.

Na-K Pump Activity (Ouabain-Sensitive [86]Rb Uptake)

The activity of the Na-K pump is estimated by measuring the uptake of ^{86}Rb in the presence and absence of the pump inhibitor ouabain. Murine cells are ouabain resistant and high concentrations of the cardiac glycoside (1-2 mM) are required to inhibit the Na-K pump. In cells from other species, ouabain concentrations in the micromolar range are effective. [86]Rb is used since rubidium substitutes for K but not Na on the Na-K pump.[12] The short half-life of [42]K (12.36 hr) makes experiments with this isotope inconvenient.

[86]Rb uptake is usually measured in 33-mm dishes. Cultures are preincubated in 1 ml of medium containing 5 mM K for varying periods of time with or without growth factors. Ten minutes before the end of the preincubation, ouabain is added to one-half of the dishes. At the end of the preincubation, ^{86}Rb is added (1-2 × 10⁶ cpm/dish). To terminate the uptake of ^{86}Rb, cultures are washed six times with 0.1 M MgCl₂ containing 10 mM Tris-HCl (pH 7.0) at 0'' as described in the section on measurement of intracellular Na and K. The uptake of ^{86}Rb is linear in Swiss 3T3 cells for at least 30 min.[3] In most experiments, the uptake is measured for 10 min. When the dishes are dry, the cells are lysed with 750 μl of 5% TCA. This can contain 15 mM LiCl if intracellular Na and K are to be measured simultaneously. An aliquot is transferred to a liquid scintillation counter to measure Cerenkov radiation using a window for [3]H. Alternatively, the aliquot can be dissolved in scintillation counting fluid and counted using a ^{32}P window.

Comments

1. The activity of the Na-K pump is estimated as the difference between ^{86}Rb uptake in the absence and in the presence of a maximal concentration of ouabain. In Swiss 3T3 cells, ouabain-sensitive ^{86}Rb uptake is usually 50-90% of total ^{86}Rb uptake. [86]Rb uptake is usually expressed in nmol/mg protein/min. The protein can be measured in parallel cultures or in the same dishes using the techniques described in the section on measuring intracellular Na and K.

2. In calculating ^{86}Rb uptake, the specific activity of the isotope is expressed as cpm/nmol of K.

3. [86]Rb efflux into cold MgCl₂ is very slow so it is possible to wash as many as 20 dishes simultaneously without adversely affecting the results of

---

[12] F. M. A. H. Schuurmans Stekhoven and S. L. Bonting, *in* "Membrane Transport" (S. L. Bonting and J. J. H. H. M. de Pont, eds.), p. 159. Elsevier/North-Holland, Amsterdam, 1971.

the experiment. The addition of ^{86}Rb should be staggered with several seconds between the addition to consecutive dishes. This permits the addition of the first MgCl$_2$ wash, which effectively ends the uptake, to occur for each dish precisely at the interval planned.

[86]Rb Efflux

In certain circumstances, it is desirable to measure the efflux of ^{86}Rb to confirm that changes in ^{86}Rb uptake caused by growth factors do not reflect an alteration in the efflux of the isotope. For this study, dishes are loaded with ^{86}Rb to equilibrium by adding $1-2 \times 10^6$ cpm of ^{86}Rb to the medium for 4 hr. After this preincubation, some dishes are washed with cold MgCl$_2$ as described above to assess the amount of ^{86}Rb at the beginning of the efflux. Other dishes are washed three times with prewarmed medium and then incubated at 37° under varying conditions. At intervals of 10 min, dishes are washed with cold MgCl$_2$ and intracellular ^{86}Rb measured as described for measuring ^{86}Rb uptake.

Comments

1. In Swiss 3T3 cells ^{86}Rb efflux is relatively slow with about 40–50% of the isotope remaining inside the cells after 1 hr.[13]
2. The results of this experiment are usually plotted as the log (fraction of initial ^{86}Rb remaining in the monolayer) vs time to calculate the ^{86}Rb half-time.

[22]Na Uptake

In these experiments, quiescent cultures of Swiss 3T3 cells are washed twice with either modified DME or an electrolyte solution. DME is modified by the isotonic replacement of NaCl with either choline chloride or sucrose to increase the specific activity of [22]Na. The electrolyte solution contains 50 mM NaCl, 200 mM sucrose, 5 mM KCl, 1.8 mM CaCl$_2$, 0.9 mM MgCl$_2$, 25 mM glucose, and 20 mM HEPES–Tris, pH 7.4. Cultures are then incubated in modified DME or electrolyte solution for 20–50 min with or without mitogens prior to the addition of 2 mM ouabain dissolved in the same medium. Twenty minutes later, $2-4 \times 10^6$ cpm of [22]NaCl is added. After 3 min, the dishes are washed rapidly five times with 0.1 M MgCl$_2$–10 mM Tris–HCl (pH 7.0). The cultures are allowed to dry and the cells lysed with 5% TCA. This can contain 15 mM LiCl if intracellular Na and K content are to be measured in the same dishes. An aliquot is

[13] S. A. Mendoza, N. M. Wigglesworth, and E. Rozengurt, *J. Cell. Physiol.* **105**, 153 (1980).

counted either in a liquid scintillation counter set at a ^{14}C window or in a gamma spectrometer.

Comments

1. The efflux of Na through the Na–K pump is extremely rapid. It is therefore essential to inhibit the activity of the pump with ouabain before measuring ^{22}Na uptake. It is also possible to block the activity of the Na–K pump by removing K from the medium.

2. The uptake of ^{22}Na is linear for 3 min in Swiss 3T3 cells.[5] By 5–10 min the uptake is no longer linear with time and a true initial rate is no longer being measured.[6]

3. In view of the fact that the uptake period is short, it is preferable to do a relatively small number of dishes (10–20) in each set of experiments so that the uptake can be terminated exactly 3 min after the addition of the ^{22}Na.

4. The electrolyte solution described above simplifies this technique because the entire experiment can be done with the dishes kept at 37° in a water bath in which the level of the water is maintained just above the level of the rack on which the dishes are placed. This eliminates the need to open and close the incubator frequently, thereby making it difficult to maintan the temperature and CO_2 concentration in the desired range. It should be noted that a portion of Na uptake is linked to the uptake of amino acids.[9] This possibility can be tested by adding a mixture of amino acids to the electrolyte solution providing a final amino acid composition identical to that of DME.

5. The amount of ^{22}Na taken up during a 3-min incubation is relatively small, so it is usually necessary to count the samples for 10–20 min. Furthermore, to obtain a true initial rate of ^{22}Na uptake, it is necessary to correct for extracellular ^{22}Na not removed by the washing process. To do this, dishes are placed on ice to cool. ^{22}Na is then added and the dishes washed immediately. The counts remaining on these dishes are subtracted from those measured after a 3-min incubation with ^{22}Na at 37° to calculate the true initial rate of ^{22}Na uptake.

6. Na uptake can also be assessed by following the increase in intracellular Na after the addition to ouabain.

Intracellular pH

The intracellular pH is determined by measuring the uptake at equilibrium of the weak acid 5,5-dimethyl[2-^{14}C]oxazolidine-2,4-dione (DMO) by the cells (see also Cassel *et al.*, this series, Vol. 147 [38]). The protonated form of the acid is highly permeant while the ionized form is impermeant. Since at equilibrium the concentration of the unionized form is the same on both sides of the plasma membrane, the distribution between intracellular and extracellular H_2O is determined by the intracellular and extracellular pH values.

In this procedure, cells are washed twice with modified DME or electrolyte solution. The DME is modified by replacing $NaHCO_3$ with 38 mM HEPES–Tris, pH 7.0. Most experiments are done with an NaCl concentration of 50 mM with the remaining NaCl replaced by either choline chloride or sucrose. The experiment can be done at a Na concentration of 140 mM, however. The electrolyte solution contains NaCl (50 mM), KCl (5 mM), $CaCl_2$ (1.8 mM), $MgCl_2$ (0.9 mM), glucose (25 mM), HEPES–Tris (30 mM; pH 7.0) with the remaining NaCl replaced by either sucrose or choline chloride. It should be noted that either solution makes it unnecessary for the cultures to be equilibrated with CO_2. This is enormously helpful in view of the need to wash the dishes individually and rapidly and to maintain the extracellular pH constant which is essential to the calculation of intracellular pH (see below). It is possible to do this procedure in a CO_2 box; however, careful attention must be paid to maintain a constant CO_2 concentration, thereby providing a constant extracellular pH.

After washing, the cultures are incubated for 1 hr in 1.925 ml of solution containing the growth factors to be tested. At that point, 75 μl of [14C]DMO is added, giving a final concentration of 150 μM; $3-4 \times 10^6$ cpm/dish. After 15 additional minutes, cultures are washed by aspirating the medium and then rapidly dipping the dish into four beakers, each containing 150 ml of 0.1 M $MgCl_2 - 10$ mM Tris–HCl (pH 7.0) placed in ice. This process is generally complete in 5–6 sec. The remaining $MgCl_2$ is then aspirated from the dish and the dish is inverted. Three to five minutes later, the $MgCl_2$ remaining on the rim of the dish is carefully removed with a tissue.

When the dishes are dry, the cells are dissolved by adding 1 ml of 0.1 M NaOH/2% Na_2CO_3/1.0% SDS. Thirty minutes later, a 200-μl aliquot is taken from each dish to estimate protein[11] and the remaining liquid is transferred to a liquid scintillation counter to measure radioactivity. The values obtained are corrected for "trapped" extracellular [14C]DMO by subtracting the counts obtained when [14C]DMO is added to cultures chilled on ice and the medium immediately aspirated and the dishes washed as described above.

Comments

1. The uptake of [14C]DMO reaches a plateau after 10 min and remains stable between 10 and 30 min (Fig. 4). Since the calculation of

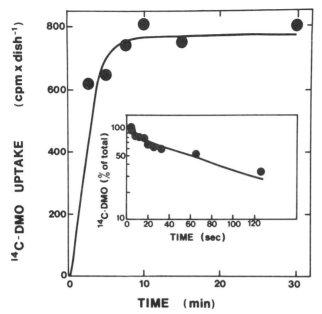

FIG. 4. Time course of [¹⁴C]DMO uptake and efflux in Swiss 3T3 cells. Confluent cultures of Swiss 3T3 cells were incubated for 1 hr at 37° in modified DME medium (pH 7.00) containing insulin (10 μg/ml), vasopressin (40 ng/ml), and platelet-derived growth factor (0.9 μg/ml). Then, 133 μM [¹⁴C]DMO containing 1.0×10^6 cpm was added. The cultures were incubated at 37° for varying lengths of time and then rapidly washed four times with 0.1 M MgCl₂. *Inset:* The cultures were treated as described except that the washing was interrupted after the addition of the second wash. The total wash time is plotted which is the sum of the average washing time (5.3 sec) and the duration of the interruption. Taken from Schuldiner and Rozengurt.[6]

intracellular pH requires that the DMO concentration be equilibrated between the intracellular and extracellular space, the uptake must be measured during the period of the plateau.

2. DMO in the un-ionized form is highly permeant. For this reason, it is essential that the washing procedure be carried out rapidly as described above. Even during a 5- to 6-second wash, it is estimated that 10–15% of intracellular DMO is lost (Fig. 4, inset). The loss of DMO during washing is a more significant problem than with any of the other Methods described in this section.

3. In general, four or five replicate dishes are used for each experimental condition. In view of the necessity for washing the dishes rapidly and individually, it is difficult to test more than seven experimental conditions in any given set of experiments.

4. The [¹⁴C]DMO is added to the dishes at 20-sec intervals to allow the washing of each to begin exactly 15 min later. Although the uptake is

usually measured after exactly 15 min, this is not absolutely essential since intracellular [^{14}C]DMO does not change between 10 and 30 min after the addition of the isotope to the medium (Fig. 4).

5. After washing five to seven dishes, the first beaker in the sequence is discarded, the three remaining beakers are rotated, and a beaker containing fresh, cold 0.1 M MgCl$_2$– 10 mM Tris–HCl (pH 7.0) is added at the end of the sequence.

6. The pH$_i$ is calculated by the formula of Waddell and Butler.[14]

$$pH_i = 6.3 + \log\left[\left(\frac{[DMO]_i}{(\text{cell H}_2\text{O})(\text{cell protein})[DMO]_e}\right)(10^{pH_e - 6.3} + 1) - 1\right]$$

where pH$_i$ is the intracellular pH; [DMO]$_i$, total cpm of DMO in the experimental dish − total cpm of DMO "trapped"; cell H$_2$0, expressed in μl/mg protein; cell protein, expressed in mg/dish; [DMO]$_e$, concentration of [^{14}C]DMO in the medium expressed as cpm/μl; and pH$_e$, pH of the medium.

7. Other laboratories have reported measurement of [pH]$_i$ using the fluorescent indicator 2′,7′-bis(carboxyethyl)-5,6-carboxyfluorescein (BCECF).[15] The theory and general methodology of this technique is identical to that described below utilizing Quin 2 to measure intracellular calcium concentration.

Ca Fluxes and Intracellular Ca

There has recently been considerable interest in the role of Ca in the regulation of various cellular processes in general and of cell proliferation in particular. Intracellular Ca exists in a number of pools including the mitochondria, endoplasmic reticulum, and plasma membrane.[16,17] Cytoplasmic free Ca concentration is maintained between 50 and 200 nM in most cell types.[18-20] Since cytosolic Ca is only a small fraction of total intracellular Ca, changes in calcium homeostasis are best assessed by measuring both cytosolic free Ca concentration and Ca fluxes across the plasma membrane.

[14] W. J. Waddell and T. C. Butler, *J. Clin. Invest.* **38,** 720 (1959).
[15] W. H. Moolenaar, L. G. J. Tertoolen, and S. W. de Laat, *J. Biol. Chem.* **259,** 7563 (1984).
[16] E. Carafoli and M. Crompton, *Curr. Top. Membr. Transp.* **10,** 151 (1978).
[17] S. K. Joseph, K. E. Coll, R. H. Cooper, J. S. Marks, and J. R. Williamson, *J. Biol. Chem.* **258,** 731 (1983).
[18] W. H. Moolenaar, L. G. J. Tertoolen, and S. W. de Laat, *J. Biol. Chem.* **259,** 8066 (1984).
[19] J. D. H. Morris, J. C. Metcalfe, G. A. Smith, T. R. Hesketh, and M. V. Taylor, *FEBS Lett.* **169,** 189 (1984).
[20] R. Y. Tsien, T. Pozzan, and T. J. Rink, *Nature (London)* **295,** 68 (1982).

^{45}Ca Uptake

Quiescent cultures of 3T3 cells are preincubated as indicated by the experimental design. Cultures are washed twice with DME at 37°, then incubated with 1 ml of DME containing 4 μCi/ml of ^{45}Ca. At the conclusion of the uptake the cultures are washed rapidly seven times at 37° with DME to which 3 mM EGTA is added. After the cultures are dry, 1 ml of 0.1 M NaOH/2% Na_2CO_3/1% SDS is added. Thirty minutes later, an aliquot is counted in the liquid scintillation counter and 200 μl is used to estimate protein.[11] Uptake is expressed as nmol/mg protein/min.

Comments

1. Since the washing is done at 37° it must be done rapidly. For this reason, each dish is washed individually. A capillary pipet is connected to the vacuum line and supported with a ring stand. This allows the addition and aspiration of the wash solution to be done virtually simultaneously. Seven washes can be completed in less than 10 sec.

2. The wash solution contains EGTA to remove Ca from an extracellular bound pool. This pool can be reduced from 40% to less than 5% of total cell ^{45}Ca by washing in this manner.[7]

3. In Swiss 3T3 cells, ^{45}Ca uptake is essentially linear for about 5 min. By 10–20 min, however, ^{45}Ca uptake slows, eventually reaching equilibrium which is determined by the intracellular exchangeable Ca.[21]

^{45}Ca Efflux

Quiescent cultures of Swiss 3T3 cells are loaded with ^{45}CaCl$_2$ (4 μCi/ml) by incubating them with the isotope for 24 hr in conditioned medium. At the onset of experiment, cultures are washed rapidly seven times with DME at 37° (see comment 1 above). The cultures are then incubated with 2 ml of DME at 37° in an atmosphere of 90% air : 10% CO_2. At appropriate times, a 200-μl sample is taken for counting and replaced by 200 μl of fresh medium. At the conclusion of the experiment, the radioactivity remaining in the cells is determined using the technique described above for measuring ^{45}Ca uptake. Efflux can be expressed as nmol/mg protein or as a percentage of the total ^{45}Ca taken up by the cells during the loading period. This equals the sum of the radioactivity which enters the medium during the experiment and the radioactivity remaining in the cells.

Comments

1. After the loading period, cultures can be washed with different solutions depending upon the experimental design. For example, if the

[21] A. Lopez-Rivas and E. Rozengurt, *Am. J. Physiol.* **247,** C156 (1984).

experiment is designed to eliminate measurement of efflux from the extra-cellular bound pool, cultures can be washed with DME containing 3 mM EGTA.[7,21] If ^{45}Ca–^{40}Ca exchange is to be assessed, some cultures can be washed with Ca-free DME containing 50 μM EGTA and efflux into that solution measured. The difference between efflux measured in this manner and the standard technique described above provides an estimate of the magnitude of ^{45}Ca–^{40}Ca exchange.

2. The frequent samples required in these studies allows only a small number of cultures to be tested simultaneously.

3. In Swiss 3T3 cells, ^{45}Ca efflux is relatively rapid. Therefore, samples are taken at intervals of 30 sec–1 min. The half-life of the "fast" compartment of intracellular Ca is 9 min.[7]

Intracellular Ca ([Ca]$_i$)

Until recently, measurement of intracellular Ca ([Ca]$_i$) was technically difficult. This situation was changed by the development of fluorescent indicators.[20,22] These indicators, of which the most widely used is Quin 2, are anions with multiple carboxylate groups (Quin 2 has four). As such, they are impermeant. They do enter cells, however, if the carboxylate groups are esterified. When the permeant ester enters the cell, endogenous esterases cleave the ester linkages, regenerating the impermeant parent compound. This is trapped and can accumulate intracellularly at high concentrations. Quin-2 fluorescence is measured most easily in suspended cells. For these experiments, therefore, cells are grown using a different technique from the one described above.

Cytodex 2 Cell Culture

Cytodex 2 is placed in phosphate-buffered saline (PBS) at 1 g/50 ml and allowed to swell for at least 4 hr. The beads are washed twice with fresh PBS and the bottles are autoclaved.

Cells are grown on Cytodex 2 beads suspended in DME containing 10% FBS in 250- or 500-ml flasks. Three hundred seventy-five milligrams of Cytodex 2 is used/250 ml of final volume. Cells are seeded at a density of 5×10^4 cells/ml of final volume. To allow the cells to adhere to the heads, the initial volume of medium used is one-third of the final volume. This suspension of cells and beads is gassed with 10% CO_2 : 90% air and maintained at 37°. Using a Techne MCS-104L microcarrier stirrer, the suspension is stirred intermittently (2 min on : 30 min off) overnight to maximize adherence. The next morning, the medium volume is increased to the final

[22] R. Y. Tsien, T. Pozzan, and T. J. Rink, *Trends Biochem. Sci. (Pers. Ed.)* **9**, 263 (1984).

volume, regassed with 10% CO_2 : 90% air, and stirred continuously at the slowest rate which prevents the beads from settling (about 25 rpm). About 1 week after seeding, the medium is changed to DME containing 1% FBS for 24 hr. Then, the cells are quiescent and ready for study.

Measurement of $[Ca]_i$

At the onset of the experiment, the quiescent cells on beads are washed twice with serum-free DME. The beads from a 500-ml flask are resuspended in 180 ml by gentle shaking and 10 ml is transferred to a plastic tube. They are incubated in the tube with 15 μM Quin 2 tetraacetoxymethyl ester for 45 min at 37°. The stock solution of 7.5 mM Quin 2 tetraacetoxymethyl ester is dissolved in DMSO. After the incubation, the beads are washed three times in medium A which contains 140 mM NaCl, 5 mM KCl, 1.8 mM $CaCl_2$, 0.9 mM $MgCl_2$, 25 mM glucose, 16 mM HEPES, 6 mM Tris, and a mixture of amino acids at the concentrations found in DME (pH 7.20). The washes are removed after the beads are allowed to settle. The beads are then transferred to a 1-cm^2 cuvette. Enough beads are used to produce a layer of 4–5 mm at the bottom of the cuvette. Medium A is added to produce a total volume of 2.0 ml.

The cuvette is placed in a fluorimeter with a thermostatted cell holder which maintains the temperature at 37° and which allows the beads to be suspended evenly by stirring. Fluorescence can be monitored continuously before and after the addition of growth factors at an excitation wavelength of 339 nm and emission wavelength of 492 nm.

The measurement described above gives relative fluorescence. This can be converted into actual readings of $[Ca]_i$ if, at the end of the experiment, first Triton X-100 and then EGTA are added to the cuvette. The addition of 0.02% Triton X-100 lyses the cells, releasing the trapped Quin 2 from the cells. Since the concentration of Ca in medium A is high, this produces maximal fluorescence of the Quin 2. If 100 mM EGTA is then added, the Ca in the solution is chelated and the fluorescence falls to the background plus the fluorescence of Quin 2 in the absence of Ca. Intracellular Ca is calculated using the formula of Tsien et al.[20]:

$$[Ca]_i = 115 \text{ n}M (F - F_{min})/(F_{max} - F)$$

where F is the fluorescence corresponding to the unknown $[Ca]_i$; F_{max} is the fluorescence after the addition of Triton X-100; F_{min} is the fluorescence after the addition of EGTA; and 115 nM is the apparent dissociation constant of Quin 2.

In quiescent Swiss 3T3 cells, $[Ca]_i$ is 207 ± 9 nM; $n = 139$ (Mendoza, Sinnett-Smith, and Rozengurt, unpublished observations).

Comments

1. It is important to prevent the cells from becoming hypoxic during the course of the experiment. For this reason beads should be kept in the main flask and stirred until the beginning of the incubation with Quin 2 tetraacetoxymethyl ester. If the beads are allowed to settle, particularly in a conical tube, the cells will deteriorate in 1–2 hr.

2. Even mixing of the beads in the cuvette is essential to produce a stable baseline of fluorescent readings. We have found that the mixing system in our Perkin-Elmer LS-5 luminescence spectrometer becomes overloaded if a larger quantity of beads than is described above is added to the cuvette. When this happens, the readings are so variable that accurate measurements of [Ca]$_i$ are impossible.

3. Fura 2 can be used as a calcium indicator in place of Quin 2. When this is done, cells on Cytodex beads are incubated with 1 μM Fura 2 tetraacetoxymethylester for 10 min at 37°. Fluorescence is measured as described above except that the excitation wavelength for Fura 2 is 335 nm and the emission wavelength is 510 nm. [Ca]$_i$ is calculated as described for Quin 2. The apparent dissociation constant for Fura 2 is 224 nM.[23]

Summary

Techniques for the measurement of cell volume, intracellular Na and K concentrations, Na–K pump activity, ^{86}Rb efflux, [22]Na uptake, intracellular pH, [45]Ca uptake and efflux, and intracellular Ca concentration have been described. The technique for washing and solubilizing the cells is similar in many of these methods. This allows several measurements to be made on the same culture. Measurement of ^{86}Rb uptake and intracellular Na and K in the same sample has been discussed above. It is also possible to measure ^{86}Rb uptake and [^{14}C]DMO distribution in a single dish. There are other possible combinations of these methods. In addition, aspects of ion transport can be measured along with the assessment of macromolecular synthesis by measuring the incorporation of labeled amino acids or thymidine into TCA-insoluble material. We have used these techniques to study the effect of such mitogens as serum,[3,5,7,24] platelet-derived growth factor,[24,25] vasopressin,[13,21,24] tumor promoters,[24,26] fibroblast-derived growth factor,[27] and others on Na–K pump activity,[3,9,25,27] Na uptake,[5,9,25] intracellular pH,[6,24,25] Ca efflux,[7,21] and [Ca]$_i$.[28]

[23] G. Grynkiewicz, M. Poenie, and R. Y. Tsien, *J. Biol. Chem.* **260**, 3440 (1985).

[24] C. P. Burns and E. Rozengurt, *Biochem. Biophys. Res. Commun.* **116**, 931 (1983).

[25] A. Lopez-Rivas, P. Stroobant, M. D. Waterfield, and E. Rozengurt, *EMBO J.* **3**, 939 (1984).

[26] P. Dicker and E. Rozengurt, *Biochem. Biophys. Res. Commun.* **100**, 433 (1981).

[27] H. R. Bourne and E. Rozengurt, *Proc. Natl. Acad. Sci. U.S.A.* **73**, 4555 (1976).

[28] E. Rozengurt and S. A. Mendoza, *J. Cell. Sci.,* Suppl. 3, 229 (1985).

Acknowledgment

This manuscript was written during the tenure of an American Cancer Society Eleanor Roosevelt International Cancer Fellowship awarded by the International Union Against Cancer to S.A.M.

[38] Growth Factor Stimulation of Sugar Uptake

By WENDELYN H. INMAN and SIDNEY P. COLOWICK*

The measurement of glucose transport activity is a relatively simple and rapid technique which can be utilized as an indicator of early growth factor effects on cultured cells. In an investigation of factors which regulate metabolic events, it is important to examine the relationship of membrane transport processes and intracellular substrates at rates which are consistent with metabolic activity. Mediation of the transport of essential nutrients, such as glucose, can be a component in the regulation of cell growth by growth factors.[1] Glucose transport in most nonepithelial cells occurs via the process of facilitated diffusion, which involves the establishment of an equilibrium across the plasma membrane.[2] This mechanism can be subcategorized into two systems: nonregulated and regulated transport.[3] Nonregulated transport is characteristic of red blood cells, hepatocytes, and brain cells.[4] Regulated transport, which gives the system the ability to adjust to cellular requirements, is observed in such cells as fibroblasts, adipocytes, and muscle fibers.[5] In the latter type the transport step becomes rate limiting in the utilization of sugar. Glucose is transported into the cell by stereospecific protein carriers that bind glucose and can be regulated independent of the glycolytic pathway.

Because it is difficult to directly measure the establishment of an equilibrium, only the uptake of glucose into the cell is actually measured, under controlled conditions in which the transport step is the rate-limiting step. The rate of uptake is measured using nonmetabolized radiolabeled glucose analogs such as 2-deoxyglucose or 3-O-methylglucose. The rate of glucose uptake is then defined by the amount of radiolabel retained by the cell. Both are dead-end metabolites. The 2-deoxyglucose is phosphorylated, but does not undergo further processing by enzymes in the glycolytic pathway. The 3-O-methylglucose, while not phosphorylated, accumulates in the cell.

* Deceased.
[1] D. Barnes and S. P. Colowick, *J. Cell. Physiol.* **89**, 633 (1976).
[2] J. Elbrink and I. Bihler, *Science* **188**, 1177 (1975).
[3] A. Klip, *Life Sci.* **31**, 2537 (1982).
[4] H. M. Kalckar, *J. Cell. Physiol.* **89**, 503 (1976).
[5] L. P. Taylor and G. D. Holman, *Biochim. Biophys. Acta* **642**, 325 (1981).

The following is a description of the procedure for determining the rate of glucose uptake into monolayer mouse Swiss 3T3 fibroblast cells in culture. A specific application of the method has been published, in which the stimulation of glucose uptake by a TGFα (EGF) and TGFβ is described.[6]

Preparation of Cells

Uptake rates of cells in different stages of growth respond differently to growth factors in culture. It is important that the investigator decide at what stage of cell growth it is desired to determine growth factor effects. Basal uptake levels differ in cells in various stages of growth.[7] Cells which are rapidly dividing (logarithmic growth) or transformed cells[8] have higher basal rates of uptake, which are less likely to exhibit responsiveness to transport-stimulating growth factors. Such rapidly growing cells may be useful tools in examining factors which depress or inhibit uptake rates. For this reason it is customary and more convenient to use quiescent (nondividing confluent or serum-starved) cells with lower basal levels of uptake to study factors which are mitogenic.

Equipment
Incubator, maintained at 37° with a 5% CO_2 atmosphere
Suction apparatus
Pastuer pipets
Volumetric pipets
Hemacytometer or Coulter counter
Water bath, maintained at 37°
Scintillation counter
Disposable plastic cell scrapers (Costar) or rubber policemen
Reagents and Solutions
PBS: Dulbecco's phosphate-buffered saline
PBS/BSA: PBS with 1 mg/ml bovine serum albumin
LSC Aquasol (New England Nuclear)
0.1% Triton X-100 solution
Uptake solution: 0.5 μCi/ml 2-deoxy [³H] glucose and 0.5 μCi/ml [¹⁴C]mannitol in PBS/BSA

Procedure

1. Standard procedures are described for fibroblasts plated in 35-mm culture dishes in a total volume of 2 ml of culture medium. Prepare cell monolayers by exposing cells to growth factors for predetermined times

[6] W. H. Inman and S. P. Colowick, *Proc. Natl. Acad. Sci. U.S.A.* **82**, 1346 (1985).
[7] S. K. Bose and B. J. Zlotnick, *Proc. Natl. Acad. Sci. U.S.A.* **70**, 2374 (1973).
[8] M. Hatanaka, *Biochim. Biophys. Acta* **355**, 77 (1974).

before assaying for glucose transport activity. Growth factor(s) should be added under sterile conditions directly to the culture medium if incubation times are greater than 12 hr. Changing medium or the addition of large volumes of fresh medium may significantly alter basal levels of uptake. For assays involving shorter incubations (less than 6 hr) medium is aspirated and growth factors are added in PBS/BSA. Cells are washed gently twice with warm PBS/BSA prior to growth factor addition to remove culture medium and serum factors. Growth factors, diluted in PBS/BSA, are then added in a total volume of 1 ml. Very few viable cells are removed by initial washes under normal conditions. After PBS/BSA incubations care should be taken to wash cells very gently in small volumes (0.5 to 1 ml) to minimize cell loss.

2. After growth factor incubations at 37°, cells are washed three times with 1 ml warm PBS/BSA. Uptake solution (1 ml) at 37° is added to washed cells. Cells are incubated at 37° for 10 min. Alternately, labeled mannitol is included in the uptake solution to correct counts retained for nonspecific hexose transport.

3. The reaction is stopped by quickly aspirating the uptake solution and washing cells three times in ice-cold PBS. The last wash is aspirated and plates are inverted to dry at room temperature approximately 1 hr. (Note: cells can be allowed to dry overnight.) After drying plates can be stored at −20°, but the assay should be completed within 3 days.

4. Cells are dispersed by the addition of 1 ml 0.1% Triton X-100 to each plate and the cells are scraped from the plates. Aliquots can be taken directly from scraped cell suspensions.

5 Remove aliquots (25 to 200 μl). Add to 10 ml LSC Aquasol and count in scintillation counter.

6. Corrections for non-carrier-mediated diffusion is calculated by subtracting counts due to [^{14}C]mannitol from total counts retained by cells. The reverse combination of labels can also be used: 2-deoxy [^{14}C]glucose and [^{3}H]mannitol.

7. Total protein is determined by the method described by Bradford.[9] The averages of triplicate data can be given in counts per minute (cpm)/mg total cell protein if desired or cpm/plate/10-min uptake. It may also be useful to determine sample cell number with a hemocytometer or Coulter counter prior to cell disruption.

Comments

This method can be adapted for use on cells in suspension culture. For example, between washes cells could easily be spun down using a tabletop

[9] M. M. Bradford, *Anal. Biochem.* **72**, 248 (1976).

centrifuge. It is then useful to determine cell number per sample. The procedure described usually requires 2–3 hr to complete exclusive of growth factor incubation time. The reagents can be prepared in advance. Uptake solution should be kept frozen (−20°) until use and is stable for several weeks.

Acknowledgments

We are grateful to Sidney Harshman for his suggestions and assistance. This work was supported by Grant AM26518 from the National Institutes of Health and Grant RDP-5F from the American Cancer Society.

[39] Assay of Growth Factor Stimulation of Fluid-Phase Endocytosis

By H. STEVEN WILEY and DANA N. MCKINLEY

During the process of endocytosis, cells necessarily incorporate a bit of the extracellular medium. This unavoidable medium incorporation is known as fluid-phase endocytosis and it can reveal some of the general features of membrane dynamics in cells. For example, during exocytosis, the membrane surface area of cells can be significantly increased. A subsequent membrane retrieval event is required to bring the cells into homeostasis. This "compensatory endocytosis" can be observed by using markers of fluid-phase endocytosis since the nonselective nature of the process necessarily reveals all membrane retrieval events.[1] Thus fluid-phase endocytosis can provide insight into the general dynamics of the cell surface. Perhaps not surprisingly, growth factors have been shown to have a pronounced effect of the rate of fluid-phase endocytosis which is probably linked to their ability to significantly alter the functional properties of the cell surface and to stimulate secretion.[2,3]

Criteria for a Purely Fluid-Phase Marker

Fluid-phase endocytosis is measured by quantitating the rate at which some noninteracting and nonpenetrating marker in the culture medium is incorporated by cells. These measurements are complicated by two sources

[1] S. Schacher, E. Holtzman, and D. C. Hood, *J. Cell Biol.* **70**, 178, (1976).
[2] H. T. Haigler, J. A. McKanna, and S. Cohen, *J. Cell Biol.* **83**, 82 (1979).
[3] H. S. Wiley and D. D. Cunningham, *J. Cell. Biochem.* **19**, 383 (1982).

of error which both cause an overestimation of uptake. First, there is the adsorption of the marker to the plasma membrane (specific or nonspecific). A true fluid-phase marker should show no interaction whatsoever with cells. Since the vesicles responsible for pinocytosis are very small (<100 nm in diameter), they have a very high surface-to-volume ratio. Hence even slight adsorption during the process of invagination and endocytosis will result in a significant overestimation of the amount of fluid incorporated by cells. The second problem is that it is essential to remove all extracellular probes before quantitating the amount incorporated by the cells. Usually the cells must be exhaustively washed and removed from their culture dishes prior to measuring their uptake.

Theoretical criteria for fluid-phase markers have been proposed for a number of years, but none has proved to be satisfactory. It has been proposed that a fluid-phase marker should show incorporation rates that are completely linear as a function of concentration.[4] However, this is actually a criterion for a nonsaturable process rather than a noninteractive one. Linear adsorption isotherms are expected for all weak and nonsaturable adsorption processes. For example, both horseradish peroxidase and the nonspecifically adsorbed polylysine conjugate of horseradish peroxidase are incorporated as a linear function of their concentration in the medium, although the absolute magnitudes of their uptake show a 900-fold difference.[5] The ability to completely rinse away a marker from the cell surface has also been argued to be a criterion for a noninteracting marker,[4] but actually only indicates than any adsorption occurring is reversible. During a fluid uptake experiment the external concentration of a marker is orders of magnitude greater than after the cells have been rinsed, so extrapolating the amount of adsorption occurring during the former by measuring the latter is inappropriate. A completely linear uptake rate as a function of time has also been argued to be characteristic of fluid-phase endocytosis,[6] but this would only be true if fluid-phase endocytosis is an irreversible process. Whether this is true or not is unclear, but is is certainly not an intrinsic property of fluid uptake per se. Unavoidably, one must rely on empirical criteria for establishing the usefulness of any single marker in a given system, a judgment that was reached a number of years ago by Lloyd and colleagues.[6] The best marker of fluid-phase endocytosis is that marker that shows the lowest rate of uptake into the cell. Its uptake may or may not represent true fluid uptake, but it will establish an *upper* limit for the process.

[4] R. M. Steinman, J. M. Shaver, and Z. A. Cohn, *J. Cell Biol.* **63**, 949 (1974).
[5] H. J. P. Ryser, I. Drummond, and W. C. Shen, *J. Cell. Physiol.* **113**, 167 (1982).
[6] J. B. Lloyd and K. E. Williams, in "Protein Turnover and Lysosome Function" (H. L. Segal and D. J. Doyle, eds.), p. 395. Academic Press, New York, 1978.

Potential Fluid-Phase Markers

^{125}I-Labeled Poly(vinylpyrrolidone)

One of the earliest fluid-phase markers was ^{125}I-labeled poly(vinylpyrrolidone) (^{125}I-PVP). The parent molecule is a synthetic water-soluble polymer ($M_r \simeq 40,000$) which is very hydrophilic and uncharged except for one primary amine group at each chain terminus. This amino group is displaced during the iodination reaction. Apparently it is quite nontoxic to cells since it has been used as a substitute for albumin in human plasma. In fibroblasts, uptake of ^{125}I-PVP at a concentration of 10 mg/ml is linear for at least 16 hr with no apparent deleterious effects. In our hands, it has a lower degree of cell adsorption than other putative fluid-phase markers. The very large size of the polymer precludes leakage across membranes and this may be an important consideration in the analysis of subcellular distribution of fluid-phase markers in cells. In addition, ^{125}I-PVP is readily precipitated by trichloroacetic acid and can be counted directly in a gamma counter. ^{125}I-PVP cannot be degraded by cells and is lost very slowly from the lysosomes during chase incubations. Thus it would seem to be an ideal fluid-phase marker.

Virtually all reports on the use of ^{125}I-PVP as a fluid phase marker are from Europe. Since it has been used as a marker of urinary clearance rates in humans, federal laws restrict its sale by Amersham, which is currently its only manufacturer. However, ^{125}I-PVP itself is very inexpensive and can be easily radioiodinated by the technique described below.

[U-^{14}C]Sucrose

Two of the most popular fluid-phase markers are [^{3}H]sucrose and [U-^{14}C]sucrose. Although the ^{14}C isotope is more expensive, it is generally preferable since there is no possibility of isotopic exchange reactions and is easier to quantitate reliably since it has a higher energy spectrum. The cell surface is impermeable to [U-^{14}C]sucrose and we know of no reports that indicate that it is adsorbed or internalized by receptors. Thus it is a potentially useful marker. One major disadvantage of [U-^{14}C]sucrose is the necessity to remove glucose impurities immediately before use since mammalian cells actively incorporate that breakdown product. Since very large quantities of radiolabeled markers must be used in fluid uptake studies it may be prohibitively expensive for larger studies. Another disadvantage of [U-^{14}C]sucrose is that its small size makes it much less suitable for a marker of fractionated subcellular organelles. Isolated vesicles can leak molecules as small as sucrose at a substantial rate even on ice.[7]

[7] G. Meissner and D. N. McKinley, *J. Membr. Biol.* **30**, 79 (1976).

Horseradish Peroxidase

The enzyme horseradish peroxidase (HRP) has been widely used in fluid-phase uptake studies. However, it should be used only if the purely fluid-phase nature of its uptake can be demonstrated by comparison with either ^{125}I-PVP or [U-^{14}C]sucrose since the structural complexity of HRP provides greater opportunities for adsorptive uptake. Some cells are known to bind HRP by way of the cellular mannose/*N*-acetylglucosamine receptors which are efficiently internalized through receptor-mediated endocytosis.[8] Since HRP is an enzyme, it can potentially be inactivated and degraded after internalization. In addition, some cells have endogenous levels of HRP-like activity which can complicate the measurement of incorporated marker. On the other hand, there are several great advantages to using HRP. It is inexpensive, nonradioactive, and can be easily handled. It can be localized within cells by a variety of methods. Enzymatic activity can be quantitated readily using a colorimetric asay.[9] Its subcellular localization can be determined directly by histochemical staining with diaminobenzidine.[10] Finally, the treatment of isolated vesicles containing HRP with diaminobenzidine will significantly increase their buoyant density. The more dense vesicles can then be isolated by density gradient centrifugation.[11]

When HRP is used as a fluid-phase marker, it is necessary to minimize its uptake by mechanisms other than fluid endocytosis. Any uptake occurring through receptors recognizing terminal mannose residues can be blocked by the addition of yeast mannans.[8] Another approach would be to measure the rate of HRP uptake in the presence of a 10-fold excess of catalytically inactive HRP (available from Sigma Chemical Co.).

Fluorescent Markers

Highly fluorescent compounds, such as Lucifer Yellow, have been proposed as suitable probes for quantitating fluid-phase endocytosis.[12] However, experience in our laboratory indicates that these compounds suffer from an unacceptable high degree of adsorption to the cell surface. This is really not surprising since their fluorescent properties depend upon their aromatic ring structures which are hydrophobic in nature. Conjugates such as FITC-Dextran with less than one fluor/polymer have applications for special purposes since they can be taken up in large amounts and used to monitor the ionic nature of their environment.[13]

[8] S. J. Sung, R. S. Nelson, and S. C. Silverstein, *J. Cell. Physiol.* **116**, 21 (1983).
[9] R. M. Steinman and Z. A. Cohn, *J. Cell Biol.* **55**, 186 (1972).
[10] R. C. Graham, Jr., and M. J. Karnovsky, *J. Histochem. Cytochem.* **14**, 291 (1966).
[11] P. J. Courtoy, J. Quintart, and P. Baudhuin, *J. Cell Biol.* **98**, 870 (1984).
[12] J. A. Swanson, B. D. Yirinec, and S. C. Silverstein, *J. Cell Biol.* **100**, 851 (1985).
[13] M. J. Geisow, *Exp. Cell Res.* **150**, 29 (1984).

Preparing Fluid-Phase Markers

Preparation of ^{125}I-PVP

This procedure is a modification of the method described by Briner.[3,14] It is extremely important to use appropriate safety precautions at each step in the protocol. The procedure uses a large amount of radioisotope (10 mCi) and requires you to add the iodine directly to a very acidic solution. Since these conditions will maximize the possibility of vaporization of the radioisotope into the air, it is essential that all steps be performed in an approved radiochemical fume hood. At least two layers of PVC gloves should be worn at all times and a survey meter should be present (such as a Ludlum model 3 survey meter with a model 44-3 detector. Available from Ludlum Measurements, Inc., Sweetwater, TX).

Materials

Poly(vinylpyrrolidone)   (Sigma,   pharmaceutical   grade;   average M_r 40,000)

10 mCi [125]I (Amersham; protein iodination grade at a concentration of 100 mCi/ml)

Sephadex G-10 preswollen in water (Pharmacia)

Dowex 1-X8-400 resin, chloride form (Sigma)

Model UVS-11 Mineralight

Quartz reaction tubes (see Fig. 1). These were made by the local glass shop from a 6 × 0.35 cm i.d. quartz tube and a screw-threaded conical glass reaction vial with a silicon rubber Teflon-lined cap (Wheaton Glass Company, part #240581). The quartz tube is fused at one end and the bottom of the reaction vial is cut off with a diamond saw. The screw-threaded top of the reaction vial is then attached to the open end of the quartz tubes with silicon rubber glue

Predialysis of the PVP. Approximately 5 g of PVP is dissolved in 50 ml of purified water and dialyzed for 2 days against several changes of water using Spectropore membrane with an M_r 16,000 cutoff. The PVP is then recovered by lyophilization with an 80% yield.

Iodination Procedure. PVP is dissolved at a concentration of 200 mg/ml in 0.2 N sulfuric acid and cooled to 0°. A total of 250 µl is placed in the bottom of each prechilled quartz tube containing a 1.5 × 15 mm magnetic stirring bar. A 50-µl aliquot of an ice-cold sodium nitrite solution (100 mg/ml) is gently layered on top followed immediately by 50 µl of [125]I. The tubes are quickly capped and inverted several times so that the magnetic stirring bar mixes the viscous solution. The tubes are placed on their

---

[14] W. H. Briner, *J. Nucl. Med.* **2**, 94 (1961).

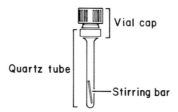

FIG. 1. Quartz reaction vial used in the synthesis of ^{125}I-labeled PVP.

side on a magnetic stirrer covered with aluminum foil. The ultraviolet light is then placed 1 cm above the tubes. The reaction is allowed to proceed for 48 hr and is terminated by the addition of 50 μl of 5 N NH$_4$OH containing 60 mg/ml potassium iodide and 100 mg/ml sodium sulfite.

Purification of ^{125}I-PVP. The reaction mixture is placed on a small, disposable column consisting of 8 ml Sephadex G-10 above 2 ml of Dowex resin washed with water. The ^{125}I-PVP elutes in a sharp peak at about 6 ml and is contained in a total volume of about 3 ml. Virtually all of the free ^{125}I is retained by the column. The ^{125}I-PVP product is then usually dialyzed against water for several days using Spectropore membrane with an M_r 3500 cutoff. This dialysis step removed a small amount of nonprecipitable radioactivity associated with the ^{125}I-PVP. The ^{125}I-PVP should then be dialyzed overnight against two changes of the medium that will be used in the uptake studies. It is then filtered through a 0.2-μm sterile filter prior to storage at $-20°$. Final yields of radioactivity incorporated into the ^{125}I-PVP range between 10 and 80%.

Evaluation of the Final Product

It is important to determine if the radioactivity incorporated by cells incubated with ^{125}I-PVP actually represents ^{125}I-PVP incorporated by fluid-phase endocytosis, or instead represents free iodine released from the polymer. One method is gel filtration of cells dissolved in formic acid. Since ^{125}I-PVP is a high-molecular-weight polymer, it will elute considerably before the free iodine. However, the simplest method to discriminate the two is by acid precipitation. Cells are usually dissolved in 10 mM CHAPS detergent. A 100-μl aliquot of the sample is added to a microfuge tube together with 100 μl of unlabeled PVP (10 mg/ml). Then 1 ml of 1% phosphotungstic acid in 0.5 N HCl is added. The precipitate is pelleted by a 5-min centrifugation at 1000 g. The tip of the tube containing the precipitate is then cut off with a razor blade and counted in a gamma counter.

Preparation of [U-^{14}C]Sucrose

Sucrose has the disadvantage that a normal breakdown product, glucose, can be incorporated by cells at relatively high rate by facilitated diffusion. Therefore, the object of repurification is to separate intact [U-^{14}C]sucrose from glucose. Radiochemical purity is often reported by the manufacturer on the basis of paper chromatography in 80% phenol. However, glucose and sucrose are not separated in that solvent system. A small amount of impurities can result in large errors. [U-^{14}C]Sucrose (560 mCi/mmol) is obtained from ICN radiochemicals in a 10 mCi/ml alcoholic solution. This material is repurified by paper chromatography as described by Besterman *et al.*[15] The [U-^{14}C]sucrose solution is spotted at the bottom of 60-cm lengths of Whatman #1 paper and dried using a hair drier without heat. The chromatogram is then developed in the descending fashion for 14–18 hr at room temperature using *n*-butanol:95% ethanol:water (104:60:20). Radioactivity on the dried chromatogram is located by autoradiography using a 5-min exposure to Kodak XAR-5 film. Since the impurities are a small percentage of the total radioactivity, an hour exposure is usually necessary to localize them. The material migrating as [U-^{14}C]sucrose is easily differentiated from the small amount of monosaccharide running ahead. The strip of paper containing the [U-^{14}C]sucrose is cut out and simply eluted from the paper directly with the desired medium. Elution by dripping from the top to the bottom of the paper is recommended to minimize the volume of the final solution. The solution should then be supplemented with 1 mM each of sucrose, glucose, and fructose to attenuate any specific or nonspecific interaction with the cells. The solution is then filter sterilized and used immediately.

Amount of Fluid-Phase Markers to Use

One of the most frequently encountered problem in conducting fluid-phase studies is the extremely small amount of fluid actually incorporated by cells. This necessitates the addition of extremely high concentrations of markers. If there is any adsorptive component in the uptake of a marker, then high concentrations of markers may increase its relative contribution. Thus the rule of thumb in determining the amount of fluid-phase marker to add is that concentration that will give an adequate amount of signal with respect to the time frame that is being examined. It is difficult to give exact guidelines since the amount of signal obtained is also dependent on the number of cells used for each point. Table I shows the approximate

15 J. M. Besterman, J. A. Airhart, R. C. Woodworth, and R. B. Low, *J. Cell Biol.* **91**, 716 (1981).

TABLE I
MINIMUM CONCENTRATIONS OF FLUID-PHASE MARKERS TO USE FOR DIFFERENT
TIME FRAMES[a]

Marker	Concentration		
	1 min	10 min	100 min
Isotopes	10^9 cpm/ml	10^8 cpm/ml	10^7 cpm/ml
Lucifer Yellow	10 mg/ml	1 mg/ml	0.1 mg/ml
Horseradish peroxidase	3 mg/ml	0.3 mg/ml	0.03 mg/ml

[a] The values shown are for a signal of approximately 5× the expected background values in 10^6 cells incubated for the indicated periods. For HRP, these values assume a 0.2 ng/ml detection limit and no endogenous peroxidase activity. For the longer incubation periods, the concentration of probe could be increased to yield more signal.

amount of marker that is appropriate in an experiment using 10^6 cells/point for different time intervals.

Protocol for Fluid-Phase Studies

Since most studies on growth factor action are carried out on cells grown on culture dishes, all of the described protocols will assume that the cells are attached to dishes. If it is necessary to conduct studies on cells grown in suspension, then specific protocols using oil flotation or exhaustive rinsings should be used.[16]

Because of the extremely low rate at which cells undergo fluid-phase endocytosis, it is usually necessary to use extremely high concentrations of markers. Thus uptake studies are usually initiated by removing the medium that the cells are growing in and replacing it entirely with medium containing an appropriate concentration of marker. Unfortunately, the simple act of replacing the medium can significantly increase the basal rate of membrane insertion and internalization.[17] Since this can give rise to apparently complex kinetic behavior (such as an apparent "diacytosis"), it is extremely important to tightly control the experimental conditions for fluid uptake studies. We have found that for rapid kinetic measurements of fluid uptake (< 10 min), the following parameters should remain constant between the medium used for culturing the cells and that used for the fluid uptake experiment: (1) temperature; (2) ion composition, (3) glucose, (4) osmolarity, (5) pH, and (6) serum. For longer experiments (> 1 hr), evapo-

[16] L. Ose, T. Ose, R. Reinertsen, and T. Berg, *Exp. Cell Res.* **126,** 109 (1980).
[17] H. S. Wiley and J. Kaplan, *Proc. Natl. Acad. Sci. U.S.A.* **81,** 7456 (1984).

ration and consequent increase in the concentration of the culture medium becomes the overriding concern. We have found that it is also useful to change the medium on the cells 30 min before initiating the uptake studies since this seems to prevent the accumulation of significant concentrations of "conditioning factors" that the cells can secrete. The following is the specific protocol that we use for assaying the rate of fluid phase endocytosis in human fibroblasts.

1. Stock cultures of human fibroblasts are grown in Dulbecco's modified Eagle's medium (DV medium) containing 10% calf serum. Cultures are plated at a density of 3×10^5 cells/60-mm plastic dish and grown to confluency ($0.9 - 1.2 \times 10^6$ cells/dish). The day before the experiment, the cells are changed to DV medium lacking bicarbonate, but containing 20 mM HEPES, penicillin at 100 U/ml, streptomycin at 100 μg/ml, glutamine at 4 mM, and 1 mg/ml of bovine serum albumin (DV–HEPES). The pH of the medium is adjusted to 7.4 at 37° before sterilization by filtration. The cells are incubated overnight in an air incubator at 37°.

2. One hour before the experiment, the medium was again changed to DV–HEPES which also contained an appropriate concentration of nonradioactive marker (such as 10 mg/ml of unlabeled PVP for studies using ^{125}I-PVP and 1 mM each of glucose, fructose, and sucrose for studies using sucrose). The cells are then transferred to a table-top incubator which consisted of a porous stainless steel platform supported in a circulating water bath enclosed in a Plexiglas box with a hinged cover. Ideally, the table-top incubator should be kept in a 37° warm room. The temperature was monitored by a flat thermistor probe (type YSI 409A, Yellow Springs Instrument Co.) installed in a 60-mm plastic dish containing an identical volume of medium as the cell cultures. It is extremely important to keep the air above the plates as humidified and as close to 37° as possible. If there is a significant temperature difference between the plates (which should be kept in contact with the 37° bath) and the surrounding air, the thermal gradient will result in a very high rate of evaporation and hence the concentration of the fluid-phase marker will change as a function of time. We always keep the lids on the plates except when adding or removing solutions.

3. Uptake of the various probes was initiated by rapidly removing the medium with an aspiration device consisting of a Pasteur pipet attached to a vacuum flask by a long hose. The plates are simply tilted in the incubator, the lid lifted, and the medium removed. The lids are immediately replaced. A prewarmed 5-ml plastic pipet is then used to add the prewarmed incubation medium to the plate. A minimum of 1 ml of medium is used for cells

grown in 35-mm plates while a minimum of 2 ml is used for 60-mm plates. A stopwatch is used to indicate the exact time at which the uptake study is initiated. If a series of time points is to be collected, then the medium is added in a staggered schedule. Usually this means that the original medium is removed from a series of plates individually and replaced with the medium containing a fluid-phase marker at 20-sec intervals. If a very rapid time course study is being conducted, the plates of cells are usually processed individually.

4. All experiments are terminated by transferring the plates of cells to an aluminum tray kept on ice. These trays can be obtained at a local restaurant supply store together with plastic busboy trays which make convenient containers for the ice. After the dishes are transferred to ice, the medium is rapidly removed and saved. The cells are then rinsed with HP saline (0.1% PVP, 130 mM NaCl, 5 mM KCl, 0.5 mM MgCl$_2$, 1 mM CaCl$_2$, 20 mM HEPES, and NaOH to give a pH of 7.2) prechilled to 0°. For experiments with [U-^{14}C]sucrose, the HP saline is supplemented with 1 mM each of glucose, fructose, and sucrose. Cells are rinsed 12 times each with 5 ml of HP saline/rinse. Each rinse is allowed to remain on the cells for approximately 5 min before being changed. Cells are then gently scraped off the plates using a rubber policeman into 2–2.5 ml of HP saline. The plates are then scraped again with 2.5 ml of HP saline to improve the yield of cells. The two samples are pooled and centrifuged in a refrigerated centrifuge for 5 min at 800 g. The supernatant from the scraped cells is carefully removed and saved for measurement. The cell pellet is gently resuspended in 1 ml of HP saline and again centrifuged at 800 g for 5 min. The total time for the rinsing and recovery of the cells is approximately 1 hr.

Quantitating the Amount of Fluid Incorporated

All results for fluid-phase studies should be in terms of volume/cell. The standard unit of uptake is the *endocytic index,* which is expressed as nl/10^6 cells/hr. Investigators are strongly encouraged to use this standard since units such as nl/μg protein are virtually impossible to interpret in the context of different cell types. Since it is usually more convenient to measure the amount of protein in a sample rather than the number of cells present, a standard curve relating cell number to protein for the particular cell type being used should be prepared. This is then used to convert the measured amount of protein recovered in the assay to the number of cells present. However, direct measurement of cell number in each sample is still the preferable method for correcting the data.

Since it is not possible to remove all of the liquid from a culture dish prior to adding the medium containing the fluid-phase marker, the actual concentration of the markers in the medium will be less than the original solution. Therefore, an aliquot of the medium which is removed from the plate of cells prior to rinsing them should be assayed to determine the actual concentration of marker to which the cells were exposed. This concentration is then used in conjunction with the amount of marker incorporated by the cells as a function of time to determine the endocytic index.

Quantitating ^{125}I-PVP Uptake

In the case of experiments in which ^{125}I-PVP is the marker, the amount incorporated by the cells can be measured in intact cell pellets using a gamma counter. Since the counting efficiency in gamma counter is dependent on the geometry of the sample, the volume of both the cell pellet and the aliquot used to determine the specific activity of the medium should be the same. For convenience, we usually pellet the cells in a microfuge tube, discard the supernatant, and then cut off the tip of the tube containing the cells and then transfer the material to a plastic counting vial. Then 0.2 ml of 1 N NaOH, 1% SDS, or concentrated formic acid is added to the vial to solubilize the cells. Be sure to add a tissue solubilizer that is compatible with the particular protein assay that is being used. An aliquot of the medium in which the cells were incubated is then serially diluted and added at an appropriate concentration in 0.2 ml of liquid in a counting vial.

Quantitating [U-^{14}C]Sucrose Uptake

The cells are pelleted in a microfuge tube and the tips containing the pellets are cut off and transferred to plastic tubes containing 0.2 ml of 1% SDS. The tubes are capped and placed in a boiling water bath for 5 min to dissolve the cells. Aliquots of the dissolved cells are removed for protein determination and the rest of the sample is transferred to a scintillation vial. Do not use strong acid or base to solubilize the cells since these can interfere with scintillation counting. Samples are added to 10 ml of Aquasol-2 (New England Nuclear). The amount of [U-^{14}C]sucrose in the sample is measured in a scintillation counter using the appropriate channel. Again, it is important that the composition and volume of the aliquot used for determining the specific activity of the medium be adjusted so that it matches that of the cell sample.

Quantitating HRP Uptake

When HRP is used as the fluid-phase marker, the rinsed cells are solubilized at a concentration of 2×10^6 cells/ml in either 0.1% SDS or 0.05% Triton X-100. Although the SDS is a far faster and more effective method for solubilizing cells, HRP is slowly inactivated in its presence unless diluted 1:10 in 1% serum saline.[9] Higher concentrations of Triton X-100 (such as 0.5%) should be used unless it interferes with the method used for the protein determination. Control cultures not exposed to HRP during incubation should also be processed similarly to measure any endogenous peroxidase activity. Aliquots of the detergent-solubilized cells are assayed for peroxidase activity using the *o*-dianisidine assay.[9] *o*-Dianisidine (3,3'-dimethyloxybenzidine) is purchased as the dihydrochloride salt and is dissolved at a concentration of 3 mg/ml in absolute methanol. The assay reagent is freshly prepared before each assay and consists of 10 mM imidazole buffer (pH 7.4), 1 mM *o*-dianisidine, and 0.05% Triton X-100. The addition of imidazole increases the sensitivity of the assay. One-tenth milliliter of the solution to be assayed is placed in a 1.25-ml cuvette of 1-cm path length together with 0.9 ml of the assay reagent. The reaction is initiated by the addition of 6 μl of 3% H_2O_2 followed by rapid mixing. The rate of development of a colored product is followed at 460 nm with a recording spectrophotometer. The reaction rate is converted to volume of HRP by comparing it to that obtained with the serially diluted incubation medium. Be sure to dilute the standards with the same solution used to extract the cells and subtract any endogenous cellular peroxidase activity from that obtained after incubation with HRP.

Quantitating Uptake of Fluorescent Markers

The amount of Lucifer Yellow or other fluorescent marker that is incorporated by cells can be determined directly in a spectrofluorometer using appropriate blanks to correct for low levels of light scattering by cellular debris. Scraped cells supernatants and the cell pellet rinse solutions are measured directly. Cell pellets are extracted overnight in 1 ml of 0.5% Triton X-100. Cell debris is removed by centrifugation immediately before the assay. The amount of Lucifer Yellow in the incubation medium is determined by measuring aliquots of medium serially diluted in 0.5% Triton X-100. Excitation and emission wavelengths are 430 and 540 nm, respectively, for Lucifer Yellow. As little as 10 ng/ml Lucifer Yellow can be accurately quantitated, although this may depend in part on the sensitivity and stray-light rejection capability of your spectrofluorometer.

Discriminating Absorbed Versus Fluid Uptake

Whichever probe is chosen to measure uptake, first it must be shown to provide adequate signal for the time frame being investigated. If it is important that uptake is occurring only by way of fluid-phase endocytosis, then one must rule out as much as possible the contribution of adsorbed, trapped of specific components in the overall "uptake" rate. Since adsorptive processes are far more effective than fluid-phase endocytosis in associating markers with cells, the tendency in an experiment is for the adsorptive component to swamp out the fluid-phase ones.

Specific interactions with the membrane leading to increased uptake (such as receptor-mediated endocytosis of facilitated diffusion) should be saturable and are indicated by the following phenomena:

1. Saturation of uptake at high concentrations of the marker
2. Inhibition of uptake by substances saturating specific uptake systems or receptors
3. Inhibition of uptake by factors preventing the actions of a specific uptake system (e.g., EGTA, where there is a specific requirement of Ca^{2+} for binding but not internalization)

Nonspecific adsorption, on the other hand, is not necessarily indicated by any of the tests described above. Since extremely low levels of adsorption are capable of invalidating fluid-phase uptake experiments, direct detection of interaction with the membrane is virtually impossible. Neither linearity of uptake as a function of probe concentration nor ease of removal of probes by rinsing proves the absence of adsorption. Therefore, since there is no way to conclusively rule out adsorptive components of uptake, the only way to generate confidence that a given probe is taken up exclusively in the fluid phase is by direct comparison with the best established markers. To date, these are ^{125}I-PVP and [U-^{14}C]sucrose. Since there is no evidence that a fluid-phase marker can be excluded from the endocytic pathway, the probe with the lowest level of uptake will most closely reflect the actual extent of the process. Any increment above that level for a different probe will reflect an adsorbed component.

In addition to the general features described above, we have found that the following criteria are a good indication of marker incorporation occurring by fluid-phase endocytosis:

1. There should be no incorporation of the probe when the cells are incubated with the marker for several hours at $0°$.

2. After incorporation and gentle homogenization of the cells, the probe should be sedimentable by centrifugation at 100,000 g for 1 hr.

3. Uptake should be strictly linear as a function of marker concentration.

Although it may be difficult to establish with an absolute certainty that what one is measuring is fluid-phase endocytosis, in most cases it is not strictly necessary since many experiments are only directed at establishing the ability of a growth factor to affect endocytosis, not in arriving at a precise quantitation of the process. Nevertheless, if the relative ability of two markers to be internalized by purely fluid-phase endocytosis is known, then the relative ability of growth factors to perturb their uptake can yield valuable information. For example, if one fluid-phase marker is taken up at five times the rate of a second, then it is safe to assume that the marker with the greater uptake has a large adsorbed component. The ratio of the incorporation of the two markers can be used as an indication of the surface : volume ratio of the vesicles engaged in the endocytosis process if adsorption is nonspecific. If, during the addition of a growth factor, the ratio of the uptake of the two markers is different, then one can interpret this result as reflecting an alteration in the types of vesicles engaged in the stimulated endocytosis rather than there simply being an overall stimulation of the endocytosis process itself.

Interpretation of Growth Factor-Stimulated Uptake Kinetics

A complete treatment on the analysis of uptake kinetics is beyond the scope of this chapter. The interested reader is referred to other works that focus on that aspect of experimental analysis.[15] Fortunately, the kinetics of fluid-phase endocytosis have usually been found to obey simple kinetics. The uptake can usually be classified as consisting of a reversible and an irreversible component (at least on the time scale of the experiment). The reversible component is usually observed as an initial nonlinearity in the uptake curve and is usually seen at short time intervals (<10 min).[15] The irreversible component is characterized as a linear rate of marker accumulation.[3,6] For those experiments in which the time scale is greater than 5 min but less than 1 hr, the reversible component can usually be ignored. In those situations, fluid uptake is usually observed as obeying simple linear kinetics with a slope of accumulation equal to the endocytic index. For studies in which fluid accumulation is investigated on the time scale of hours, a nondegradable marker such as [U-^{14}C]sucrose or ^{125}I-PVP should be used since internalized enzymes such as HRP are slowly degraded.[4] Since it has been shown than even nondegradable markers are eventually lost from the cell, it is best to restrict the time course of the actual experiment to less than 6 hr. In addition, evaporation and consequent concentration of the medium can occur during long incubation periods.

When examining the effect of growth factors on fluid-phase endocytosis, one is usually interested not only in the kinetics of the *induction* of increased fluid uptake, but also in the relationship between receptor occupancy and the *magnitude* of the response. As mentioned above, owing to the extremely small amount of liquid actually incorporated by fluid-phase endocytosis, it is extremely difficult and/or expensive to accurately quantitate the process at time intervals of less than 5 min. Since many of the rapid responses of cells to growth factors occur within 5 min, an investigation of the induction kinetics of the process should not be undertaken lightly. For example, to investigate the effect of EGF on fluid-phase endocytosis in human fibroblasts, we used 65 μCi/ml of ^{125}I-PVP as the marker ($\sim 1 \times 10^8$ cpm/ml). Yet the unstimulated rate of incorporation observed in 1×10^6 cells amounted to only 80 cpm/min. The addition of EGF raised this rate transiently to approximately 200 cpm/min. Clearly these types of studies are feasible only if very high concentrations of fluid-phase markers are used and if the markers have extraordinarily low adsorptivity.

All studies that investigate the effect of growth factors on fluid-phase endocytosis should report both the absolute effect of the growth factor on the process and the relative effect. Usually the absolute effects are presented as linear kinetic plots using time of exposure to the fluid-phase marker on the x axis and the volume incorporated (nl/10^6 cells) on the y axis. Both the control (untreated) and the experimental (exposed to growth factor) results should be shown on the same graph. The relative effect is expressed in terms of percentage of control values at that time interval and can either be presented as a graph in the case of kinetic studies or in a table for long-range incubations. It is important to perform all experimental manipulations on the control cells as well as the experimental ones since, as mentioned above, simple acts such as changing the medium can have significant effects on membrane dynamics.

Another useful way of presenting data on growth factor-stimulated fluid-phase endocytosis is in the form of differential rates.[3] The difference in the amount of fluid incorporated between any two time intervals is equivalent to the rate of fluid accumulation during that interval. The effect of growth factors on the rate of fluid uptake is more informative than their effect on the absolute amount of fluid accumulated at any given time point since the latter is both a product of the rate of accumulation at the time of measurement and the integral of all of the rates existing since the start of the experiment. The stimulated rate of fluid uptake is calculated as

$$V_u = [I_e(t + i) - I_e(t) - I_c(t + i) + I_c(t)]/i \tag{1}$$

where V_u is the differential velocity of uptake, I_e is the amount of fluid incorporated in the experimental cultures (treated with growth factors), I_c is the amount of fluid incorporated in the control cultures, (t) refers to the time of measurement, and i is the length of the time interval being measured. The results of this type of calculation should be presented as a bar graph with the x axis displaying the time intervals and the y axis displaying the differential rate.[3] It is particularly easy to identify both the kinetics and magnitude of growth factor-induced fluid-phase endocytosis when this type of data presentation is used.

Author Index

Numbers in parentheses are footnote reference numbers and indicate that an author's work is referred to although the name is not cited in the text.

Subject Index

A

A431 cells, 84
  binding of EGF receptor monoclonal
    antibody to, 71–72
  for biochemical studies of epidermal
    growth factor, 49
  in EGF receptor phosphotyrosine kinase
    assay, 143
  EGF receptor-related protein, purifica-
    tion, 87–88
  EGF receptors, 63–64
    isolation, 52–53
    internalization, effect of EGF receptor
      monoclonal antibody, 77–78
    number of, 49, 64, 88
    properties, 64–65
    secreted form, identification, 61–63
  extracellular receptor-related protein,
    biosynthesis, 62
  growth
    effects of EGF and EGF receptor
      monoclonal antibody, 76–77
    effects of EGF receptor monoclonal
      antibody, 74
    in serum-containing medium, 89–90
    in serum-free medium, 90–91
  growth responses to EGF, 88–89
  in vitro growth, 88–92
  maintenance, 66–67
  membrane preparations, TGFα/EGF
    receptor proteins, identification,
    152–153
  membranes, time course of synthetic
    peptide phosphorylation by, 358
  mixed vesicle preparation from cultures
    of, 376–378
  purification of radioiodinated TGFα
    using binding and elution from, 98
  response to epidermal growth factor, 73
  solubilization of EGF receptors in, 51–
    52
  source, 89
  uses, 88
  variant
    exhibiting altered growth response to
      EGF, isolation, 91–92
    to monitor purification and/or activity
      of TGF, 343
  TGFα receptors, 144
A431R-1 cells
  in soft agar assay, 343
  source, 344
A549 cells
  as indicator cell in soft agar assay, 344
  source, 344
  TGFβ binding, 172–173
  TGFβ receptor parameters on, 173
A673 cells, TGFβ receptor parameters on,
    173
Acromegaly, 225
Affinity cross-linking, 147, 184–185
African Green monkey kidney cells. See
    BSC-1 cells
AKR-2B cells
  in soft agar assay, 343
  TGFβ binding to, 101–102
Albumin, in bone matrix, 281
α₂HS-glycoprotein, in bone matrix, 281
Aminomethyl-resin, preparation, 131–134
Anchorage-independent growth, 112–113,
    174. See also Soft agar assay; Trans-
    formed phenotype
Aspartate carbamoyltransferase, induction
    by mitogen, 340
Autocrine growth control, 341

B

BALB/c 3T3 cells
  affinity labeling with radioiodinated
    TGFβ, 189, 191–192
  DNA synthesis, effect of IGF-I on, 206–
    207
  TGFα/EGF receptor proteins, identifica-
    tion, 151–153
β-bungarotoxin, 207
BGP. See Osteocalcin

E

Endochondral ossification, 313
Endocytic index, 411
Endocytosis
  compensatory, 402
  fluid-phase, 402
    absolute effect of growth factor, vs.
      relative effect, 416
    and dynamics of cell surface, 402
    growth factor effects on, 402
    growth factor-stimulated, differential
      rates, 416–417
    growth factor-stimulated uptake kinet-
      ics, interpretation, 415–417
    irreversible component, 415
    marker
      amount used in assay, 408–409
      criteria for, 402–403
      fluorescent, 405
      potential candidates, 404–405
      preparation, 406–408
    quantitation of amount of fluid incor-
      porated, 411–413
    quantitation of fluorescent marker
      uptake, 413
    quantitation of HRP uptake, 413
    quantitation of ^{125}I-PVP uptake,
      412
    quantitation of [U-^{14}C]sucrose uptake,
      412
    reversible component, 415
    studies
      discrimination between absorbed vs.
        fluid uptake, 414–415
      protocol for, 409–411
Endothelial cell growth factor, in bone
  matrix, 310
Epidermal growth factor
  biological activity, 11
  in control of cell growth, 49
  distribution, 3
  effect on chondrocytes, 319
  effect on human foreskin fibroblast
    growth, 8, 10
  guinea pig, 4
  human, 103, 127
    amino acid composition, 3
    characterization, 48
    forms, 8–9, 11

gene, 22–41
  construction, 34–39
  design of *Hga*I linker derivative,
    36–37
  purification, 34–39
  synthetic, characterization, 41
  synthetic, expression, 40–41
  synthetic, ligation and purification
    procedure, 37–39
  synthetic, molecular cloning, 39–41
  synthetic, transference to α-factor
    yeast expression vector, 40–41
  synthetic fragments, preparative 5'-
    phosphorylation, 37
hEGF-A, 8–9
minor form, from urine, 8–9
monoclonal antibody, 5, 11–22
  binding domains on hEGF molecule,
    22
  competitive binding to hEGF, 18–
    19, 21–22
  dissociation constant, determination
    of, 18
  effect on biological activities of
    hEGF, 19–22
  immunoglobulin class, determination
    of, 16–17
  production, 13–17
  production, cell lines, 14
  production, generation of hybrid-
    omas, 14–16
  production, immunization of mice,
    14
  production, media, 13–14
  properties, 17–21, 21–22
  purification, 17
  purification by DEAE Affi-Gel Blue
    chromatography, 5–6
  radioiodination, 18–19
  specificity, determination of, 17–18
  storage, 17
  yield, 17
monoclonal antibody affinity chromatography, 5–11
PAGE, 9
purification, 4
  monoclonal antibody affinity chro-
    matography, 3–11
  premonoclonal antibody affinity
    column steps, 6–8